Milestones in Drug Therapy
MDT

Hepatitis Prevention and Treatment

Edited by J.M. Colacino and B.A. Heinz

Springer Basel AG

Editors
Joseph M. Colacino
PTC Therapeutics
Drug Discovery
100 Corporate Court
South Plainfield, NJ 07080
USA

Beverly A. Heinz
Eli Lilly and Company
Discovery Research
Lilly Corporate Center
Indianapolis, IN 46285-0438
USA

Library of Congress Cataloging-in-Publication Data

Hepatitis prevention and treatment / Joseph M. Colacino, Beverly A. Heinz, editors.
p. cm. -- (Milestones in drug therapy)
Includes biblographical references and index.
ISBN 978-3-0348-9617-7 (alk. paper)
1. Hepatitis, Viral. I. Colacino, Joseph M., 1953- II. Heinz, Beverly A. III. Series.

RC848.H43H47 2004
616.3'623--dc22 2004048866

Bibliographic information published by Die Deutsche Bibliothek
Die Deutsche Bibliothek lists this publication in the Deutsche Nationalbibliografie; detailed bibliographic data is available in the internet at http://dnb.ddb.de

ISBN 978-3-0348-9617-7 ISBN 978-3-0348-7903-3 (eBook)
DOI 10.1007/ 978-3-0348-7903-3

Originally published by Birkhäuser Verlag, Basel - Boston - Berlin in 2004
Printed on acid-free paper produced from chlorine-free pulp. TFC ∞
Cover illustration: HBV (top) and HCV (bottom) genomes. With the friendly permission of J.M. Colacino and L. Condreay.

ISBN 3-7643-5956-0

9 8 7 6 5 4 3 2 1 www. birkhauser.ch

Contents

Contents

List of contributors

Eleanor J. Barnes, Nuffield Department of Medicine, University of Oxford, Peter Medawar Building for Pathogen Research, South Parks Rd, Oxford OX1 3SY, UK

Lawrence M. Blatt, InterMune Inc, 3280 Bayshore Boulevard, Brisbane, CA 94005, USA; e-mail: lblatt@intermune.com

Richard E. Boehme, Idenix Pharmaceuticals, 60 Hampshire St., Cambridge, MA 02139, USA; e-mail: richard.boehme@idenix.com

Nathaniel A. Brown, Idenix Pharmaceuticals, 60 Hampshire St., Cambridge, MA 02139, USA; e-mail: nat.brown@idenix.com

Lynn D. Condreay, GlaxoSmithKline, Department of Virology, P.O.Box 13398, Research Triangle Park, NC 27709-3398, USA; e-mail: Lynn.D.Condreay@gsk.com

Anouk Dev, Duke Clinical Reserach Institute and Division of Gastroenterology, Duke University Medical Centre, P.O. Box 17969, Durham, NC 27715, USA

Craig S. Gibbs, Gilead Sciences, 333 Lakeside Drive, Foster City, CA 94404, USA; e-mail: Craig_Gibbs@gilead.com

Sarah A. Harris, GlaxoSmithKline, Department of Virology, P.O.Box 13398, Research Triangle Park, NC 27709-3398, USA; e-mail: Sarah.A.Harris@gsk.com

Gururaj Kalkeri, Vertex Pharmaceuticals, Inc., Infectious Diseases Biology, 130 Waverly Street, Cambridge, MA 02139, USA; e-mail: gururaj_kalkeri@vrtx.com

Paul Klenerman, Nuffield Department of Medicine, University of Oxford, Peter Medawar Building for Pathogen Research, South Parks Rd, Oxford OX1 3SY, UK; e-mail: klenerman@ndm.ox.ac.uk

Ann D. Kwong, Vertex Pharmaceuticals, Inc., Infectious Diseases Biology, 130 Waverly Street, Cambridge, MA 02139, USA; e-mail: ann_kwong@vrtx.com

Stephen Locarnini, Victorian Infectious Diseases Reference Laboratory, Locked Bag 815, Carlton South, Victoria 3053, Australia; e-mail: stephenlocarnini@compuserve.com

Guangxiang (George) Luo, Department of Microbiology, Immunology and Molecular Genetics, University of Kentucky College of Medicine, 800 Rose Street, MN475, UKMC, Lexington, Kentucky, 40536, USA; e-mail: gluo0@uky.edu

James M. McGill, Bone & Inflammation Research, Lilly Research Laboratories, Indianapolis, IN 46285, USA, and Division of Gastroenterology and

Hepatology, Indiana University School of Medicine, Indianapolis, IN, USA; e-mail: jmcgill@lilly.com

John G. McHutchison, Duke Clinical Reserach Institute and Division of Gastroenterology, Duke University Medical Centre, P.O. Box 17969, Durham, NC 27715, USA; e-mail: mchut001@mc.duke.edu

Tomasz I. Michalak, Molecular Virology and Hepatology Research, Faculty of Medicine, Health Sciences Centre, Memorial University, St. John's, Newfoundland, Canada A1B 3V6; e-mail: timich@mun.ca

Keyur Patel, Duke Clinical Reserach Institute and Division of Gastroenterology, Duke University Medical Centre, P.O. Box 17969, Durham, NC 27715, USA

Raymund R. Razonable, Bone & Inflammation Research, Lilly Research Laboratories, Indianapolis, USA; and Division of Infectious Diseases and Internal Medicine, Mayo Clinic College of Medicine, Rochester, MN, USA; e-mail: Razonable.raymund@mayo.edu

Nasser Semmo, Nuffield Department of Medicine, University of Oxford, Peter Medawar Building for Pathogen Research, South Parks Rd, Oxford OX1 3SY, UK

Tim Shaw, Victorian Infectious Diseases Reference Laboratory, Locked Bag 815, Carlton South, Victoria 3053, Australia; e-mail: Tim.Shaw@mh.org.au

Myron Tong, The Liver Center, Huntington Medical Research Institutes, Pasadena, CA 94005, USA

Preface

Despite remarkable advances in our understanding of their replicative strategies, the human hepatotrophic viruses hepatitis B and C remain major causes of viral hepatitis worldwide and represent important medical needs that are not fully met. Hepatitis B virus (HBV) and hepatitis C virus (HCV) are genetically unrelated and belong to different virus families; thus they must be targeted by separate approaches.

HBV is the prototype virus of the Hepadnaviridae family. It is a small DNA virus that is most prevalent in Asia and in developing countries, chronically infecting nearly 400 million people worldwide. Approximately 10–15% of individuals who become infected with HBV go on to become chronic carriers of this virus. Chronic HBV infection can result in morbidity and mortality that includes severe liver disease, cirrhosis and hepatocellular carcinoma. Importantly, chronically infected individuals serve as sources of this bloodborne and sexually transmitted virus. Although there is a safe and effective vaccine for the prevention of HBV infection, the large pool of chronic carriers must be treated with effective antiviral agents in order to ameliorate or cure disease and reduce horizontal and vertical (*in utero*) transmission of the virus. Currently there are three approved therapies for chronic HBV infection including interferon α, the nucleoside analog lamivudine (3TC), and the prodrug of the nucleotide analog PMEA known as Hepsera (adefovir dipivoxil). Because of the limitations of these therapies, the search for more effective antiviral agents continues and a number of agents are in preclinical or clinical development.

HCV is a small, positive-strand RNA virus that is a member of the Flaviviridae family and that infects an estimated 170–350 million people worldwide. In contrast to HBV, the vast majority (85%) of people who become infected with HCV go on to develop a chronic infection and, of these, approximately 20% develop severe sequelae including liver dysfunction, cirrhosis and, in many cases, hepatocellular carcinoma. Chronic infection with HCV is currently the leading reason for liver transplantation in the United States. There is no vaccine available for the prophylaxis of HCV infection and the only approved therapy, a combination of interferon plus ribavirin, is not effective in over half of the patients and is associated with severe adverse effects.

This monograph brings together in one volume the most recent information on all aspects of these two fascinating and medically important viruses. In planning and developing this book, we sought contributors who are active in their fields and well known for their significant contributions to our understanding of HBV and HCV and the diseases they establish. We have selected authors who can survey their areas authoritatively and from a personal perspective. Our contributors include scientists and clinicians who have made singularly important advances in our understanding of the molecular biology of HBV and HCV, have increased our understanding of the host immune response to viral infection, have been integral to the discovery and development of agents that hold promise for the treatment of chronic HBV and HCV infection, and who have been instrumental in the design and implementation of clinical trials for the evaluation of agents that are now approved for use against HBV and HCV or are in development.

The first two chapters offer an interesting look at the history and epidemiology of the different causes of viral hepatitis with, of course, an emphasis on HBV and HCV. A time is certainly within the memory of many of our readers when there was no HBV, but rather "Australia antigen," and when there was no HCV, but rather "non-A, non-B hepatitis." Since those earlier times, rapid advances in the characterization of the etiologic agents that cause these diseases have been made and our understanding of the molecular virology of HBV and HCV has increased dramatically. Accordingly, the next two chapters are concerned with the molecular virology of HBV and HCV and offer state-of-the-art discussions of the latest findings concerning the virus genomes, the structure and function of viral proteins, and virus-replication strategies. Our understanding of the immunology of HBV and HCV has increased remarkably over the last decade, and the next two chapters are concerned with the immunologic basis of the host immune response to infection. The topics covered include the antibody, cellular, and innate responses to viral infection, viral resistance to interferon, and persistence of viral infection. The next three chapters are concerned with the current standard-of-care for HBV- and HCV-infected patients. The efficacy, adverse events, limitations of approved therapies, and the viral response to therapy are topics that are covered in these chapters. Finally, the last two chapters contain discussions of HBV and HCV therapies under development. Here are presented surveys of the more classical as well as novel approaches to viral and cellular targets that are being explored recently. Importantly, the limitations inherent to these therapies and the prospects for future approaches are summarized.

We have intended that this book will be relevant for working virologists in industry and academia, clinical investigators and physicians, graduate students and post-doctoral fellows. It is hoped that this up-to-date and comprehensive single source of information regarding HBV and HCV will be a useful introduction to researchers not yet well acquainted with these viruses and a convenient source of review and state-of-the-art information for those scientists already established in the field. The book directs the reader to useful primary sources as well as current review articles for a more in-depth treatment of specific areas of interest. It is certain that the fields of HBV and HCV will continue to advance as we learn even more about the molecular virology and immunology of these viruses and as new experimental therapies become approved. Therefore, in the years to come, it is hoped that this book will serve as a convenient source of knowledge that will provide a solid foundation for an understanding and appreciation of the basic discoveries and key therapeutic advances of the future.

J.M. Colacino
B.A. Heinz

May 2004

Hepatitis Prevention and Treatment
Edited by J.M. Colacino and B.A. Heinz

History of viral hepatitis

Gururaj Kalkeri and Ann D. Kwong

Vertex Pharmaceuticals, Inc., Infectious Diseases Biology, 130 Waverly Street, Cambridge, MA 02139, USA

Introduction

Multiple etiologies of jaundice

Hepatitis is a medical condition attributed to inflammation of the liver. Jaundice is the clinical manifestation of liver malfunction, caused by faulty excretion of bile pigments from damaged hepatocytes which is characterized by a yellowish discoloration of the skin and conjunctivae of the eye and the excretion of dark colored-urine [1]. Hepatitis can be caused by a variety of reasons, including genetic disease, autoimmune disorders, drug overdoses (acetaminophen, aspirin, anesthesia agents) [2], excessive alcohol consumption and viral (hepatitis A-E viruses, yellow fever virus, cytomegalovirus, Epstein-Barr virus) as well as non-viral etiological agents (syphilis, leptospirosis, Q fever). Depending on the duration of illness, it can be further classified as 'acute' or 'chronic' hepatitis.

Jaundice in ancient civilizations

The importance of the liver for the vital functioning of the body was recognized as early as the Mesopotamian civilization [3], where detailed liver examination was used for forecasting illnesses. Hippocrates documented outbreaks of jaundice on the Greek island of Thassos [4]. He described people with fever, vomiting, jaundice and some fatalities. 'Epidemic jaundice' was also observed in the Middle Ages, mostly during times of war, due to crowded and unsanitary conditions.

Jaundice in the 18th and 19th centuries

In 1885, Lurman reported an outbreak of jaundice affecting 15% of a group of 1289 Bremen shipyard workers [5]. This outbreak was thought to be associated with the use of smallpox vaccine containing glycerinated human lymph. In the United States, the first known epidemics of jaundice occurred in the war of 1812 [6]. During the civil war in 1888, approximately 70 000 cases of jaundice in Federal troops were reported by Smart [7]. In World War I, the allied forces also suffered epidemics of jaundice [8, 9].

Iatrogenically transmitted jaundice

In the early part of the 20th century, percutaneously transmitted hepatitis was observed in patients attending clinics where non-discarding and improperly sterilized needles were used [10–12]. Initially, jaundice in these patients was attributed to commonly used drugs such as arsenic, bismuth and salvarsan. Unfortunately, the infectious nature of jaundice outbreaks was only recognized when a group of children and British troops inoculated with the mumps vaccine started developing jaundice and hepatitis [13].

Transmission of hepatitis

Identification of a causal relationship

Cockayne [14] was the first to note the epidemic nature of jaundice. In a scientific report published in 1912, he mentioned the occurrence of epidemics of jaundice in ancient Greece, Rome, China and Minorca. His scrupulous and systemic study of the occurrence, spread and infectious nature of these jaundice outbreaks led him to suspect that a viral infection was involved [14, 15]. Findlay and colleagues [16] studied an "epidemic *(of)* Catarrhal jaundice" after an outbreak in Surrey and described the causative agent as an "Ultramicroscopic virus that is pathogenic only to man". In 1937, Findlay and MacCallum noticed the contagious nature of the jaundice observed among the yellow fever recipients and used the terms 'common infective hepatic jaundice' and 'infective jaundice' [17]. In the 1940s, hepatitis was thought to be mainly transmitted through blood transfusion [13] and was called 'catarrhal jaundice.' Though experimental infection studies were conducted in chimpanzees to study the infectious nature of hepatitis, they were largely inconclusive.

At the time, scientists believed that viral involvement in the clinical manifestation of hepatitis could only be proven using transmission studies with bacteria-free filtrates. Scientists therefore decided to experimentally infect human volunteers in order to identify and observe the course of infection of the putative infectious agents. In these transmission studies, scientists tried to transmit the disease to naive human subjects by injecting or feeding filtrates of various body fluids and extracts (stool extracts, duodenal fluid, serum) from infected patients with hepatitis in order to identify the causal agent of the disease and to observe the course of the illness. The use of mentally retarded children with no ability to give informed consent for such human studies is most troubling, though these studies yielded significant information about the nature of the infectious agents. Today, an Institutional Review Board would rightly refuse permission to infect an individual with an infectious agent associated with significant morbidity and mortality, but without any existing therapy for treating the illness.

Identification of different routes of human transmission

Experimental transmission studies using infected duodenal juice were conducted in human volunteers in Germany in 1942 [18] and in the Middle East [19]. In the United States, scientists at Yale successfully transmitted hepatitis using the oral route by feeding 12 volunteers with patient serum or filtrates of stool extracts [20]. In order to test for a parenteral route of transmission, preicteric serum from patients who had developed hepatitis after oral ingestion of infected material was injected into 11 recipients; six of these infected patients developed hepatitis [21]. Since the transmission studies proved infectability of human volunteers using filtered samples, physicians recognized the probable cause of catarrhal jaundice as a viral agent [22]. Infectious hepatitis was also reported in animal-care personnel handling newly imported chimpanzees [23, 24], indicating the cross-species transmission of hepatitis infections from primates to humans.

Identification of hepatitis viruses by human transmission studies

Observation of the clinical course of the experimentally induced infections enabled distinction between the enterically transmitted, short-incubation, epidemic hepatitis A virus (HAV) and parenterally transmitted hepatitis with longer incubation hepatitis B virus (HBV). MacCallum is credited as the first to use the terms hepatitis A and B in 1947 [25]. MacCallum and Krugman, further subdivided "infectious hepatitis" (Tab. 1) into (1) fecal-oral transmitted, short incubation, epidemic hepatitis (now known as hepatitis A and E viruses) and (2) parenterally transmitted hepatitis (now known as hepatitis B, C and D viruses) [26]. Identification and characterization of the etiological agents further demonstrated that HAV and hepatitis E virus (HEV) as the key etiological agents of enterically transmitted acute hepatitis [27] and HBV was the major cause of post-transfusion hepatitis [28].

Studies conducted in human inmate 'volunteers' by MacCallum in 1940 in England [12, 29] were able to differentiate between different types of viral hepatitis based upon the primary route of infection and incubation period [17]. In the 1950s and 1960s, Krugman and colleagues [30–32] performed similar studies in the United States in mentally retarded 'volunteer' children at the Willowbrook State School in New York (Tab. 1). Krugman and Giles tested 25 000 serum specimens from more than 700 hepatitis patients. A patient (Mir)

Table 1. Milestones in hepatitis history

1947	Identification of type A and type B hepatitis by MacCallum
1950–1960s	Willowbrook studies by Krugman in human volunteers
1963	Identification of Australia antigen (Au) of hepatitis B virus by Blumberg
1967–1968	Association between Australia antigen and development of hepatitis B virus infections
1970	Discovery of Dane particles of hepatitis B virus
1972	Federal laws passed for mandatory testing of donor blood for presence of hepatitis B virus markers
1973–1974	Discovery of hepatitis A virus by Feinstone
1977	Discovery of hepatitis D virus by Rizzetto
1983–1986	Discovery of hepatitis E virus by Balayan
	Subunit vaccine for prevention of hepatitis B virus infections approved for use
1989	Discovery of hepatitis C virus by Houghton and colleagues
1995–1996	Discovery of hepatitis G virus and GB virus C isolate
1996	Approval of hepatitis A virus vaccine
1998	Discovery of TT virus
2000	Discovery of SEN virus

came down with two episodes of hepatitis and the strains of virus observed during both episodes were termed 'MS-1' (Mir serum-1) and 'MS-2' (Mir serum-2) respectively. Seven out of eight children injected with MS-1 serum developed hepatitis after a short incubation period (now known to be HAV) and seven out of nine children injected with MS-2 came down with hepatitis with a longer incubation period (now known to be HBV) [30]. Australia (Au)- or hepatitis-associated antigen was present in the MS-2 strain of serum hepatitis (SH), but not in MS-1 infectious hepatitis (IH) [33]. Both these studies helped classify and characterize two distinct forms of viral hepatitis – *infectious/epidemic* or *type A hepatitis* with a shorter incubation period (2–6 weeks) and *serum/homologous serum jaundice* or *type B hepatitis* with a longer incubation period ranging from 6 weeks to 6 months [30].

Enterically transmitted hepatitis

HAV and HEV are the most important etiological agents of enterically transmitted acute hepatitis.

HAV

As recently as in 1973, human inmate 'volunteers' of the Joliet state penitentiary in Illinois were inoculated with the MS-1 strain of hepatitis. Using electron microscopy, Feinstone et al. [27] detected HAV particles in the feces of the infected prisoners (Tab. 1). This led to the development of diagnostic assays for HAV infections. Soon the global prevalence of HAV in Europe, USA, Africa, Asia and the Middle East was demonstrated to range between 29 and 97% of the population [34]. In 1979, a breakthrough was achieved when scientists found a way to cultivate HAV (albeit replicating slowly) in cultured cells in the laboratory [35], obviating the need to infect more humans experimentally. Cell-culture-adapted strains of the virus were selected [36], which eventually paved the way for the development of a successful vaccine for the prevention of HAV infections.

HAV-associated disease and vaccine development
HAV has been shown to cause an acute self-limiting disease with a low mortality rate (2/1000 cases) [37]. HAV is highly contagious, spreading predominantly through a contaminated fecal–oral route, and is characterized by an acute influenza-like onset of symptoms with myalgia, headache, fever and malaise [38]. HAV is commonly shed during the incubation period and early symptomatic phase [39]. Diagnosis of HAV is made by the detection of anti-HAV immunoglobulins in the patient blood. In 1996 Merck got approval for an inactivated HAV vaccine followed by SmithKlineBeecham with another HAV vaccine (Tab. 1). Both vaccines are effective in the prevention of HAV infections.

HEV

Epidemiology of HEV infections
HEV is responsible for a significant proportion of acute viral hepatitis cases observed in young and middle-age adults in Asia and India [40]. The disease occurs in epidemic and endemic forms and is mainly transmitted due to the ingestion of fecal-contaminated drinking water [41]. This form of hepatitis was initially recorded in Delhi in 1955, when 29 000 cases of icteric hepatitis were identified [42]. Two more incidences were recorded in December 1975 and January 1976 in the city of Ahmedabad in India [43]. Both of these HEV outbreaks were traced to contaminated water supplies and HAV and HBV were ruled out as the etiological agent [40]. Similar outbreaks were reported during 1955–1956 in the Kirgiz republic of the former Soviet Union (10 812 cases), during 1986–1988 in China (119 280 cases) and during 1980–1986 in Africa (5300 cases) [41]. The infectious nature of this type of hepatitis was demonstrated inadvertently by scientists at the Institute of Poliomyelitis and Viral Encephalitis of Moscow who developed hepatitis after accidentally ingesting acute-phase stool samples from a waterborne epidemic case of hepatitis [44] (Tab. 1). After becoming infected, scientists recovered virus-like particles 27–30 nm in size from their own feces.

Additional confirmation of an infectious agent came from the transmission of HEV to non-human primates. Two cynomolgus macaques were intravenously inoculated with a 10% (w/v) suspension of feces from a human volunteer who was positive for virus-like particles [44]. Elevated alanine aminotransferase (ALT) activity was noticed between 24–36 days and animals excreted 27–30 nm virus-like particles. Similar transmissions of putative HEV samples to animals were also reported by other groups [45]. Taking into consideration the waterborne transmission of this form of hepatitis, a broader term, "Enterically transmitted non A, non B hepatitis", was later coined [46, 47].

Molecular cloning of HEV

The next task in understanding HEV was to isolate and clone the virus from infected patients. However due to probable degradation associated with isolation of viral nucleic acid from fecal material, this was not possible. This led researchers to look in the gall bladder for the virus [41]. Bile samples from HEV-infected cynomolgus macaques during the acute phase of the infection yielded virus particles ranging from 32–34 nm which were similar to the virus particles isolated earlier from the human fecal material. The Centers for Disease Control [48] performed the molecular cloning of HEV in association with Genelabs (Redwood City, CA, USA) and Reyes and colleagues [49] identified the first virus specific clone using differential hybridization (Tab. 2). These studies demonstrated that the HEV genome consists of a RNA genome encoding both structural and nonstructural proteins through the use of discontinuous and partially overlapping open reading frames (Tab. 2). There is significant sequence variation between the North American, Mexican and Burmese isolates [41].

HEV-associated disease and incidence of infection

HEV is the etiological agent of an acute self-limiting disease, epidemiologically resembling HAV in its enteric transmission and occurring in epidemic as well as in sporadic infections (Tab. 2) [39]. Outbreaks in India, Pakistan, Russia, China, Peru and Mexico have demonstrated a high mortality rate in pregnant women ranging up to 20% [50]. Diagnosis of HEV can be made by the presence of anti-HEV antibodies in the patient sera using enzyme immunoassays.

Hepatitis F virus

In the mid 1980s, another form of enterically transmitted hepatitis virus was reported from different parts of the world including India [51], England [52] and Italy [53]. The infectious nature of this virus in animals was demonstrated by transmission to cynomolgus macaques and tamarins and the recovery of 27–34 nm virus-like particles [45]. In 1994, rhesus monkeys were infected with the stool extracts of from five French sporadic hepatitis cases and virus-like particles were also identified [54]. Antigens of this virus were observed in 66% of the coded hepatitis cases, indicating a widespread presence. The infectious agent was termed 'hepatitis F virus' due to the French origin of the ini-

Table 2. Comparison of different hepatitis viruses

Agent	Discovery	Family	Genome and virus morphology	Incubation and transmission	Disease and mortality	Treatment	Prevention
HAV	Feinstone et al. (1973) [27]	Picorna	Genome: 7.5 kb (+) sense single-stranded RNA Virus: 27 nM non-enveloped, spherical	Incubation: 15–40 days Transmission: fecal, oral	Disease: self-resolving acute hepatitis. Mortality 0.1–0.2%	Not necessary. Usually self-limiting.	HAV vaccine and anti-HAV IgGs.
HBV	Blumberg et al. (1963) [28]	Hepadna	Genome: 3.2 kb partial double-stranded DNA Virus: 42 nm, enveloped, filamentous and spherical	Incubation: 50–180 days Transmission: parental	Disease: chronic hepatitis and hepatocellular carcinoma. Mortality 1–2%	Interferon, lamivudine liver transplantation.	HBV vaccine and anti-HBV IgGs.
HCV	Choo et al. (1989) [116]	Flavi viridae Hepaci virus	Genome: 9.6 kb (+) sense single-strand RNA Virus: 40–60, enveloped	Incubation: 30–150 days Transmission: parental	Disease: chronic hepatitis, hepatocellular carcinoma. Mortality 1–2%	Interferon, ribavirin. Liver transplantation.	None
HDV	Rizetto et al. (1977) [127]	Virusoids	Genome: 1.7 kb Circular, single-strand RNA Virusoid: 36 nm	Incubation: 21–90 days Transmission: parental	Disease: acute delta hepatitis and fulminant hepatitis. Mortality 2–20%	Interferon, liver transplantation.	HBV vaccine
HEV	Balayan et al. (1983) [44]	Caliciviridae	Genome: 7/5 kb single stranded RNA Virus: 32–34 nm	Incubation: 63 days Transmission: fecal, oral	Disease: self-limiting acute disease. Mortality –2% in general, 20% in pregnant women.	Usually not necessary. Endstage therapy: liver transplantation.	Good sanitation, clean water.

Continued on next page)

Table 2. (Continued)

Agent	Discovery	Family	Genome and virus morphology	Incubation and transmission	Disease and mortality	Treatment	Prevention
HGV	Simons et al. (1995) [152] Linnen et al. (1996) [158]	Flavi viridae	Genome: 9.5 kb RNA Virus: 50–100 nm	Incubation: unknown Transmission: unkown	Disease: lack of strong evidence for association with hepatitis.	Unknown	Unknown
TTV	Okamoto et al. 1998 [173]	Circo virida	Genome: 3.9 kb Circular single-stranded DNA. Virus: 30–50 nm	Incubation: unknown Transmission: unkown	Disease: lack of strong evidence for association with hepatitis.	Unknown	Unknown
SEN	Primi et al. (2000) [192]	Circo virida	Genome: 3.2–3.6 kb circular single-stranded DNA	Incubation: unknown Transmission: unkown	Disease: lack of strong evidence for association with hepatitis.	Unknown	Unknown

tial hepatitis cases. However, no additional cases of hepatitis due to this virus have been reported.

Parenterally transmitted hepatitis

HCV and HBV are the major causes of parenterally transmitted hepatitis cases. Prior to the initiation of routine screening of donated blood, hepatitis was a common serious consequence of blood-transfusion therapy [55]. Diagnosis of hepatitis before the advent of more sophisticated diagnostic tools was based mostly on the clinical symptoms such as fatigue, malaise, anorexia and jaundice. In the early part of the 1950s, serum markers for liver damage such as transaminases and aminotransferases [56, 57] were discovered, facilitating a more scientific and objective evaluation of clinical hepatitis. The advantage of using these serum markers for diagnosis over the traditional methods of diagnosis became evident with a few published studies that reported the incidence of transfusion-associated hepatitis to be around 11–65% [58–61] which was much more than previous estimates.

HBV

Frequency of detection of the Au antigen in different populations

In 1963 at the National Institutes of Health (NIH), Baruch Blumberg was searching for inherited polymorphic serum proteins in the blood samples of a group of hemophiliac patients from Mount Sinai Hospital in New York. Blumberg hypothesized that sera from multiply-transfused patients might have antibodies to novel antigens. An antigen precipitin band that stained with azocarbamine (indicating that the major band was a protein) but not with Sudan black was observed in these hemophiliac patients. Interestingly, serum from the patients only reacted with one of the 24 sera in the test panel. Since the antigen specimen was from an Australian aborigine, it was named the Australia antigen (Au) [28, 62]. Blumberg and colleagues screened approximately 20 000 serum specimens of blood in the collection of the Division of Clinical Research of the Institute of Cancer Research and noted that the Au antigen was very stable and samples collected and stored for 10 years still gave a positive result in the immunodiffusion assay. Further analysis of the serum-survey results demonstrated a relatively rare occurrence of Au antigen in the normal population of the United States (1 in 1000), and a higher frequency in people originating from other geographic regions such as the tropical and Asian populations (e.g. 6% in Filipinos from Cebu, 1% in Japanese and 5–15% in Pacific islander populations) [63]. Serum from transfused and leukemic patients, revealed a high frequency of Au antigen rather than anti-Au antibody. Several hypotheses were made with these preliminary observations [65]. These included (1) a common susceptibility factor between patients with leukemia and 'Au antigen positivity' and (2) a relationship of Au antigen with an infectious virus that might be the cause of leukemia.

Association of the Au antigen with hepatitis
In 1966, a Down's syndrome patient named James Blair, who had initially tested negative for Au antigen, was found to have seroconverted and tested positive. Liver chemistry tests performed after the detection of Au antigen demonstrated the development of 'chronic anicteric hepatitis'. This interesting observation opened the possible association between Au antigen and hepatitis [64]. Blumberg and colleagues [63, 65] reported that most of the disease associations could be explained by the association of Au with a virus (Tab. 1). Initially, these observations were accepted with a certain amount of skepticism by the scientific community, due to the multiple viral candidates associated with transfusion hepatitis of unknown etiology. Dr Kazuo Okochi, of the University of Tokyo, who was also pursuing a similar line of enquiry, found that Au antigen could be transmitted by transfusion. Hepatitis developed in some patients receiving the Au antigen and some of the transfused patients developed anti-Au antibodies [66, 67]. These findings were further supported by the observations of Prince [68]. In addition to this circumstantial evidence, it was also noticed that with high probability donor blood containing Au antigen was more likely to transmit hepatitis than donor blood without the antigen [69]. These observations led to the screening of donor sera for the presence of Au antigen and the exclusion of Au-positive donor blood samples from blood transfusions. As a result, Senior and colleagues [70] observed a dramatic reduction in post-transfusion hepatitis cases from 18% (before the screening for Au antigen) to 6% after the screening of donor blood for Au antigen. These studies played a pivotal role in the widespread use of blood-screening tests for Au antigen before transfusion.

HBV structure, proteins and vaccines
Three distinct morphological forms of HBV have been observed in patient sera: (1) 17–25 nm-diameter small pleiomorphic, spherical, non-infectious particles with HBV surface antigen (HBsAg); (2) filamentous, tubular forms containing HBsAg (Au antigen) and (3) double-shelled 42 nm particles with a 27-nm electron-dense inner core (Dane particles) (Tab. 2).

Ultra-structural examination of Au antigen using electron microscopy and isolation techniques developed by Alter and Blumberg revealed sausage-like particles 20 nm in size that were aggregated by anti-Au serum. Later, Au antigen was shown to be an incomplete form of the envelope of HBV which is only found in the serum of HBV patients [68]. Thus it was named HBsAg. Another antigen HBV envelope antigen (HBeAg) was identified by Magnius and Espmark [71] in 1972 and proved to be a reliable marker for the presence of intact virions and for infectivity.

In 1970, Dane and colleagues [72] used immunoelectron microscopy to detect a 42-nm double-shelled particle, consisting of an outer HBsAg and an inner core or nucleocapsid containing HBV core antigen (HBcAg). These particles were named Dane particles after their discoverer. DNA-dependent DNA polymerase activity was detected in the core of Dane particles isolated from the serum of an HBV-infected patient [73]. Lack of a requirement for a primer or template for the polymerase activity and precipitation of the enzyme with antibodies against the HBcAg further demonstrated the unique characteristics of this viral DNA polymerase. In 1974 Robinson and colleagues [74, 75] identified the hepatitis B viral genome as small circular partially double-stranded DNA with 3200 nucleotides (Tab. 2). In 1997 a cell line (HepG2.2.15 cells) was created which produces HBV virions, HBsAg and naked core particles, which proved to be a useful tool in understanding the biology of the virus [76].

HBV blood screening
The discovery of HBV led to the development of sensitive diagnostic methods such as gel immunodiffusion tests, counter electrophoresis (CEP), passive haemagglutination and radioimmunoassay. These tests eventually resulted in the widespread use of HBV detection methods for screening at blood banks around the world. In 1972 the importance of blood-screening assays in reducing the incidence of transfusion-associated hepatitis was realized by the US government and screening of blood donors for the presence of HBsAg was made mandatory [77]. By 1975, legislation had been enacted to screen the blood supply with 3rd generation radioimmunoassays. During the 1970s, five studies were initiated in the USA to evaluate the incidence and risk factors of transfusion-associated hepatitis. These studies revealed that the incidence of HBV was between 9.1 and 18.3% of the total number of hepatitis cases observed in the USA [55, 78, 79]. European prospective studies conducted in the 1960s and 1970s estimated the incidence of HBV to be approximately 3% in the Netherlands [80] and 15% in Greece [55, 81]. By the late 1970s, commercial blood donors were shown to more likely harbor the HBsAg [82] and HBeAg [83] than the volunteer donors. Thus declaration of the origin of the blood (whether from volunteer blood donors or commercially paid donors) was made mandatory [84].

HBV-associated diseases and vaccine development
HBV has been shown to affect 400 million patients worldwide [85]. HBV infections result in 5–10% chronic hepatitis cases in adults, with a propensity to cause cirrhosis and hepatocellular carcinoma and a variety of extrahepatic manifestations such as glomerulonephritis [86].

In 1980 Szmuness and colleagues at Merck demonstrated that the subunit vaccine developed by Merck provides 90% protection for people at risk without serious side effects [201]. In 1986 the US Food and Drug Administration (FDA) approved a yeast subunit vaccine containing HBsAg for use in human beings (Tab. 1).

Hepatitis C virus (HCV)

Post-transfusion non A, non B hepatitis (PT-NANBH)

After the discovery of HAV and HBV, widespread blood screening was employed to screen donor blood for transfusions. These screening strategies were expected to eliminate or reduce the risks of fresh cases of hepatitis. However, it was observed that 10% of the patients who received the blood transfusions still developed hepatitis even after ruling out HAV and HBV infections [87]. In addition, half of the patients with post-transfusion hepatitis developed chronic hepatitis and 20% of those patients progressed to cirrhosis [88]. These observations supported a hypothesis for the existence of hepatitis viruses still to be discovered [87, 89, 90]. Thus the name post-transfusion non-A, non-B hepatitis (PT-NANBH) was coined for the unidentified hepatitis virus. Infection with NANBH was observed to be clinically and epidemiologically distinct from HAV and HBV infections [91].

NANBH transmissions studies in chimpanzees

Chimpanzees were found to be susceptible for NANBH infections and proved to be useful tools in the initial characterization of the NANBH agent [92–94]. These attempts involved the use of convalescent human and chimpanzee serum and acute and chronic phase sera with counter-immunoelectrophoresis, radioimmunoassay, and immunodiffusion assays. Most of these transmission attempts were unsuccessful, most likely due to low virus titers in the infectious serum/plasma, liver homogenates and blood fractions [95]. For example, most of the stocks of chimpanzee infectious serum had titers of less than 1×10^3 chimpanzee infectious dose (CID)/ml [96–98].

Tubule-forming agents (TFAs)

In a pioneering study, chimpanzees were infected with the two strains of NANBH serum obtained from patients suffering with acute (strain H) and chronic (strain F) infections [99]. Ultrastructural changes were observed in the hepatocytes of the infected chimpanzees. These changes consisted of double-walled tubular structures in the cytoplasm of hepatocytes from the chimpanzee which received the F strain, and intranuclear particles 20–27 nm in size in the hepatocytes from the chimpanzee that was infected with the H strain. Other researchers further confirmed the presence of such alterations in the livers of the infected chimpanzees [100–102]. The structures were called tubule-forming agents (TFAs) and were thought to be the NANBH agent. The infectivity of NANBH serum in chimpanzees was destroyed by treatment with chloroform [103, 104] or formalin [98, 105], suggesting that the NANBH agent was a lipid- containing or enveloped virus. TFAs passed through a 80-nm filter [106] and partially through a 50-nm filter [107], indicating that TFAs were approximately 40–60 nm in size. These studies ruled out the possibility that NANBH was a viroid or a prion-like agent (both viroids and prions are proteinaceous in nature) [108, 109].

Detection of reverse transcriptase (RT) activity

In the 1980s, there was considerable confusion about the agent associated with NANBH. Brechot and colleagues [110] detected HBV DNA in NANBH patient hepatocytes, and in 1982 a NANBH-infected serum showed reactivity with HBV monoclonal antibodies [111]. Simultaneously, another group of researchers searching for the same NANBH agent identified particle-associated RT activity in human sera and in cultured NANBH-infected chimpanzee liver cells [112–114]. RT activity was also detected in four human NANBH serum, two plasma products which were known to transmit NANBH disease to chimpanzees [112] and in 12 of 12 human acute and chronic phase NANBH human sera, but not in normal healthy human sera.

However, the US Center for Disease Control and Prevention (CDC) was unable to reproduce the detection of RT activity using a panel of well-established cases of PT-NANBH. RT activity also was not detected in concentrated acute-phase human plasma fractions containing 1×10^6 CID/ml of NANBH TFA [115]. These findings, together with the fact that the TFA was less than 80 nm and the failure to detect any homology between RNA extracted from NANBH chimpanzee livers or hepadna viruses and RNA from retroviruses, appeared to rule out the possibility of NANBH being a retrovirus.

Physical characterization of NANBH

The CDC decided to further probe the physical characteristics of the putative PT-NANBH agent using polycarbonate membranes. TFA was concentrated from chronic-phase chimpanzee plasma and the pellet, mixed with Herpes simplex virus 6 (HSV-6) as a control for filtration experiments, was passed through 450-, 200- and 80-nm filters sequentially. Inoculation of a naïve chimpanzee with the 80-nm filtrate resulted in an increase in ALT levels which peaked after 65 days and in ultrastructural changes in hepatocytes associated with the presence of the TFA agent [115]. This disease profile correlated with the TFA transmission studies conducted at CDC.

Molecular cloning of NANBH

Although the physicochemical and biological properties of NANBH were known by this time, it was essential to understand NANBH at the molecular level. The biggest obstacle in achieving this goal was the lack of high-titer plasma pools or liver homogenates. The CDC initiated efforts to solve this problem by (1) adaptation of the virus to the chimpanzees by serial passaging, (2) induction of immunosuppression to increase the TFA replication in the infected chimpanzees and (3) titration of selected acute-phase and chronic-phase livers from infected chimpanzees [115]. The purpose of these efforts was to facilitate (1) visualization of the NANBH virus by electron microscopy, (2) purification of the virus for molecular cloning, (3) development of sensitive detection assays and (4) development of a cell-culture system to grow the virus.

At the CDC, two chimpanzees named Don and Rodney were studied intensively for the purpose of collecting high-titer plasma samples containing the

TFA agent [115]. Retrospective studies at the CDC had shown that chronic-phase plasma contained more TFA than acute-phase plasma samples and correlated well with the disease severity in the infected chimpanzees. A large pool of plasma was collected from Don over a period of 4 years following the acute-phase infection. These pooled plasma samples of 3200 ml were further tested in chimpanzees and determined to have a titer of at least 1×10^6 CID/ml. The pooled plasma samples were resuspended and concentrated by ultracentrifugation, and the total DNA and RNA in the samples were extracted [115]. Molecular cloning of the PT-NANBH was done at Chiron Corporation using a variety of molecular biology techniques (Tab. 1). cDNA was prepared from nucleic acid isolated from the concentrated TFA samples, cloned into a lambda gt11 β-galactosidase fusion expression vector, and screened with human NANBH-positive serum [116]. Two clones of 155 and 353 bp in length reacted with sera from NANBH patients and NANBH-infected chimpanzees, but not with the uninfected patients and chimpanzees. This agent was called hepatitis C virus (HCV) [116]. Further studies revealed that the TFA agent was a 9.6 kb, single-stranded RNA with a positive-sense orientation (i.e. with the same polarity as mRNA) and a similar genomic organization as other flaviviruses [117, 118] (Tab. 2). Sucrose-gradient fractionation studies revealed the presence of particles at two densities –1.06 g/ml from infectious sera and 1.12–1.17 g/ml from non-infectious sera [119, 120]. Later it was demonstrated that the particles in the lighter fractions were associated with low-density lipoproteins (LDLs) [119, 121] and the higher-density fractions were associated with immunoglobulins [122]. In 1995, Li and colleagues [123] visualized icosahedral viral particles of 60–75 nm in concentrated infectious human serum and liver samples by electron microscopy; these results were confirmed by other researchers [124, 125].

HCV-associated disease and incidence of infection

Infection with HCV affects 170 million people worldwide, and is associated with chronic hepatitis, cirrhosis and hepatocellular cancer. Active HCV infections are diagnosed by the presence of HCV RNA extracted from the serum of infected patients, which will include newly infected and chronically infected patients. Approximately 70% of all patients who are HCV-antibody-positive are HCV-RNA-negative, indicating that they were once infected but the infection did not become chronic. The incidence of HCV infection is estimated to be 8.6–17.1% of the US population [55, 78, 126]. A putative role for HCV in autoimmune disorders and non-Hodgkin's lymphoma is also under investigation.

Hepatitis D virus (HDV)

Detection of HDV

HDV was initially identified in 1977 by Rizzetto and coworkers in Italy [127] (Tab. 1) as a new nuclear antigen in the hepatocytes of HBV-infected patients that was commonly associated with severe, acute or chronic hepatitis [128]. Due to its resemblance to HBcAg, it was initially thought of as HBcAg. Surprisingly, testing of the stored serum samples using anti-HDV assay demonstrated a worldwide prevalence of HDV [129]. Ponzetto and colleagues [130, 131] were able to detect the HDV antibodies in normal human immunoglobulin prepared during World War II, in serum samples collected in 1971, and liver samples from Santa Marta hepatitis collected between 1937 and 1977 in South America [132].

Though HDV was initially thought of as yet another unidentified HBV variant, it was later shown to be a new transmissible agent and actually an RNA-containing virus enveloped by HBsAg [129]. This newly discovered antigen was termed delta antigen (HDAg). HDAg is usually detected in the patient's liver or serum samples in large HDAg (27 000 kDa) and small HDAg (24 000 kDa) molecular-weight forms [133, 134]. HDAg is now known to be a component of the delta virion [134], and the two forms of HDAg are detected in the patients during HDV infection.

HDV disease association

Initially, HDV infection was thought to be associated with severe or progressive liver disease [135, 136], but recently a considerable variation in the natural history of delta hepatitis has been observed which ranges from fulminant acute hepatitis to a healthy carrier state [137–139]. Co-infection with HDV appears to influence the clinical outcome of HBV infection by resulting in a more acute HBV infection, whereas superinfection of HDV is associated with chronic HBV infections [129, 138]. Patients with evidence of HBV replication might have more severe chronic liver disease and progress more rapidly to end-stage liver disease [140–142]. This might be due to the helper function of HBV permitting increased levels of HDV replication [143].

Incidence of HDV infection

Approximately 15 million patients are infected with HDV worldwide [137]. Annually, 7500 cases of acute HDV co-infection or superinfection, 70 000 HDV carriers and 1000 deaths from chronic or fulminant hepatitis are associated with HDV infection [144].

Novel viruses found in non-A non-E hepatitis

Around 5% cases of chronic liver disease and more than half the cases of hepatocellular carcinoma are not associated with detectable hepatitis A-E virus infection [145]. An estimated 68 000 cases of post-transfusion hepatitis cases in the US alone cannot be attributed to any known hepatitis virus infection [146]. These observations have sparked efforts to identify new viruses associated with hepatitis. In the last few years, several candidate viruses have been identified, but a firm association between these viruses and clinical hepatitis has yet to be established. These putative viruses include TT virus (TTV), SEN virus (SENV), hepatitis G virus (HGV) and hepatitis GBV-C (HGVB).

Hepatitis GB viruses

Detection of GB infectious agent-associated hepatitis
In 1967, Deinhardt and colleagues [147] were investigating the presence of an unknown agent in the serum of a 34-year-old surgeon with the initials GB who had developed acute hepatitis and jaundice for 4 weeks. The infectivity of this GB serum was demonstrated by the induction of hepatitis in naive tamarins inoculated with the serum.

Not much progress was achieved in understanding the GB virus before the 1990s, when HCV and HEV were discovered by Choo and colleagues [116] and Reyes and colleagues [49] respectively. Subsequent development of sensitive detection methods for these viruses made it feasible to rule out HCV- and HEV-positive infection cases from the pool of largely undiagnosed non-A, non-B hepatitis cases. As a result, 12 out of 98 cases from an NIH non-A, non-B transfusion series [148], 8% of blood donors with elevated ALT levels [149] and 7% of post- transfusion patients in France [150] could not be attributed to any known hepatitis viruses. Thus, it was noticed that not all the non-A, non-B hepatitis could be ascribed to any of the known etiological agents [151].

Molecular cloning of the GB infectious agent
Simons and colleagues [152] working in the Abbott virus discovery group were interested in the discovery of the still uncharacterized non-A, non-E hepatitis viruses. They employed a subtractive PCR technique called representational difference analysis (RDA) to selectively amplify unique nucleic acid sequences present only in the GB-infected tamarin plasma [152] (Tab. 1). Two distinct RNA genomes were detected in the GB-infected tamarin plasma with limited homogeneity to each other or to HCV; thus they were termed GBV-A and GBV-B.

Incidence of GBV-A and GBV-B and detection of GBV-C
In order to determine the seroprevalence of these viruses in the non-A–E hepatitis patients, antibodies to recombinant proteins of GBV-A and GBV-B expressed in *Escherichia coli* were used to screen patient plasma [153]. Since neither GBV-A nor GBV-B RNA could be isolated from these patients, Simons and colleagues [154] designed degenerate primers encompassing a putative helicase gene (NS3) from GBV-A, GBV-B or HCV-1 in hemi-nested PCR to rule out the cross-contamination with a known virus. Interestingly, though a product of the expected size was derived from a patient in West Africa, it had only 59, 53.7 and 47.9% similarity at the nucleotide level and 64.2, 57.3 and 50.4% at the protein level with GBV-A, HCV-1 and GBV-B, respectively [154]. Due to the phylogenetic similarity of this virus sequence to GBV-A and

GBV-B, it was termed GBV-C [154]. Leary et al. [155] later published the near full-length sequence of GBV-C.

Physical characterization of the GB agent

Physical characterization of the GB agent was done by inoculation of the filtrates in marmosets. Fifty- and 25-nm filtrates induced hepatitis in 100% and 80% of the inoculated marmosets, respectively [156]. The enveloped nature of the GB virus was determined by the loss of infectivity of the virus by treatment with 50% ether at 4 °C for 18 h [156]. Electron-microscopy studies demonstrated the size of the virus particles to be around 20–22 nm in infectious particles [157] (Tab. 2).

HGV

Molecular cloning of HGV

Linnen and colleagues [158] also attempted to identify cryptogenic post-transfusion and community-acquired hepatitis. A patient, PNF2161, identified as 'non-A, non-B hepatitis' [159], was used for the cloning of HGV using λgt11 immunoscreening. A clone-specific PCR assay demonstrated the presence of a unique sequence that was exogenous to *E. coli*, *Saccharomyces cerevisiae*, human patients and healthy control subjects. Eventually, a clone of 9392 nt in length was detected that showed significant similarity with HBV-A and HBV-B. This new virus was termed HGV (Tab. 2). Interestingly, HGV demonstrated 85 and 100% similarity with the GBV-C virus PCR product detected earlier by Simons and colleagues in 1995 [152] at the nucleotide and protein levels, respectively. Comparison of GBV-C and HGV demonstrated that both of them are separate isolates of the same virus [160] sharing 95% sequence homology.

HGV incidence and disease

A PCR test for the detection of HGV RNA in the blood and antibody against the E2 protein are used to screen for HGV-positive patients. A worldwide prevalence of HGV viremia (active infection with virus found in the blood) or infection with HGV ranging from 1.2 to 13% in the healthy blood donors has been observed [161]. HGV has been detected in cryptogenic chronic hepatitis (6%), cryptogenic cirrhosis (7%), chronic hepatitis C (11%), acute seronegative hepatitis (12%), end-stage hepatitis C virus before (24%) and after (28%) liver transplantation, hepatocellular cancer (25%) and fulminant hepatic failure (50%) [162–168].

However, definitive data regarding the relationship between the clinical course of hepatitis and HGV still remains to be seen. Fiordalisi et al. [169] studied the histological features in liver samples from patients infected with GBV-C alone. Among six chronic hepatitis patients positive for GBV-C, chronic active hepatitis was seen in one patient and chronic persistent hepatitis in five others. A wide spectrum of liver disease was observed in 67 GBV-C patients, including steatosis, fibrosis and cirrhosis [170]. HGV RNA was detected in 23% (three out of 13 patients) of transfusion-associated hepatitis cases without any identifiable viral causes [171]. HGV RNA was cleared in two of the patients, whereas it persisted until death in another patient. Alter and colleagues [172] probed the role of HGV in well-defined community-acquired acute hepatitis cases and observed that none of the HGV patients developed chronic hepatitis. Thus, more data is still awaited to conclusively demonstrate HGV as a cause of clinical hepatitis.

TT virus (TTV)

Detection of TTV and molecular cloning

Okamoto and colleagues in Japan discovered this single-stranded DNA virus in "non A to G post-transfusion hepatitis" patients in December 1997 ([173] and reviewed in [174]) (Tab. 1). This novel virus is named after the initials of the first patient from whom it was isolated. TTV is proposed to be a circoviridae [175] and has a 3750–3900-bp genome consisting of circular single-stranded antisense DNA [176] (Tab. 2).

TTV is present in bile, stools, breast milk and saliva [177, 178] and may be transmitted through enteric and parenteral routes [179]. The presence of TTV has been reported in the bone marrow and peripheral blood mononuclear cells [180, 181], bone marrow, lymph node, muscle, thyroid, lung, liver, spleen, pancreas and kidney [182]. TTV infection seems to be prevalent worldwide with infection rates of 13.3–93% [179, 183–187].

TTV-associated disease

TTV was successfully transmitted to chimpanzees [175] without any changes in ALT levels or liver histology, casting doubt over the role of HGV in clinical hepatitis. An overall TTV prevalence of 19.6% was observed in cases of HBV, HCV, human immunodeficiency virus (HIV), non A-E chronic hepatitis or fulminant hepatic failure, and the presence of TTV infection did not seem to affect the clinical course of the liver disease [188]. Contrary to this finding, TTV was associated with varying degrees of liver damage in transfusion-associated hepatitis patients [189, 190]. Thus, a definitive correlation between liver disease and TTV remains to be clarified [187, 191].

SEN virus (SENV)

Primi and colleagues identified a novel virus from an HIV-infected drug user and named the virus after the patient (reviewed in [174, 192]) (Tab. 1). Both parenteral [193] and vertical transmission [194] of the virus are reported. Although eight strains of SENV have been reported so far [195], most of the clinical studies are focused on strains D and H due to their link with non-A, non-E hepatitis. These strains of SENV were detected in 32% of fulminant hepatitis, 17% of acute hepatitis, 27% of chronic hepatitis, 31% of cirrhotic and 33% of autoimmune hepatitis cases [196]. SENV incidence was significantly higher (30%) in patients receiving transfusions than in non-transfused controls (3%) [193]. However, more data are needed to prove a direct causal relationship of SENV with hepatitis.

Morbidity and mortality of chronic viral hepatitis

Chronic viral hepatitis is the second most frequent cause of cirrhosis after alcohol abuse in the USA and in Western Europe. In 2001 10 615 cases of HAV infection (incidence of 3.8/100 000), 7844 cases of HBV infection (incidence of 2.8/100 000) and 3977 cases of HCV infection (incidence of 0.6/100 000) were reported in the USA [197]. HAV, HBV and HCV are responsible for 31.3, 4.9 and 1.8% of infections in the USA, respectively [48, 198, 199]. Chronic HBV infections are observed in 5% of the world population and cause the most common malignancies [200]. In the USA alone, there are approximately 5000 deaths from chronic liver disease per year due to HBV, and 8000–10 000 deaths per year overall due to chronic liver disease.

Conclusion

Since the discovery of the first viral cause of hepatitis in 1963 (see Tab. 1), significant progress has been made in understanding the clinical, epidemiological and viral characteristics of hepatitis (see Tab. 2). Understanding the causal relationship between a virus and hepatitis, coupled with the development of tools for detecting such viruses in the blood supply, has led to a dramatic drop in the incidence of transfusion-related HBV and HCV infection. The development of effective prophylactic vaccines for HAV for use in endemic areas of infection and for HBV infection for the general population has also led to a decrease in the incidence of viral hepatitis. For patients with chronic HBV and HCV infections and liver disease, a concerted effort has been made to develop better therapeutic and preventive measures for viral hepatitis, including antiviral therapy and surgical interventions such as liver transplantations. There is still a need to focus on the pathogenesis, prevention and design of rational drug therapy for hepatitis B and C viruses to fulfill the unmet needs. The incidence of a significant number of hepatitis cases with an infectious etiology without any relationship to any known viral agents stresses the need for more research in the areas of epidemiology, virus isolation and characterization.

References

1 Hollinger F (1996) Hepatitis B virus. In: DM Knipe, BN Fields, PM Howley et al. (eds): *Field's Virology.* Lippincott-Raven Publishers, Philadelphia, 2739–2807
2 Zimmerman HJ (1978) Drug-induced liver disease. *Drugs* 16: 25–45
3 Lyons A (1978) Ancient Civilizations. In: RJ II Petrucelli, AS Lyons (eds): *Medicine an Illustrated history.* HN Abrams, New York, 57–149
4 Zuckerman AJ (1972) Laboratory investigation into the etiology of human viral hepatitis. *Brit Med Bull* 28: 134–137
5 Lurman A (1885) Eine Icterusepidemia. *Berliner Klinische Wochenscrift* 2: 20–23
6 Blumer GB (1923) Infectious jaundice in the United States. *JAMA* 81: 353–358
7 Smart C (1888) *Medical history, part III.* The medical and surgical history of the war of the rebellion, Vol I. Government Printing Office, Washington, D.C., *Medical Volume III*, 874–879
8 Willcox W (1916) The epidemic jaundice of campaigns. *Brit Med J* 1: 671–675
9 Martin C (1917) Concerning the pathology and etiology of the infectious jaundice common at the Dardenelles. *Brit Med J* 1: 445
10 Flaum A, Malmrose H, Persson E (1926) Eine Nosocomiale Iktarus Epidemic. *Acta Med Scand* 16 (Suppl): 544–553
11 Hartfall SJ, Garland HG, Goldie W (1937) Gold treatment of arthritis: a review of 900 cases. *Lancet* 2: 784–788
12 MacCallum F (1947) Homologous serum Jaundice. *Lancet* ii: 691–692
13 Beeson P (1943) Jaundice occurring one to four months after transfusion of blood or plasma: report of seven cases. *JAMA* 121: 1332–1334
14 Cockayne EA (1912) Catarrhal jaundice, sporadic and epidemic, and its relation to acute yellow atrophy of the liver. *Q J Med* 6: 1–29
15 Cuthbert JA (2001) Hepatitis A: old and new. *Clin Microbiol Rev* 14: 38–58
16 Findlay GM, Dunlop JL, Brown HC (1931) Observations on epidemic catarrhal jaundice. *Trans R Soc Trop Med Hyg* 25: 7–24
17 Purcell RH (1993) The discovery of the hepatitis viruses. *Gastroenterology* 104: 955–963
18 Voegt H (1942) Zur aetiologie der hepatitis epidemica. *Muench Med Wochenschr* 89: 76–79
19 Cameron JDS (1943) Infective Hepatitis. *Q J Med* 12: 139–155
20 Havens WP Jr, Ward R, Drill VA, Paul JR (1944) Experimental production of hepatitis by feeding icterogenic materials. *Proc Soc Exp Biol Med* 57: 206–208
21 Paul JR, Havens WP, Sabin AB, Philip CB (1945) Transmission experiments in serum jaundice and infectious hepatitis. *JAMA* 128: 911–925
22 Paul JR, Havens WP Jr (1946) Recent advances in the study of infectious hepatitis and serum jaundice. *Trans Assoc Am Phys* 59: 133–141
23 Hillis WD (1961) An outbreak of infectious hepatitis among chimpanzee handlers at a United States Air Force Base. *Am J Hygien* 73: 316–328
24 Hillis WD (1963) Viral hepatitis associated with sub-human primates. *Transfusion* 59: 445–454
25 Anonymous (1947) Homologous serum hepatitis. *Lancet* ii: 691–692
26 Zuckerman AJ (1993) Viral hepatitis. *Transfus Med* 3: 7–19
27 Feinstone SM, Kapikian AZ, Purceli RH (1973) Hepatitis A: detection by immune electron microscopy of a virus like antigen associated with acute illness. *Science* 182: 1026–1028
28 Blumberg BS, Alter HJ, Visnich S (1965) A "New" Antigen in leukemia sera. *JAMA* 191: 541–546
29 MacCallum FO, Bradley WH (1944) Transmission of infective hepatitis to human volunteers. *Lancet* 2: 228
30 Krugman S, Giles JP, Hammond J (1967) Infectious hepatitis. Evidence for two distinctive clinical, epidemiological, and immunological types of infection. *JAMA* 200: 365–373
31 Krugman S (1971) Experiments at the Willow brook State School. *Lancet* 1: 966–967
32 Krugman S, Giles JP (1973) Viral hepatitis, type B (MS-2-strain). Further observations on natural history and prevention. *N Engl J Med* 288: 755–760
33 Krugman S, Giles JP (1970) Viral hepatitis. New light on an old disease. *JAMA* 212: 1019–1029
34 Szmuness W, Dienstag JL, Purcell RH, Prince AM, Stevens CE, Levine RW (1977) Hepatitis type A and hemodialysis: a seroepidemiologic study in 15 U.S. centers. *Ann Intern Med* 87: 8–12
35 Frosner GG, Deinhardt F, Scheid R, Gauss-Muller V, Holmes N, Messelberger V, Siegl G, Alexander JJ (1979) Propagation of human hepatitis A virus in a hepatoma cell line. *Infection* 7: 303–305

36 Gust ID, Lehmann NI, Crowe S, McCrorie M., Locarnini SA, Lucas CR (1985) The origin of the HM175 strain of hepatitis A virus. *J Infect Dis* 151: 365–367
37 Lemon SM (1985) Type A viral hepatitis. New developments in an old disease. *N Engl J Med* 313: 1059–1067
38 Krugman S (1970) Viral hepatitis: new clinical, epidemiological and immunological concepts. *Calif Med* 113: 57–59
39 Hsu HH, Feinstone S, Hoofnagle J (1990) Acute viral hepatitis. In: G Mandell, J Bennett, R Dolin (eds): *Principles and Practice of Infectious diseases*. Churchill Livingstone, New York, 1136–1153
40 Wong DC, Purcell RH, Sreenivasan MA, Prasad SR, Pavri KM (1980) Epidemic and endemic hepatitis in India: evidence for a non-A, non-B hepatitis virus etiology. *Lancet* 2: 876–879
41 Bradley DW, Krawczynski K, Purdy MA (1993) Epidemiology, Natural history and experimental models of hepatitis E virus. In: AJ Zuckerman, HC Thomas (eds): *Viral Hepatitis Scientific Basis and Clinical Management*. Churchill Livingstone, Edinburgh, 379–384
42 Viswanathan, R (1957) Infectious hepatitis in Delhi (1955–56): A critical study; epidemiology. *Ind J. Med Res* 45: 1–30
43 Sreenivasan MA, Banerjee K, Pandya PG, Kotak RR, Pandya PM, Desai NJ, Vaghela LH (1978) Epidemiological investigations of an outbreak of infectious hepatitis in Ahmedabad city during 1975–76. *Ind J. Med Res* 67: 197–206
44 Balayan MS, Andjaparidze AG, Savinskaya SS, Ketiladze ES, Braginsky DM, Savinov PP, Poleschuk VF (1983) Evidence for a virus in non-A, non-B hepatitis transmitted via the fecal-oral route. *Intervirology* 20: 23–31
45 Bradley DW, Krawczynski K, Cook EH Jr, McCaustland KA, Humphrey CD, Spelbring JE, Myint H, Maynard JE (1987) Enterically transmitted non-A, non-B hepatitis: serial passage of disease in cynomolgus macaques and tamarins and recovery of disease-associated 27- to 34-nm virus like particles. *Proc Natl Acad Sci USA* 84: 6277–6281
46 Bradley DW, McCaustland KA, Cook EH, Schable CA, Ebert JW, Maynard JE (1988) Aetiological agent of enterically transmitted non-A, non-B hepatitis. *J Gen Virol* 69: 731–738
47 Bradley DW (1990) Enterically transmitted non-A, non-B hepatitis. *Brit Med Bull* 46: 442–461
48 Centers for Disease Control and Prevention (1999) Prevention of hepatitis A through active or passive immunization: recommendations of the Advisory Committee on Immunization Practices (ACIP). *MMWR* 48 (RR12) 1–31
49 Reyes GR, Purdy MA, Kim JP, Luk KC, Young LM, Fry KE, Bradley DW (1990) Isolation of a cDNA from the virus responsible for enterically transmitted non-A, non-B hepatitis. *Science* 247: 1335–1339
50 Khuroo MS (1980) Study of an epidemic of non-A, non-B hepatitis. Possibility of another human hepatitis virus distinct from post-transfusion non-A, non-B type. *Am J Med* 68: 818–824
51 Khuroo MS, Duermeyer W, Zargar SA, Ahanger MA, Shah MA (1983) Acute sporadic non-A, non-B hepatitis in India. *Am J Epidemiol* 118: 360–364
52 Acheson DW, Leen CL, Tariq WU, Mandal BK (1987) Non-parenterally transmitted non-A, non-B hepatitis in children in the UK. *Lancet* 2: 798–799
53 Caredda F, Antinori S, Pastecchia C, Coppin P, Moroni M (1988) Sporadic non-A, non-B hepatitis. *Lancet* 1: 58
54 Deka N, Sharma MD, Mukerjee R (1994) Isolation of the novel agent from human stool samples that is associated with sporadic non-A, non-B hepatitis. *J Virol* 68: 7810–7815
55 Alter HJ, Seeff L (1993) Transfusion associated hepatitis, In: HC Thomas (eds): *Viral hepatitis: Scientific Basis and Clinical management*, AJ Zuckerman. Churchill Livingstone, Edinburgh, 467–499
56 Wroblewski F, Ladue JS (1955) Serum glutamic oxalacetic transaminase activity as an index of liver cell injury: a preliminary report. *Ann Intern Med* 43: 345–360
57 Wroblewski F, Ladue JS (1956) Serum glutamic oxalacetic aminopherase (transaminase) in hepatitis. *J Am Med Assoc* 160: 1130–1134
58 Shimizu Y, Kitamoto O (1963) The incidence of viral hepatitis after blood transfusions. *Gastroenterology* 44: 740–744
59 Hampers CL, Prager D, Senior JR (1964) Post-Transfusion Anicteric Hepatitis. *N Engl J Med* 271: 749–754
60 Creutzfeldt W, Severidt HJ, Schmitt H, Gallasch E, Arndt HJ, Brachmann H, Schmidt G, Tschaepe U (1966) Incidence and course of icteric and anicteric transfusion hepatitis. *Ger Med Mon* 11: 469–476

61 Walsh JH, Purcell RH, Morrow AG, Chanock RM, Schmidt PJ (1970) Post transfusion hepatitis after open-heart operations. Incidence after the administration of blood from commercial and volunteer donor populations. *JAMA* 211: 261–265
62 Blumberg BS (1964) Polymorphisms of the Serum Proteins and the Development of Iso-Precipitins in Transfused Patients. *Bull N Y Acad Med* 40: 377–386
63 Blumberg BS, Gerstley BSJ, Hungerford DA, London WT, Sutnick A (1967) A serum antigen (Australia antigen) in Down's syndrome, leukemia, and hepatitis. *Ann Intern Med* 66: 924–931
64 London WT, Sutnick AI, Blumberg BS (1969) Australia antigen and acute viral hepatitis. *Ann Intern Med* 70: 55–59
65 Blumberg BS (1976) Australia antigen and the biology of Hepatitis B. In *Nobel Lecture* December 3, 1976
66 Okochi K, Murakami S (1968) Observations on Australia antigen in Japanese. *Vox Sang* 15: 374–385
67 Okochi K, Murakami S, Ninomiya K, Kaneko M (1970) Australia antigen, transfusion and hepatitis. *Vox Sang* 18: 289–300
68 Prince AM (1968) An antigen detected in the blood during the incubation period of serum hepatitis. *Proc Natl Acad Sci USA* 60: 814–821
69 Gocke DJ, Kavey NB (1969) Hepatitis antigen. Correlation with disease and infectivity of blood-donors. *Lancet* 1: 1055–1059
70 Senior JR, Sutnick AI, Goeser E, London WT, Dahlke MB, Blumberg BS (1974) Reduction of post-transfusion hepatitis by exclusion of Australia antigen from donor blood in an urban public hospital. *Am J Med Sci* 267: 171–177
71 Magnius LO, Espmark A (1972) A new antigen complex co-occurring with Australia antigen. *Acta Pathol Microbiol Scand B Microbiol Immunol* 80: 335–337
72 Dane DS, Cameron CH, Briggs M (1970) Virus-like particles in serum of patients with Australia-antigen-associated hepatitis. *Lancet* 1: 695–698
73 Kaplan PM, Greenman RL, Gerin JL, Purcell RH, Robinson WS (1973) DNA polymerase associated with human hepatitis B antigen. *J Virol* 12: 995–1005
74 Robinson WS, Clayton DA, Greenman RL (1974) DNA of a human hepatitis B virus candidate. *J Virol* 14: 384–391
75 Robinson WS, Greenman RL (1974) DNA polymerase in the core of the human hepatitis B virus candidate. *J Virol* 13: 1231–1236
76 Sells MA, Chen ML, Acs G (1987) Production of hepatitis B virus particles in Hep G2 cells transfected with cloned hepatitis B virus DNA. *Proc Natl Acad Sci USA* 84: 1005–1009
77 Federal Report 37, 106, Title 24, part 73.775, 10937
78 Aach RD, Lander JJ, Sherman LA et al. (1978) Transfusion transmitted viruses: Interim analysis of hepatitis among transfused and non transfused patients. In *Viral hepatitis*, CS Vyas GN, Ahmed R (eds): Franklin Institute press, Philadelphia, 383–396
79 Alter MJ (1993) Community acquired viral hepatitis B and C in the United States. *Gut* 34: S17–19
80 Brandt KH, Meulendijk PN, Poulie NJ, Schalm L, Schultz MJ, Zana HC (1963) De warde van de transaminase, bepalingin donorbloed. *Geneesk* 107: 2313–2332
81 Hadziyannis (1978) *Post transfusion hepatitis*. PhD Thesis. Athens
82 Cherubian CE, Prince AM (1971) Serum hepatitis specific antigen (SH) in commercial and volunteer sources of blood. *Transfusion* 11: 25–27
83 Tabor E, Gerety RJ (1980) Hepatitis B e antigen in volunteer and paid blood donors. *Transfusion* 20: 192–198
84 Federal Register 43 (1978), 43: 2142
85 Lee WM (1997) Hepatitis B virus infection. *N Engl J Med* 337: 1733–1745
86 Lai KN, Lai FM (1991) Clinical features and the natural course of hepatitis B virus-related glomerulopathy in adults. *Kidney Int Suppl* 35: S40–S45
87 Feinstone SM, Kapikian AZ, Purcell RH, Alter HJ, Holland PV (1975) Transfusion-associated hepatitis not due to viral hepatitis type A or B. *N Engl J Med* 292: 767–770
88 Fetelson M (2002) Background of hepatitis C virus. In: M Fetelson (ed): *Hepatitis C Virus from laboratory to Clinic*. Cambridge University Press, Cambridge, 3–17
89 Prince AM, Brotman B, Grady GF, Kuhns WJ, Hazzi C, Levine RW, Millian SJ (1974) Long-incubation post-transfusion hepatitis without serological evidence of exposure to hepatitis-B virus. *Lancet* 2: 241–246
90 Alter HJ, Holland PV, Morrow AG, Purcell RH, Feinstone SM, Moritsugu Y (1975) Clinical and

serological analysis of transfusion-associated hepatitis. *Lancet* 2: 838–841
91 Krugman S, Goecke DS (1978) Acute Viral Hepatitis. In: LH Smith Jr (ed): *Viral Hepatitis*. Saunders, Philadelphia, 30–47
92 Tabor E, Gerety RJ, Drucker JA, Seeff LB, Hoofnagle JH, Jackson DR, April M, Barker LF, Pineda-Tamondong G (1978) Transmission of non-A, non-B hepatitis from man to chimpanzee. *Lancet* 1: 463–466
93 Bradley DW, Cook EH, Maynard JE, McCaustland KA, Ebert JW, Dolana GH, Petzel RA, Kantor RJ, Heilbrunn A, Fields HA, Murphy BL (1979) Experimental infection of chimpanzees with antihemophilic (factor VIII) materials: recovery of virus-like particles associated with non-A, non-B hepatitis. *J Med Virol* 3: 253–269
94 Wyke RJ, Tsiquaye KN, Thornton A, White Y, Portmann B, Das PK, Zuckerman AJ, Williams R (1979) Transmission of non-A non-B hepatitis to chimpanzees by factor-IX concentrates after fatal complications in patients with chronic liver disease. *Lancet* 1: 520–524
95 Bradley DW (1984) Transmission, etiology, and pathogenesis of viral hepatitis non-A, non-B in non-human primates. In: FV Chisari (ed): *Advances in hepatitis research*. Maason, New York, 268–280
96 Bradley DW, Maynard JE (1983) Non-A, non-B hepatitis: research progress and current perspectives. *Dev Biol Stand* 54: 463–473
97 Feinstone SM, Alter HJ, Dienes HP, Shimizu Y, Popper H, Blackmore D, Sly D, London WT, Purcell RH (1981) Non-A, non-B hepatitis in chimpanzees and marmosets. *J Infect Dis* 144: 588–598
98 Yoshizawa H, Itoh Y, Iwakiri S, Kitajima K, Tanaka A, Tachibana T, Nakamura T, Miyakawa Y, Mayumi M (1982) Non-A, non-B (type 1) hepatitis agent capable of inducing tubular ultra structures in the hepatocyte cytoplasm of chimpanzees: inactivation by formalin and heat. *Gastroenterology* 82: 502–506
99 Shimizu YK, Feinstone SM, Purcell RH, Alter HJ, London WT (1979) Non-A, non-B hepatitis: ultrastructural evidence for two agents in experimentally infected chimpanzees. *Science* 205: 197–200
100 Bradley DW, Maynard JE, Cook EH, Ebert JW, Gravelle CR, Tsiquaye KN, Kessler H, Zuckerman AJ, Miller MF, Ling C, Overby LR (1980) Non-A/non-B hepatitis in experimentally infected chimpanzees: cross-challenge and electron microscopic studies. *J Med Virol* 6: 185–201
101 Pfeifer U, Thomssen R, Legler K, Bottcher U, Gerlich W, Weinmann E, Klinge O (1980) Experimental non-A, non-B hepatitis: four types of cytoplasmic alteration in hepatocytes of infected chimpanzees. *Virchows Arch B Cell Pathol Incl Mol Pathol* 33: 233–243
102 Tsiquaye KN, Bird RG, Tovey G, Wyke RJ, Williams R, Zuckerman AJ (1980) Further evidence of cellular changes associated with non-A, non-B hepatitis. *J Med Virol* 5: 63–71
103 Feinstone SM, Mihalik KB, Kamimura T, Alter HJ, London WT, Purcell RH (1983) Inactivation of hepatitis B virus and non-A, non-B hepatitis by chloroform. *Infect Immunity* 41: 816–821
104 Bradley DW, Maynard JE (1983) Posttransfusion non-A, non-B hepatitis: physicochemical properties of two distinct agents. *J Infect Dis* 148: 254–265
105 Tabor E, Gerety RJ (1980) Inactivation of an agent of human non-A, non-B hepatitis by formalin. *J Infect Dis* 142: 767–770
106 Bradley DW, McCaustland KA, Cook EH, Schable CA, Ebert JW, Maynard JE (1985) Post transfusion non-A, non-B hepatitis in chimpanzees. Physico-chemical evidence that the tubule-forming agent is a small, enveloped virus. *Gastroenterology* 88: 773–779
107 He LF, Alling D, Popkin T, Shapiro M, Alter HJ, Purcell RH (1987) Determining the size of non-A, non-B hepatitis virus by filtration. *J Infect Dis* 156: 636–640
108 Merz PA, Somerville RA, Wisniewski HM, Manuelidis L, Manuelidis EE (1983) Scrapie-associated fibrils in Creutzfeldt-Jakob disease. *Nature* 306: 474–476
109 Prusiner SB, Groth DF, Bolton DC, Kent SB, Hood LE (1984) Purification and structural studies of a major scrapie prion protein. *Cell* 38: 127–134
110 Brechot C, Hadchouel M, Scotto J, Fonck M, Potet F, Vyas GN, Tiollais P (1981) State of hepatitis B virus DNA in hepatocytes of patients with hepatitis B surface antigen-positive and -negative liver diseases. *Proc Natl Acad Sci USA* 78: 3906–3910
111 Shafritz DA, Lieberman HM, Isselbacher KJ, Wands JR (1982) Monoclonal radioimmunoassays for hepatitis B surface antigen: demonstration of hepatitis B virus DNA or related sequences in serum and viral epitopes in immune complexes. *Proc Natl Acad Sci USA* 79: 5675–5679
112 Seto B, Coleman WG Jr, Iwarson S, Gerety RJ (1984) Detection of reverse transcriptase activity

in association with the non-A, non-B hepatitis agent(s). *Lancet* 2: 941–943
113 Prince AM, Huima T, Williams BA, Bardina L, Brotman B (1984) Isolation of a virus from chimpanzee liver cell cultures inoculated with sera containing the agent of non-A, non-B hepatitis. *Lancet* 2: 1071–1075
114 Iwarson S, Schaff Z, Mitchell F, Gerety RJ (1985) Non-A, non-B hepatitis virus. *Lancet* 1: 171
115 Bradley D (2000) Studies of Non-A, Non-B Hepatitis and characterization of the Hepatitis C virus in Chimpanzees. In: CH Hagedorn, C Rice (eds): *The Hepatitis C Viruses*. Springer, 1–23
116 Choo QL, Kuo G, Weiner AJ, Overby LR, Bradley DW, Houghton M (1989) Isolation of a cDNA clone derived from a blood-borne non-A, non-B viral hepatitis genome. *Science* 244: 359–362
117 Houghton M, Weiner A, Han J, Kuo G, Choo QL (1991) Molecular biology of the hepatitis C viruses: implications for diagnosis, development and control of viral disease. *Hepatology* 14: 381–388
118 Weiner AJ, Brauer MJ, Rosenblatt J, Richman KH, Tung J, Crawford K, Bonino F, Saracco G, Choo QL, Houghton M et al. (1991) Variable and hypervariable domains are found in the regions of HCV corresponding to the flavivirus envelope and NS1 proteins and the pestivirus envelope glycoproteins. *Virology* 180: 842–848
119 Miyamoto H, Okamoto H, Sato K, Tanaka T, Mishiro S (1992) Extraordinarily low density of hepatitis C virus estimated by sucrose density gradient centrifugation and the polymerase chain reaction. *J Gen Virol* 73: 715–718
120 Hijikata M, Shimizu YK, Kato H, Iwamoto A, Shih JW, Alter HJ, Purcell RH, Yoshikura H (1993) Equilibrium centrifugation studies of hepatitis C virus: evidence for circulating immune complexes. *J Virol* 67: 1953–1958
121 Thomssen R, Bonk S, Propfe C, Heermann KH, Kochel HG, Uy A (1992) Association of hepatitis C virus in human sera with beta-lipoprotein. *Med Microbiol Immunol (Berl)* 181: 293–300
122 Yoshikura H, Hijikata M, Nakajima N, Shimizu YK (1996) Replication of hepatitis C virus. *J Viral Hepat* 3: 3–10
123 Li X, Jeffers LJ, Shao L, Reddy KR, de Medina M, Scheffel J, Moore B, Schiff ER (1995) Identification of hepatitis C virus by immunoelectron microscopy. *J Viral Hepat* 2: 227–234
124 Kaito M, Watanabe S, Tsukiyama-Kohara K, Yamaguchi K, Kobayashi Y, Konishi M, Yokoi M, Ishida S, Suzuki S, Kohara M (1994) Hepatitis C virus particle detected by immunoelectron microscopic study. *J Gen Virol* 75: 1755–1760
125 Prince AM, Huima-Byron T, Parker TS, Levine DM (1996) Visualization of hepatitis C virions and putative defective interfering particles isolated from low-density lipoproteins. *J Viral Hepat* 3: 11–17
126 Alter HJ, Purcell RH, Holland PV, Morrow AG (1978) Non-A. non-B hepatitis: a review and interim report of an ongoing prospective study. In: Cohen SN, GN Vyas, R Schmid (eds): *Viral Hepatitis*. Franklin Institute Press, Philadelphia, 359–381
127 Rizzetto M, Canese MG, Arico S, Crivelli O, Trepo C, Bonino F, Verme G (1977) Immunofluorescence detection of new antigen-antibody system (delta/anti-delta) associated to hepatitis B virus in liver and in serum of HBsAg carriers. *Gut* 18: 997–1003
128 Rizzetto M, Verme G (1985) Delta hepatitis-present status. *J Hepatol* 1: 187–193
129 Rizzetto M, Hoyer B, Canese MG, Shih JW, Purcell RH, Gerin JL (1980) Delta Agent: association of delta antigen with hepatitis B surface antigen and RNA in serum of delta-infected chimpanzees. *Proc Natl Acad Sci USA* 77: 6124–6128
130 Ponzetto A, Hoofnagle JH, Seeff LB (1984) Antibody to the hepatitis B virus-associated delta-agent in immune serum globulins. *Gastroenterology* 87: 1213–1216
131 Ponzetto A, Forzani B, Parravicini PP, Hele C, Zanetti A, Rizzetto M (1985) Epidemiology of hepatitis delta virus (HDV) infection. *Eur J Epidemiol* 1: 257–263
132 Buitrago B, Popper H, Hadler SC, Thung SN, Gerber MA, Purcell RH, Maynard JE (1986) Specific histologic features of Santa Marta hepatitis: a severe form of hepatitis delta-virus infection in northern South America. *Hepatology* 6: 1285–1291
133 Zyzik E, Ponzetto A, Forzani B, Hele C, Heermann K, Gerlich WH (1987) Proteins of hepatitis virus in serum and liver. In: *Hepadnaviruses*, Robinson W, Koike K, Will H (eds): Alan R Liss, New York, 565–577
134 Bonino F, Heermann KH, Rizzetto M, Gerlich WH (1986) Hepatitis delta virus: protein composition of delta antigen and its hepatitis B virus-derived envelope. *J Virol* 58: 945–950
135 Rizzetto M, Verme G, Recchia S, Bonino F, Farci P, Arico S, Calzia R, Picciotto A, Colombo M, Popper H (1983) Chronic hepatitis in carriers of hepatitis B surface antigen, with intrahepatic

expression of the delta antigen. An active and progressive disease unresponsive to immunosuppressive treatment. *Ann Intern Med* 98: 437–441
136 Sagnelli E, Piccinino F, Pasquale G, di Costanzo MG, Franzese R, Peinetti P (1984) Delta agent infection: an unfavourable event in HBsAg positive chronic hepatitis. *Liver* 4: 170–176
137 Rizzetto MDM, Durazzo M (1991) Hepatitis Delta virus (HDV) infections. Epidemiological and clinical heterogeneity. *J Hepatol* 13: S116–S118
138 Hoofnagle JH (1986) Type D hepatitis and hepatitis Delta virus. In: HC Thomas, EA Jones (eds): *Recent Advances in Hepatology*. Churchill Livingstone, Edinburgh, 73–92
139 Hoofnagle JH (1989) Type D (Delta) hepatitis. *JAMA* 261: 1321–1325
140 Smedile A, Rosina F, Saracco G, Chiaberge E, Lattore V, Fabiano A, Brunetto MR, Verme G, Rizzetto M, Bonino F (1991) Hepatitis B virus replication modulates pathogenesis of hepatitis D virus in chronic hepatitis D. *Hepatol* 13: 413–416
141 Smedile A, Rosina F, Chiaberge E, Lattore V, Saracco G, Brunetto MR, Bonino F, Verme G, Rizzetto M (1991) Presence and significance of hepatitis B virus replication in chronic type D hepatitis. *Prog Clin Biol Res* 364: 185–195
142 Smedile A, Chiaberge E, Rosina F et al. (1991) HBV replication influences the course of chronic hepatitis B virus replication in chronic type D hepatitis. In: SM Lemon, FB Hollinger, HS Margolis (eds): *Viral Hepatitis and Liver Disease*. Williams and Wilkins, Baltimore, 480–483
143 Conjeevaram HS, Di Bisceglie AM (1993) Natural history of hepatitis D virus. In: HC Thomas, AJ Zuckerman (eds): *Viral Hepatitis: Scientific Basis and Clinical Basis*. Churchill Livingstone, Edinburgh, 341–349
144 Bisceglie AD (1993) Epidemiology and diagnosis of hepatitis D virus. In: AJ Zuckerman, HC Thomas (eds): *Viral Hepatitis Scientific Basis and Clinical Management*. Churchill Livingstone, Edinburgh, 351–361
145 Miyakawa Y, Mayumi M (1997) Hepatitis G virus–a true hepatitis virus or an accidental tourist? *N Engl J Med* 336: 795–796
146 Hollinger FB, Mushahwar I (2000) New candidate hepatitis viruses: In: *Summary of plenary session VIII. Tenth International Symposium on Viral Hepatitis and Liver Disease*. International Medical Press, London
147 Deinhardt F, Holmes AW, Capps RB, Popper H (1967) Studies on the transmission of human viral hepatitis to marmoset monkeys. I. Transmission of disease, serial passages, and description of liver lesions. *J Exp Med* 125: 673–688
148 Alter MJ, Mast EE (1994) The epidemiology of viral hepatitis in the United States. *Gastroenterol Clin North Am* 23: 437–455
149 Marcellin P, Martinot-Peignoux M, Gabriel F, Branger M, Degott C, Elias A, Xu LZ, Larzul D, Erlinger S, Benhamou JP (1993) Chronic non-B, non-C hepatitis among blood donors assessed with HCV third generation tests and polymerase chain reaction. *J Hepatol* 19: 167–170
150 Thiers V, Lunel F, Valla D, Azar N, Fretz C, Frangeul L, Huraux JM, Opolon P, Brechot C (1993) Post-transfusional anti-HCV-negative non-A non-B hepatitis (II) serological and polymerase chain reaction analysis for hepatitis C and hepatitis B viruses. *J Hepatol* 18: 34–39
151 Simons JN, Desai SM, Mushahwar IK (2000) The GB viruses. In: CH Hagedorn, CM Rice (eds): *The Hepatitis C Viruses*. Springer, 342–375
152 Simons JN, Leary TP, Dawson GJ, Pilot-Matias TJ, Muerhoff AS, Schlauder GG, Desai SM, Mushahwar IK (1995) Isolation of novel virus-like sequences associated with human hepatitis. *Nat Med* 1: 564–569
153 Pilot-Matias TJ, Muerhoff AS, Simons JN, Leary TP, Buijk SL, Chalmers ML, Erker JC, Dawson GJ, Desai SM, Mushahwar IK (1996) Identification of antigenic regions in the GB hepatitis viruses GBV-A, GBV-B, and GBV-C. *J Med Virol* 48: 329–338
154 Simons JN, Pilot-Matias TJ, Leary TP, Dawson GJ, Desai SM, Schlauder GG, Muerhoff AS, Erker JC, Buijk SL, Chalmers ML et al. (1995) Identification of two flavivirus-ike genomes in the GB hepatitis agent. *Proc Natl Acad Sci USA* 92: 3401–3405
155 Leary TP, Muerhoff AS, Simons JN, Pilot-Matias TJ, Erker JC, Chalmers ML, Schlauder GG, Dawson GJ, Desai SM, Mushahwar IK (1996) Sequence and genomic organization of GBV-C: a novel member of the flaviviridae associated with human non-A-E hepatitis. *J Med Virol* 48: 60–67
156 Parks WP, Melnick JL (1969) Characterization of marmoset hepatitis virus. *J Infect Dis* 120: 548–559
157 Almeida JD, Deinhardt F, Holmes AW, Peterson DA, Wolfe L, Zuckerman AJ (1976) Morphology of the GB hepatitis agent. *Nature* 261: 608–609

158 Linnen J, Wages J Jr, Zhang-Keck ZY, Fry KE, Krawczynski KZ, Alter H, Koonin E, Gallagher M, Alter M, Hadziyannis S et al. (1996) Molecular cloning and disease association of hepatitis G virus: a transfusion-transmissible agent. *Science* 271: 505–508
159 Alter MJ, Margolis HS, Krawczynski K, Judson FN, Mares A, Alexander WJ, Hu PY, Miller JK, Gerber MA, Sampliner RE et al. (1992) The natural history of community-acquired hepatitis C in the United States. The Sentinel Counties Chronic non-A, non-B Hepatitis Study Team. *N Engl J Med* 327: 1899–1905
160 Zuckerman AJ (1996) Alphabet of hepatitis viruses. *Lancet* 347: 558–559
161 Moaven LD, Hyland CA, Young IF, Bowden DS, McCaw R, Mison L, Locarnini SA (1996) Prevalence of hepatitis G virus in Queensland blood donors. *Med J Aust* 165: 369–371
162 Aikawa T, Sugai Y, Okamoto H (1996) Hepatitis G infection in drug abusers with chronic hepatitis C. *N Engl J Med* 334: 195–196
163 Masuko K, Mitsui T, Iwano K, Yamazaki C, Okuda K, Meguro T, Murayama N, Noue T, Tsuda F, Okamoto H, Miyakawa Y, Mayumi M (1996) Infection with hepatitis GB virus C in patients on maintenance hemodialysis. *N Engl J Med* 334: 1485–1490
164 Tanaka E, Alter HJ, Nakatsuji Y, Shih JW, Kim JP, Matsumoto A, Kobayashi M, Kiyosawa K (1996) Effect of hepatitis G virus infection on chronic hepatitis C. *Ann Intern Med* 125: 740–743
165 Yoshiba M, Okamoto H, Mishiro S (1995) Detection of the GBV-C hepatitis virus genome in serum from patients with fulminant hepatitis of unknown aetiology. *Lancet* 346: 1131–1132
166 Berenguer M, Terrault NA, Piatak M, Yun A, Kim JP, Lau JY, Lake JR, Roberts JR, Ascher NL, Ferrell L, Wright TL (1996) Hepatitis G virus infection in patients with hepatitis C virus infection undergoing liver transplantation. *Gastroenterology* 111: 1569–1575
167 Bralet MP, Roudot-Thoraval F, Pawlotsky JM, Bastie A, Tran Van Nhieu J, Duval J, Dhumeaux D, Zafrani ES (1997) Histopathologic impact of GB virus C infection on chronic hepatitis C. *Gastroenterology* 112: 188–192
168 Pessoa MG, Wright TL (1996) Hepatitis G: a virus in search of a disease. *Hepatology* 24: 461–463
169 Fiordalisi G, Zanella I, Mantero G, Bettinardi A, Stellini R, Paraninfo G, Cadeo G, Primi D (1996) High prevalence of GB virus C infection in a group of Italian patients with hepatitis of unknown etiology. *J Infect Dis* 174: 181–183
170 Colombatto P, Randone A, Civitico G, Monti Gorin J, Dolci L, Medaina N, Oliveri F, Verme G, Marchiaro G, Pagni R et al. (1996) Hepatitis G virus RNA in the serum of patients with elevated gamma glutamyl transpeptidase and alkaline phosphatase: a specific liver disease? [corrected]. *J Viral Hepat* 3: 301–306
171 Alter HJ, Nakatsuji Y, Melpolder J, Wages J, Wesley R, Shih JW, Kim JP (1997) The incidence of transfusion-associated hepatitis G virus infection and its relation to liver disease. *N Engl J Med* 336: 747–754
172 Alter MJ, Gallagher M, Morris TT, Moyer LA, Meeks EL, Krawczynski K, Kim JP, Margolis HS (1997) Acute non-A-E hepatitis in the United States and the role of hepatitis G virus infection. Sentinel Counties Viral Hepatitis Study Team. *N Engl J Med* 336: 741–746
173 Okamoto H, Nishizawa T, Kato N, Ukita M, Ikeda H, Iizuka H, Miyakawa Y, Mayumi M (1998) Molecular cloning and characterization of a novel DNA virus (TTV) associated with posttransfusion hepatitis of unknown origin. *Hepatol Res* 10: 1–16
174 Menon KN (2002) Non-A to E hepatitis. *Curr Opin Infect Dis* 15: 529–534
175 Mushahwar IK, Erker JC, Muerhoff AS, Leary TP, Simons JN, Birkenmeyer LG, Chalmers ML, Pilot-Matias TJ, Desai SM (1999) Molecular and biophysical characterization of TT virus: evidence for a new virus family infecting humans. *Proc Natl Acad Sci USA* 96: 3177–3182
176 Kamahora T, Hino S, Miyata H (2000) Three spliced mRNAs of TT virus transcribed from a plasmid containing the entire genome in COS1 cells. *J Virol* 74: 9980–9986
177 Ross RS, Viazov S, Runde V, Schaeffer UW, Roggendorf M (1999) Detection of TT virus DNA in specimens other than blood. *J Clin Virol* 13: 181–184
178 Itoh M, Shimomura H, Fujioka S, Miyake M, Tsuji H, Ikeda F, Tsuij T (2001) High prevalence of TT virus in human blie juice samples: importance of secretion through bile into feces. *Dig Dis Sci* 46: 457–462
179 Kondili LA, Pisani G, Beneduce F, Morace G, Gentili G, Ballati G, Rapicetta M (2001) Prevalence of TT virus in healthy children and thalassemic pediatric and young adult patients. *J Pediatr Gastroenterol Nutr* 33: 629–632
180 Kanda Y, Tanaka Y, Kami M, Saito T, Asai T, Izutsu K, Yuji K, Ogawa S, Honda H, Mitani K et al. (1999) TT virus in bone marrow transplant recipients. *Blood* 93: 2485–2490

181 Zhong S, Yeo W, Tang M, Liu C, Lin XR, Ho WM, Hui P, Johnson PJ (2002) Frequent detection of the replicative form of TT virus DNA in peripheral blood mononuclear cells and bone marrow cells in cancer patients. *J Med Virol* 66: 428–434
182 Okamoto H, Nishizawa T, Takahashi M, Asabe S, Tsuda F, Yoshikawa A (2001) Heterogeneous distribution of TT virus of distinct genotypes in multiple tissues from infected humans. *Virology* 288: 358–368
183 Yan J, Chen LL, Luo YH, Mao YF, He M (2001) High frequencies of HGV and TTV infections in blood donors in Hangzhou. *World. J Gastroenterol* 7: 637–641
184 Ali S, Fevery J, Peerlinck K, Verslype C, Schelstraete R, Gyselinck F, Emonds MP, Vermylen J, Yap SH (2002) TTV infection and its relation to serum transaminases in apparently healthy blood donors and in patients with clotting disorders who have been investigated previously for hepatitis C virus and GBV-C/HGV infection in Belgium. *J Med Virol* 66: 561–566
185 Saback FL, Gomes SA, Niel C (2002) High frequency of mixed TT virus infections in healthy adults and children detected by a simplified heteroduplex mobility assay. *J Virol Meth* 101: 117–125
186 Krekulova L, Rehak V, Killoran P, Madrigal N, Riley LW (2001) Genotypic distribution of TT virus (TTV) in a Czech population: evidence for sexual transmission of the virus. *J Clin Virol* 23: 31–41
187 Kadayifci A, Guney C, Uygun A, Kubar A, Bagci S, Dagalp K (2001) Similar frequency of TT virus infection in patients with liver enzyme elevations and healthy subjects. *Int J Clin Pract* 55: 434–436
188 Masia G, Ingianni A, Demelia L, Faa G, Manconi P, Pilleri E, Ciancio G, Rizzetto AM, Coppola RC (2001) TT virus infection in Italy: prevalence and genotypes in healthy subjects, viral liver diseases and asymptomatic infections by parenterally transmitted viruses. *J Viral Hepat* 8: 384–390
189 Tanaka Y, Hayashi J, Ariyama I, Furusyo N, Etoh Y, Kashiwagi S (2000) Seroepidemiology of TT virus infection and relationship between genotype and liver damage. *Dig Dis Sci* 45: 2214–2220
190 Comar M, Ansaldi F, Morandi L, Dal Molin G, Foschini PM, Croce SL, Bonin S, Stanta G, Tiribelli C, Campello C (2002) *In situ* polymerase chain reaction detection of transfusion-transmitted virus in liver biopsy. *J Viral Hepat* 9: 123–127
191 Chemin I, Zoulim F, Merle P, Arkhis A, Chevallier M, Kay A, Cova L, Chevallier P, Mandrand B, Trepo C (2001) High incidence of hepatitis B infections among chronic hepatitis cases of unknown aetiology. *J Hepatol* 34: 447–454
192 Primi D, Sortini A (2000) Identification and characterization of SEN virus, a family of novel DNA viruses. (Abstract). *Antivir Ther* 5: G7
193 Umemura T, Yeo AE, Sottini A, Moratto D, Tanaka Y, Wang RY, Shih JW, Donahue P, Primi D, Alter HJ (2001) SEN virus infection and its relationship to transfusion-associated hepatitis. *Hepatology* 33: 1303–1311
194 Pirovano S, Bellinzoni M, Matteelli A, Ballerini C, Albertini A, Imberti L (2002) High prevalence of a variant of SENV in intravenous drug user HIV-infected patients. *J Med Virol* 68: 18–23
195 Tanaka Y, Primi D, Wang RY, Umemura T, Yeo AE, Mizokami M, Alter HJ, Shih JW (2001) Genomic and molecular evolutionary analysis of a newly identified infectious agent (SEN virus) and its relationship to the TT virus family. *J Infect Dis* 183: 359–367
196 Shibata M, Wang RY, Yoshiba M, Shih JW, Alter HJ, Mitamura K (2001) The presence of a newly identified infectious agent (SEN virus) in patients with liver diseases and in blood donors in Japan. *J Infect Dis* 184: 400–404
197 *Hepatitis Surveillance Report* (2003) *CDC report number 58*, US Department of Health and Human Services Centers for Disease Control and Prevention, Atlanta, GA
198 McQuillan G (1999) Prevalence of hepatitis B virus infection in the united states: The national Health and Nutrition and Examination Surveys, 1976 through 1994. *AJPH* 89: 14–18
199 Alter MJ, Kruszan-Moran D, Nainan OV, McQuillan GM, Gao F, Moyer LA, Kaslow RA, Margolis HS (1999) The prevalence of hepatitis C virus infection in the United States, 1988 through 1994. *N Engl J Med* 341: 556–562
200 Beasley RP, Hwang LY, Lin CC, Leu ML, Stevens CE, Szmuness W, Chen KP (1982) Incidence of hepatitis B virus infections in preschool children in Taiwan. *J Infect Dis* 146: 198–204
201 Szmuness W, Stevens CE, Horley EJ, Zang EA, Oleszko WR, William DC, Sadovsky R, Morrison JM, Kellner A (1980) Hepatitis B vaccine: demonstration of efficacy in a controlled clinical trial in a high risk population in the United States. *N Engl J Med* 303: 833–841

Hepatitis Prevention and Treatment
Edited by J.M. Colacino and B.A. Heinz

Epidemiology of chronic hepatitis viruses: hepatitis B virus and hepatitis C virus

Lawrence M. Blatt[1] and Myron Tong[2]

[1] *InterMune Inc, Brisbane, CA 94005, USA*
[2] *The Liver Center, Huntington Medical Research Institutes, Pasadena, CA 94005, USA*

Introduction

Chromic infection with one of the prevalent hepatotrophic viruses, namely hepatitis B virus (HBV) or hepatitis C virus (HCV), is a major public-health concern worldwide due to the significant morbidity and mortality they cause. The prevalence of chronic hepatic viral infections is approximately 10 times higher than that of human immunodeficiency virus (HIV). According to the World Health Organization (WHO) report entitled *AIDS Epidemic Update December 2003*, approximately 34–46 million people are currently infected with HIV [1]. More specifically, an estimated 2 billion people worldwide have been infected with HBV, and approximately 300 million individuals have chronic liver disease due to infection with HBV [2]. For HCV, approximately 175 million people are chronically infected, with an estimated 3–4 million new infections occurring each year [3]. Given that HBV and HCV share no genomic sequence identity (HBV and HCV are DNA and RNA viruses, respectively) and that the only commonality between the two viruses at a molecular level is hepatotrophy, it is interesting to note that both viral infections can lead to liver inflammation, fibrosis, cirrhosis and hepatocellular carcinoma (HCC) [4, 5]. Thus, for HBV-infected individuals, an estimated 15–40% of patients with chronic infection will develop liver cirrhosis leading to liver failure and possibly HCC [6]. For chronic HCV infection, an estimated 20–40% of patients progress to liver fibrosis and cirrhosis [7]. For these patients, a history of liver decompensation and serum albumin abnormalities serve as independent risk factors that are associated with development of HCC [8].

One major difference between HBV and HCV infection is the percentage of those patients infected with either virus who go on to have chronic liver disease. For HBV approximately, 15% of patients who are infected go on to chronic liver disease whereas for HCV-infected individuals approximately 80% of infected individuals go on to chronic liver disease.

Mode of transmission

HBV DNA has been detected in blood, semen, saliva, vaginal fluids and breast milk; transmission of HBV can be vertical or horizontal [9]. For HBV, the mode of transmission depends on geographic location, with areas of high HBV prevalence, such as Asia and sub-Saharan Africa, having perinatal transmission and unsafe injections as major causes of new disease [10]. With the advent of sensitive blood-screening methodologies for HBV, unsafe blood transfusions are only a minor source of transmission in the developed world but do play a significant role as a source of transmitted virus in developing countries [10]. In areas of lower HBV sero-prevalence such as North America and Western Europe, transmission is primarily though illicit intravenous drug abuse and through unsafe sexual contact with infected partners [11]. The most common risk factors associated with acute hepatitis B in the United States include heterosexual contact, male homosexual contact and injection drug use. However, the largest pool of chronic HBV-infected individuals in the United States is Asian-Americans, especially those that have recently emigrated from the Asian continent. In the United States, HBV infection is more often encountered in the African-American than in the Caucasian or Hispanic populations.

In recent years, transmission of HBV has been greatly reduced due to the availability of a recombinant subunit vaccine that has demonstrated significant success in lowering the incidence in perinatal transmission, especially in key endemic areas such as Taiwan and Korea [12, 13].

As in the case of HBV, HCV can be found in blood and in bodily secretions and fluids such as semen [14], saliva [15–17] and breast milk [18] although transmission of HCV is primarily through blood-to- blood contact. In 1992, Alter and co-workers [19] published a study that examined the demographic characteristic of patients in the United States with HCV infection. Among the most important findings of the study was the determination that although it had previously been assumed that most transmission of non-A, non-B hepatitis (i.e. HCV) was the result of exposure to contaminated blood products through blood transfusion, a large proportion of the cases in the Alter study had no known risk factors for acquiring HCV. More specifically, the Alter study demonstrated that 36% of the patients infected with HCV had a history of using injected illegal drugs, 29% had no known exposure but could be classified as low socioeconomic class, 14% had no known risk factors, 13% had exposure through sexual or household contacts, 4% had a history of post-transfusion hepatitis, 2% had occupational exposure as healthcare workers and 1% were dialysis patients. Additional studies have identified other risk factors for HCV infection, including incarceration [20], acupuncture and tattooing [21], exposure to immune globulin [22], nosocomial exposure [23] and receiving injections in developing countries [24]. Unlike HBV, perinatal transmission and sexual transmission of HCV is uncommon [25, 26]. In addition, no vaccine has yet been developed that can protect individuals from HCV infection [27]. The reasons for lack of an effective vaccine may be due to several factors including the high rate of HCV RNA mutations, especially in the envelope regions, the wide range of genetic diversity of HCV strains or genotypes (HCV exists as a family of distinct genotypes that share approximately 70% sequence identity), and the fact that HCV constantly evolves within a patient to form a heterogeneous population of quasispecies (HCV quasispecies share approximately 90% sequence identity) [27].

Thus, both HBV and HCV are blood-borne infections that can be transmitted by a number of routes including blood transfusion (rarely in developed countries) and unsafe injection practices both in the healthcare setting and through illicit drug abuse. A recent report from the WHO estimated that unsafe injections account for approximately 35% of new cases of HBV and 55% of new cases of HCV in the developing world [28]. Thus, programs to educate healthcare workers and improve injection practices in developing countries are badly needed.

Incidence and prevalence

Worldwide, it is estimated that 300 million and 175 million individuals are chronically infected with HBV and HCV, respectively. HBV prevalence rates are variable depending on geographic location, but range from 0.1 to 20% depending on the population and age of the individual at the time of transmission [29]. The incidence and prevalence of HBV are highly variable geographically. Areas of high HBV seropositivity include sub-Saharan Africa, Pacific Rim countries with the exception of Japan, south east Europe and the Amazon basin [10]. In those areas where HBV is endemic, incidence rates can be as high as 20% leading to a substantial number of cases of primary HCC each year. Areas with low HBV seropositivity rates include North America, Japan and Western Europe. In the United States, the incidence of cases of acute hepatitis B has decreased significantly from a peak of 13.8 cases per 100 000 in 1987 to 3.3 cases per 100 000 in 1998 [30]. However, due to the chronicity of this insidious disease, the burden of chronic hepatitis B in the United States remains unchanged. In the United States, the number of individuals who are hepatitis B surface antigen (HBsAg) positive is estimated to be approximately 1.25 million, and chronic hepatitis B infection continues to be a major cause of the cirrhosis and of HCC that are encountered in our health system today.

DNA sequencing of the HBV genome reveals that there are eight distinct HBV genotypes; HBV genotypes A–H [31]. These eight HBV genotypes have important clinical implications and vary with respect to disease progression and response to anti-HBV therapies [32]. More specifically, infection with HBV genotype C has been associated with a greater severity in the progression of liver disease, and infection with HBV genotype B or D leads to a greater rate of progression to HCC [33]. In contrast, infection with HBV genotype A (check- inconsistent with previous statement) is associated a good clinical prognosis. Because conflicting data have been reported, the role of HBV genotypes in relationship to response to anti-HBV therapy is not well understood [34, 35].

Worldwide, annual new cases of HCV are estimated to be between 3 and 4 million each year [36], with a global prevalence estimated at 3% [37]. The distribution of HCV infection by age group varies by geographic region. In regions such as the United States and Australia, HCV infection is most frequently seen in persons who are 30–49 years of age. In Italy and Japan, most individuals who are infected with HCV are older than 49. In contrast, HCV is endemic in Egypt and can be detected in over 5% of the population; however, no age-specific pattern of HCV infection can be seen [37]. In the United States, studies conducted by the Centers for Disease Control and Prevention (CDC) estimated that 3.9 million Americans have been infected and approximately 2.7 million have chronic hepatitis C [38]. This places the prevalence of antibody to HCV in the United States at 1.8%. With the advent of blood-screening methodologies in 1990, the annual rate of new cases of chronic hepatitis C infections in the United States has declined from a high of 180 000 per year [36] to a current yearly rate of approximately 36 000 [39]. The annual death rate associated with HCV infection in the United States is now estimated at 8000 to 10 000 [36]. Due to the fact that infections in the United States occurred relatively recently, and that progression to liver cirrhosis and HCC can take 20–30 years [4], it is estimated that the HCV-related disease burden in the United States will increase dramatically over the next decade. More specifically, it is estimated that the number of patients with hepatic decompensation will increase 106%, the incidence of HCV-related HCC will increase by approximately 80%, and liver-related deaths will increase 180% by the year 2013 [40]. In Western Europe, a similar prevalence of HCV seropositivity has been demonstrated. For example, in a recent study [41], the number of cases of chronic HCV in France was estimated to be 500 000 with a prevalence of HCV seropositivity of approximately 1.1%. In Japan, the prevalence of HCV seropositivity is estimated to be 3.6% [42], with 90% of Japanese patients having chronic liver disease attributable to HCV infection.

Comparative sequence data from the HCV core, E1 and NS5 regions in the genomes of different HCV isolates have demonstrated that HCV exists as distinct genotypes that share approximately 70% sequence identity [43]. Several studies have shown that HCV genotypes have clinical significance with respect to disease progression and treatment outcomes [44, 45]. More specifically, hepatitis C patients chronically infected with HCV genotypes other than type 1 respond to interferon α and the combination of interferon α plus ribavirin at a significantly higher rate than do patients infected with HCV genotype 1. A large study of over 6500 chronic -hepatitis C patients across the United States demonstrated that HCV genotype 1 was seen in 73% of patients; 14% had genotype 2, followed by genotypes 3 (8%), mixed (4%), and genotypes 4, 5 and 6 with a frequency of <1% [46]. In other regions of the world such as Japan, the rate of infection with HCV genotype 1 can be as high as 89% [47]. In contrast, certain regions of eastern Europe have HCV genotype 1 prevalence rates as low as 60%, with HCV genotype 2 and 3 making up for the remaining numbers of patients [48]. Rare HCV genotypes can be seen in certain populations, presumably due to founder effects (a founder effect occurs when an individual with a genetically unique viral genome in an isolated geographic region is responsible for infecting subsequent individuals with the genetically unique viral genome). For example, HCV genotype 4 is most prevalent in patients from Egypt, while HCV genotype 6 is found predominately in patients from south east Asia [48].

As infection with either HBV or HCV can lead to primary HCC, it is important to understand the growing prevalence of HCC as a result of these viral infections. A recent report [49] estimates that the number of new cases of HCC worldwide is now approximately 500 000. This is a significant public-health burden, since most cases of HCC are untreatable and lead to rapid mortality [50]. In a study by Montalto et al. [49] it was estimated that 52.3% of new HCC cases are due to infection with HBV and approximately 25% of new cases of HCC are attributed to HCV infection worldwide. The percentage of HCC due to either HBV or HCV is geographically dependent, with more cases of HCC due to HBV in Asia (except Japan) and more cases of HCC due to HCV in the United States, Japan and southern Europe.

Natural history of the disease

HBV natural history

Chronic infection by HBV results in different clinical outcomes. Most infected individuals, approximately 85%, remain asymptomatic carriers of HBV, while others develop chronic hepatitis manifested by clinical symptoms and signs and abnormal liver tests. Thereafter, a proportion, 15–40%, of chronic hepatitis B patients may progress to more serious and life-threatening conditions such as cirrhosis, leading to eventual decompensation and to HCC. The reasons for the differences in the clinical outcomes may be a result of the host's immune response to HBV infection, the HBV genotype, and the level of viral activity, but these factors remain to be elucidated. In addition, predictors of progression to HCC and to death from chronic liver disease in patients with chronic hepatitis B need to be identified.

HCV natural history

After initial exposure to HCV, the infected patient will experience a transient rise in liver enzymes, [the most frequently monitored liver test is serum alanine aminotransferase (ALT)] indicating that inflammatory processes are occurring [51]. This elevation in liver enzymes will occur at least 4 weeks after the initial exposure to the virus and may last for up to 2 months [52]. Prior to the rise in liver enzymes, HCV RNA can be detected in the patient's serum using PCR analysis [52]. However, this acute stage of the disease usually goes undetected since 75% of patients with acute viral hepatitis due to HCV infection are asymptomatic [53]. The remaining 25% of these patients develop jaundice or other symptoms consistent with hepatitis.

Acute HCV infection is a benign disease; however, as many as 80% of acute HCV patients progress to chronic liver disease as evidenced by persistent elevation of serum ALT levels and by the continual presence of circulating HCV RNA [51, 54]. The natural progression of chronic HCV infection over a 10–20-year period leads to cirrhosis in 20–50% of patients [54], and the progression of HCV infection to HCC has been well documented [4, 55]. It is important to note that the survival for patients diagnosed with HCC is only 0.9–12.8 months from initial diagnosis [56]. Treatment of HCC with chemotherapeutic agents has not proven effective and only 10% of patients will benefit from surgery due to extensive tumor invasion of the liver [57]. Given the aggressive nature of primary HCC, the only viable treatment alternative to surgery is liver transplantation [58]. Although liver transplantation will replace the failed liver of the chronic HCV patient, most patients will reinfect the transplanted liver with HCV within first 6 months of transplantation. This is further complicated by the fact that liver-transplant patients must be on immuno-suppressant drugs to ameliorate the rejection process. Reinfection of HCV post-liver transplant may be due to extrahepatic reservoirs of HCV which may include T lymphocytes and other immune cells. In fact, HCV has become the most common reason for liver transplantation in the United States [40].

Progression to cirrhosis for HBV and HCV: disease manifestations

Upon progression to cirrhosis, patients with chronic HBV or HCV infection present with clinical features that are common to clinical cirrhosis regardless of the initial cause. These clinical features may include bleeding esophageal varices, ascites, jaundice, and encephalopathy [59].

In 1986, D'Amico et al. [59] described the clinical manifestations and survival rates in 1155 patients with both alcoholic and viral-associated cirrhosis. Of these patients, 435 (37%) had compensated disease, 70% of whom were asymptomatic at the beginning of the study. The remaining 720 patients (63%) had decompensated liver disease and 78% of these presented with a history of ascites, 31% with jaundice, 17% with bleeding, and 16% with encephalopathy. HCC was observed in six (0.5%) patients with compensated disease and in 30 (2.6%) patients with decompensated disease. Over the course of 6 years, patients with compensated cirrhosis developed clinical features of decompensated disease at a rate of 10% per year. In most cases, ascites was the first presentation of decompensation. In addition, HCC developed in 59 patients who initially presented with compensated disease by the end of the 6-year study.

With respect to survival, the D'Amico study [59] indicated that the 5-year survival rate for all patients on the study was only 40%. The 6-year survival rate for the patients who initially had compensated cirrhosis was 54% while the 6-year survival rate for patients who initially presented with decompensated disease was only 21%. There were no significant differences in the survival rates between the patients who had alcoholic cirrhosis and the patients with viral-related cirrhosis. The major causes of death for the patients in the D'Amico study [59] were as follows: liver failure in 49%, hepatocellular carcinoma in 22% and bleeding in 13%.

Impact of therapy

In recent years treatment of both HBV and HCV infection with interferons has emerged as an important therapeutic alternative. Interferons are cytokines that are induced as a consequence of viral and microbial infections, and have pleiotropic biological effects that play a role in modulating innate and adaptive immunity. Human interferons have been classified according to the cell-surface receptor that they bind. Type 1 interferons bind to the interferon α heterodimeric receptor IFNAR1/IFNAR2 and include the 21 non-allelic subtypes of interferon α, interferon β, interferon ω and interferon τ [60]. Interferon γ binds a unique cell-surface receptor, is classified as type 2 interferon, and displays antiviral, antifibrotic and immunomodulating activity that effects the TH1 response [61].

For hepatitis B, three approved drugs, interferon α adefovir and lamivudine, are currently available for the treatment of HBV infection. Interferon α therapy results in a complete response [HBV DNA negative, hepatitis B envelope antigen (HBeAg) negative] in only 15–37% [62, 63] of patients and is associated with some potentially severe side effects [63]. Lamivudine, a nucleoside analogue, is a very potent inhibitor of HBV DNA synthesis, but unlike treatment with interferon, does not eliminate the HBV from infected cells [64]. Rather, chronic administration of lamivudine controls viral replication and results in improved liver histology in >50% of patients. Sustained suppression of HBV DNA and HBeAg seroconversion occurs in only 16–17% of patients following 1 year of lamivudine treatment [65, 66] and cessation of therapy often results in a reactivation of HBV replication [67]. In addition, recent reports have documented lamivudine resistance in greater than 50% of chronically treated patients [68].

References

1 World Health Organization (2003) AIDS Epidemic Update December 2003 website, www.who.int/hiv/pub/epidemiology/epi2003/en/
2 World Health Organization (2000) Hepatitis B The World Health Fact Sheet 204 (revised October 2000); www.who.int/inf-fs/en/fact204.html

3 World Health Organization (2000) Hepatitis C The World Health Fact Sheet Fact 164 (revised October 2000); www.who.int/mediacentre/factsheets/fs164/en/print.html
4 Tong MJ, el-Farra NS, Reikes AR, Co RL (1995) Clinical outcomes after transfusion-associated hepatitis C. *N Engl J Med* 332(22): 1463–1466
5 Huang MA, Lok AS (2003) Natural history of hepatitis B and outcomes after liver transplantation. *Clin Liver Dis* 7(3): 521–536
6 Lok AS (2002) Chronic hepatitis B. *N Engl J Med* 346(22): 1682–1683
7 Shiffman ML (2003) Natural history and risk factors for progression of hepatitis C virus disease and development of hepatocellular cancer before liver transplantation. *Liver Transpl* 9(11): S14–S20
8 Bonis PA, Tong MJ, Blatt LM, Conrad A, Griffith JL (1999) A predictive model for the development of hepatocellular carcinoma, liver failure, or liver transplantation for patients presenting to clinic with chronic hepatitis C. *Am J Gastroenterol* 94(6): 1605–1612
9 Gust ID (1996) Epidemiology of hepatitis B infection in the Western Pacific and South East Asia. *Gut* 38 Suppl 2: S18–S23
10 Lavanchy D (2004) Hepatitis B virus epidemiology, disease burden, treatment, and current and emerging prevention and control measures. *J Viral Hepat* 11(2): 97–107
11 World Health Organization (2000) Hepatitis B Transmission 2000 www.who.int/emc-documents/hepatitis/docs/whocdscsrlyo20022/disease/transmission.html
12 Williams A (2002) Reduction in the hepatitis B related burden of disease–measuring the success of universal immunisation programs. *Commun Dis Intell* 26(3): 458–460
13 Kao JH, Chen DS (2002) Global control of hepatitis B virus infection. *Lancet Infect Dis* 2(7): 395–403
14 Kato N, Yokosuka O, Omata M, Hosoda K, Ohto M (1990) Detection of hepatitis C virus ribonucleic acid in the serum by amplification with polymerase chain reaction. *J Clin Invest* 86(5): 1764–1767
15 Pekler VA, Robbins WA, Nyamathi A, Yashina TL, Leak B, Robins TA (2003) Use of versant TMA and bDNA 3.0 assays to detect and quantify hepatitis C virus in semen. *J Clin Lab Anal* 17(6): 264–270
16 Hermida M, Ferreiro MC, Barral S, Laredo R, Castro A, Diz Dios P (2002) Detection of HCV RNA in saliva of patients with hepatitis C virus infection by using a highly sensitive test. *J Virol Meth* 101(1–2): 29–35
17 Fabris P, Infantolino D, Biasin MR, Marchelle G, Venza E, Terribile Wiel Marin V, Benedetti P, Tositti G, Manfrin V, de Lalla F (1999) High prevalence of HCV-RNA in the saliva cell fraction of patients with chronic hepatitis C but no evidence of HCV transmission among sexual partners. *Infection* 27(2): 86–91
18 Ogasawara S, Kage M, Kosai K, Shimamatsu K, Kojiro M (1993) Hepatitis C virus RNA in saliva and breastmilk of hepatitis C carrier mothers. *Lancet* 341(8844): 561
19 Alter MJ, Margolis HS, Krawczynski K, Judson FN, Mares A, Alexander WJ, Hu PY, Miller JK, Gerber MA, Sampliner RE (1992) The natural history of community-acquired hepatitis C in the United States. The Sentinel Counties Chronic non-A, non-B Hepatitis Study Team. *N Engl J Med* 327(27): 1899–1905
20 Pallas JR, Farinas-Alvarez C, Prieto D, Delgado-Rodriguez M (1999) Coinfections by HIV, hepatitis B and hepatitis C in imprisoned injecting drug users. *Eur J Epidemiol* 15(8): 699–704
21 Sun CA, Chen HC, Lu CF, You SL, Mau YC, Ho MS, Lin SH, Chen CJ (1999) Transmission of hepatitis C virus in Taiwan: prevalence and risk factors based on a nationwide survey. *J Med Virol* 59(3): 290–296
22 Bresee JS, Mast EE, Coleman PJ, Baron MJ, Schonberger LB, Alter MJ, Jonas MM, Yu MY, Renzi PM, Schneider LC (1996) Hepatitis C virus infection associated with administration of intravenous immune globulin. A cohort study. *JAMA* 276(19): 1563–1567
23 Grethe S, Gemsa F, Monazahian M, Bohme I, Uy A, Thomssen R (1999) Molecular epidemiology of an outbreak of HCV in a hemodialysis unit: Direct sequencing of HCV-HVR1 as an appropriate tool for phylogenetic analysis. *J Med Virol* 60(2): 152–158
24 Kane A, Lloyd J, Zaffran M, Simonsen L, Kane M (1999) Transmission of hepatitis B, hepatitis C and human immunodeficiency viruses through unsafe injections in the developing world: model based regional estimates. *Bull WHO* 77(10): 801–807
25 Polywka S, Schroter M, Feucht HH, Zollner B, Laufs R (1999) Low risk of vertical transmission of hepatitis C virus by breast milk. *Clin Infect Dis* 29(5): 1327–1329

26 Hammer GP, Kellogg TA, McFarland WC, Wong E, Louie B, Williams I, Dilley J, Page-Shafer K, Klausner JD (2003) Low incidence and prevalence of hepatitis C virus infection among sexually active non-intravenous drug-using adults, San Francisco, 1997–2000. *Sex Transm Dis* 30(12): 919–924
27 Poordad FF, Tran T, Martin P (2003) Developments in hepatitis C therapy during 2000–2002. *Expert Opin Emerg Drugs* 8(1): 9–25
28 World Health Organization Report (2001) Improving access to injection equipment to improve injection safety November 2001; www.who.int/medicines/organization/par/edl/policy_analysis.pdf
29 World Health Organization Report (2002) Hepatitis Bincidence and Epidemiology; www.who.int/emcdocuments/hepatitis/docs/whocdscsrlyo20022/surveillance/incidence_epidemiology.html
30 Goldstein ST, Alter MJ, Williams IT et al. (2002) Incidence and risk factors for acute hepatitis B in the United States, 1982–1998: implications for vaccination programs. *J Infect Dis* 185: 713–719
31 Bartholomeusz A, Schaefer S (2004) Hepatitis B virus genotypes: comparison of genotyping methods. *Rev Med Virol* 14(1): 3–16
32 Halfon P, Pol S, Bourliere M, Cacoub P (2002) Hepatitis B virus genotypes: clinical, epidemiological and therapeutic implications. *Gastroenterol Clin Biol* 26(11): 1005–1012
33 Kao JH (2002) Hepatitis B viral genotypes: clinical relevance and molecular characteristics. *J Gastroenterol Hepatol* 17(6): 643–650
34 Chien RN, Yeh CT, Tsai SL, Chu CM, Liaw YF (2003) Determinants for sustained HBeAg response to lamivudine therapy. *Hepatology* 38(5): 1267–1273
35 Ciancio A, Smedile A, Rizzetto M, Lagget M, Gerin J, Korba B (2004) Identification of HBV DNA sequences that are predictive of response to lamivudine therapy. *Hepatology* 39(1): 64–73
36 Alter MJ (1997) Epidemiology of hepatitis C. *Hepatology* 26(3, Suppl 1): 62S–65S
37 Wasley A, Alter MJ (2000) Epidemiology of hepatitis C: geographic differences and temporal trends. *Semin Liver Dis* 20(1): 1–16
38 Centers for Disease Control and Prevention (1998) Recommendations for prevention and control of hepatitis C virus (HCV) infection and HCV-related chronic disease. *MMWR* 47(RR-19): 1–39
39 Williams I (1999) Epidemiology of hepatitis C in the United States. *Am J Med* 27107(6B): 2S–9S
40 Davis GL, Albright JE, Cook SF, Rosenberg DM (2003) Projecting future complications of chronic hepatitis C in the United States. *Liver Transpl* 9(4): 331–338
41 Desenclos JC (2000) Epidemiology of hepatitis C. *Rev Prat* 50(10): 1066–1070
42 Lavanchy D, Gavinio P. Hepatitis C (2000) Can. *J Gastroenterol* 14 Suppl B: 67B–76B
43 Simmonds P, Smith DB, McOmish F, Yap PL, Kolberg J, Urdea MS et al. (1994) Identification of genotypes of hepatitis C virus by sequence comparisons in the core, E1 and NS-5 regions. *J Gen Virol* 75: 1053–1061
44 Hoofnagle JH, Di Bisceglie AM (1997) The treatment of chronic viral hepatitis. *N Engl J Med* 336: 347–356
45 Zein NN, Rakela J, Krawitt EL, Reddy KR, Tominaga T, Persing DH et al. (1996) Hepatitis C virus genotypes in the United States: epidemiology, pathogenicity, and response to interferon therapy. *Ann Intern Med* 125: 634–639
46 Blatt LM, Mutchnick MG, Tong MJ, Klion FM, Lebovics E, Freilich B, Bach N, Smith C, Herrera J, Tobias H et al. (2000) Assessment of hepatitis C virus RNA and genotype from 6807 patients with chronic hepatitis C in the United States. *J Viral Hepat* 7(3): 196–202
47 Takada N, Takase S, Takada A, Date T (1993) Differences in the hepatitis C virus genotypes in different countries. *J Hepatol* 17: 277–283
48 McOmish F, Yap PL, Dow BC, Follett EA, Seed C, Keller AJ, Cobain TJ, Krusius T, Kolho E, Naukkarinen R et al. (1994) Geographical distribution of hepatitis C virus genotypes in blood donors: an international collaborative survey. *J Clin Microbiol* 32(4): 884–892
49 Montalto G, Cervello M, Giannitrapani L, Dantona F, Terranova A, Castagnetta LA (2002) Epidemiology, risk factors, and natural history of hepatocellular carcinoma. *Ann N Y Acad Sci* 963: 13–20
50 Ikeda M, Maeda S, Shibata J, Muta R, Ashihara H, Tanaka M, Fujiyama S, Tomita K (2004) Transcatheter arterial chemotherapy with and without embolization in patients with hepatocellular carcinoma. *Oncology* 66(1): 24–31
51 Alter HJ (1989) Chronic consequences of non-A, non-B hepatitis. In: LB Seeff, JH Lewis (eds): *Current Perspectives in Hepatology*. Plenum Medical Book Co, New York, 83–89

52 Farci P, Alter H, Wong D, Miller RH, Shih JW, Jett B, Purcell RH (1991) Long-term study of Hepatitis C Virus replication in non-A, non-B hepatitis. *N Engl J Med* 325: 98–104
53 Alter HJ (1990) The hepatitis C virus and its relationship to the clinical spectrum of NANB hepatitis. *J Gastroenterol Hepatol* 1: 78–94
54 Sherlock S (1992) Viral hepatitis. *Lancet* 339: 802
55 Liang TJ, Jeffers LJ, Reddy R et al. (1993) Viral pathogenesis of hepatocellular carcinoma in the United States. *Hepatology* 18: 1326–1333
56 Tsukuma H, Hiyama T, Tanaka S, Nakao M, Yabuuchi T, Kirtamura T, Nakanishi K, Fujimoto I, Inoue A, Yamazaki H, Kawashima T (1993) Risk Factors for hepatocellular carcinoma among patients with chronic liver disease. *N Engl J Med* 328: 25;1797–1801
57 Trinchet JC, Beaugrand M (1994) Hepatocellular carcinoma treatment with arterial chemoembolization. *Presse Medicine* 23: 831–833
58 Pichlmayr R, Weimann A, Ringe B (1994) Indications for liver transplantation in hepatobiliary malignancy. *Hepatology* 20: 33S–40S
59 D'Amico G, Morabito A, Pagliaro L, Marubini E, and the liver study group of "V. Cervello" Hospital (1986) Survival and Prognostic Indicators in Compensated and Decompensated Cirrhosis. *Dig Dis Sci* 31: 5: 468–475
60 Pestka S (2000) The human interferon alpha species and receptors. *Biopolymers* 55(4): 254–287
61 Bach EA, Aguet M, Schreiber RD (1997) The IFN gamma receptor: a paradigm for cytokine receptor signaling. *Annu Rev Immunol* 15: 563–591
62 Perrillo RP, Schiff ER, Davis GL, Bodenheimer HC, Lindsay K, Payne J, Dienstag JL, O'Brien C, Tamburro C, Jacobson IM et al. (1990) A randomized controlled trial of interferon alfa-2b alone and after prednisone withdrawal for the treatment of chronic hepatitis B. *N Engl J Med* 323: 295–301
63 Lok ASF, Wu PC, Lai CL, Lau JYN, Leung EKY, Wong LS, Ma OC, Lauder IJ, Ng CP, Chung HT (1992) A controlled trial of interferon with or without prednisone priming for chronic hepatitis B. *Gastroenterology* 102: 2091–2097
64 Mason WS, Cullen J, Moraleda G, Saputelli J, Aldrich CE, Miller DS, Tennant B et al. (1998) Lamivudine therapy of WHV-infected woodchucks. *Virology* 245: 18–32
65 Lai CL, Chien RN, Leung NW, Chang TT, Guan R, Tai DI, Ng KY, Wu PC, Dent JC, Barber J et al. (1998) A one-year trial of lamivudine for chronic hepatitis B. *N Engl J Med* 339: 61–68
66 Dienstag JL, Schiff ER, Wright TL, Perrillo RP, Hann HW, Goodman Z, Crowther L, Condreay LD, Woessner M, Rubin M, Brown NA (1999) Lamivudine as initial treatment for chronic hepatitis B in the United States. *N Engl J Med* 341: 1256–1263
67 Honkoop P, de Man RA, Niesters HG, Zondervan PE, Schalm SW (2000) Acute exacerbation of chronic hepatitis B virus infection after withdrawal of lamivudine therapy. *Hepatology* 32: 635–639
68 Lau DT, Khokhar MF, Doo E, Gnany MG, Herion D, Park Y, Kleiner DE et al. (2000) Long-term therapy of chronic hepatitis B with lamivudine. *Hepatology* 32: 828–834

Hepatitis Prevention and Treatment
Edited by J.M. Colacino and B.A. Heinz

Molecular virology of hepatitis B virus

Lynn D. Condreay and Sarah A. Harris

GlaxoSmithKline, Department of Virology, P.O. Box 13398, Research Triangle Park, NC 27709-3398, USA

Introduction

Hepatitis B virus (HBV) is the prototype virus of the Hepadnaviridae family that infects humans. Hepadnaviruses are widespread, infecting many different mammalian and avian species [1–5] and displaying both host [6–10] and tissue [11, 12] tropism. Each virus type typically infects only very closely related species and in each case the primary target organ is the liver with infection and replication detected in the hepatocytes [1, 6–8]. These viruses are classified according to host species and degree of homology between the viral genomes. Greater homology is detected between members of the same genus than between members of diverse groups. The mammalian viruses [1, 2, 5, 8, 13–21] are members of the genus Orthohepadnavirus while the avian viruses [3, 4, 22, 23] are grouped in the genus Avihepadnavirus. This host and tissue tropism reflects both receptor limitations and the impact of liver-specific transcription factors on viral replication [9, 12, 24]. Infection of other tissues (bile duct epithelial, mononuclear, pancreatic, and renal cells) has been reported but the significance of these extrahepatic sites is unclear [25–34]. Hepatic tissue tropism is consistent with the disease associated with hepadnavirus infection. However, the specific disease sequelae vary for each species. In the case of HBV infection, the disease may present as acute or chronic hepatitis with manifestations leading to fibrosis, cirrhosis, hepatocellular carcinoma, and hepatic failure occurring in patients infected for decades [35].

Phylogenetic analyses of the hepadnaviruses are ongoing. HBV has been classified historically into four the subtypes adr, adw, ayr, and ayw [36–40], based on the antigenic determinants within the viral envelope proteins with additional serotypes defined within each subtype. More recently, work has focussed on classification by genotype [38, 39, 41–44] or even subgenotype groups [45–47] as defined by degree of homology across the entire viral genome. Eight genotypes (A–H) have been identified for human HBV [41, 48]. Although some virus types within a genotype display some variation in nucleotide sequence, this is substantially less than the ≥8% divergence between the genotypes. Genotypes A–G have clearly defined geographic and ethnic distributions [44, 49, 50]. Genotype A is detected in patients infected with HBV in North America, northwest Europe, and central Africa while genotypes B and C are detected in samples from southeast Asia, China, and Japan. Genotype D is predominantly found in the Mediterranean region of southern Europe, the Middle East, and India; genotype E is detected in patients in Africa; and genotype F is found in people in Central and South America and Polynesia, and in Native Americans. The distribution of genotype G is less well defined but this genotype has been found in the United States, France, and Canada. Limited information on ethnic and geographic distribution exists for genotype H which has only been detected in the past 2 years in Central America [41].

It is difficult to define the role viral genotype plays in the course of disease because of the linkage of genotype with geography, ethnicity, diet, and lifestyle. Some comparisons are being made between genotypes with overlapping distribution with respect to impact on disease, viral variant evolution, and response to antiviral therapies [47, 51–55]. Genotype B is associated with earlier hepatitis B e antigen (HBeAg) seroconversions than is genotype C in Chinese patients chronically infected with HBV [51]. However, progression to advanced fibrosis and ultimately hepatocellular carcinoma may not differ between patients infected with genotypes B and C [55]. Reports differ as to the association of genotype with disease progression and severity (reviewed in [53]). Clearly, there are key differences with respect to the impact of the viral sequence and genotype on the emergence of HBeAg-negative HBV infections, which may be associated with a more severe disease course. The most commonly detected naturally occurring variants of HBV affecting HBeAg levels in patients are precore stop codon mutants and basal core promoter mutants. Precore stop-codon mutants are most frequently detected in genotypes D, B, and C but also in some clades of genotype F [47]. This is explained at a molecular level by the finding that the precore stop codon mutation, $G_{1896}A$, stabilizes a structural component of the viral pregenome (the stem-loop structure of epsilon, discussed below) required for replication by base-pairing with T_{1858} in genotypes D, B and C, and in some clades of F. T_{1858} is not generally detected in wild-type genotype A HBV resulting in a multistep mutational pathway for precore stop-codon-mutant emergence in this genotype.

The hepadnaviral genome

Until recently [56], there have been no simple *in vitro* infection systems to study HBV infection. *In vitro* HBV infection with infectious serum was only described for cultured human, tupaia, and chimpanzee hepatocytes [10, 21, 57, 58]. Infection of limited cell lines could be obtained if the virus was purified and protease-treated [59, 60]. Component systems evaluating isolated aspects of the HBV life cycle exist but simple and robust infection systems for HBV have proven problematic. As such, direct studies into the molecular biology of HBV infection and replication have been restricted. In contrast, more readily available and more reproducible infection systems have been developed and used with other members of this family [7, 61, 62]. Therefore, much of the research into understanding the molecular biology of HBV infection and the viral life cycle has been performed in surrogate systems with other Hepadnaviridae members such as the avihepadnavirus, duck hepatitis B virus, and the orthohepadnavirus: woodchuck hepatitis virus.

All members of the Hepadnaviridae family have a small DNA genome of approximately 3000 base pairs (Fig. 1). The relaxed circular viral genome is

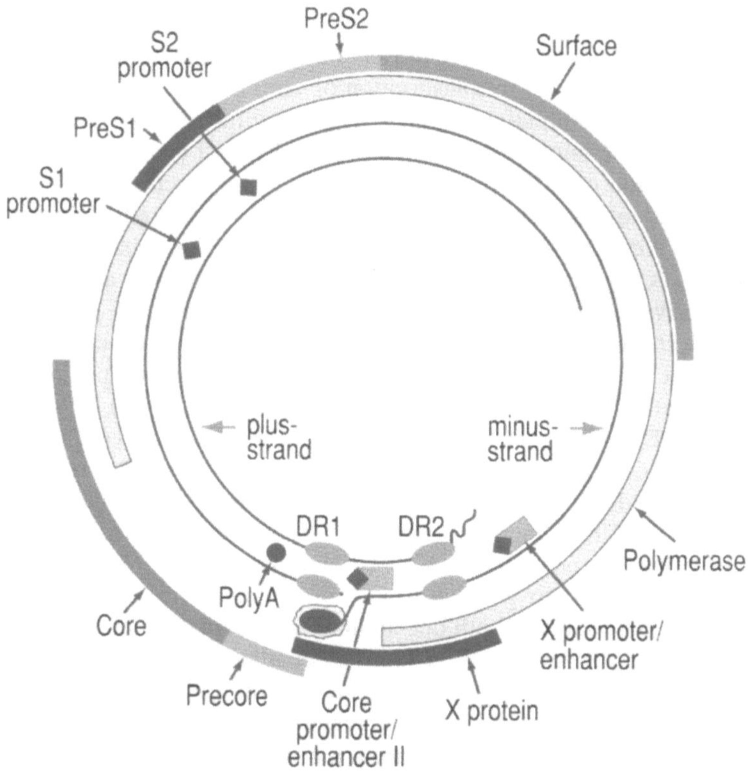

Figure 1. Organization of the hepadnavirus genome. The inner, gapped, circles represent the relaxed circular DNA genome of HBV (plus- and minus-strands). The minus-strand DNA has a short terminal redundancy. The plus-strand is incomplete. The terminal protein covalently attached to the 5' end of the minus-strand DNA and the short RNA primer for plus-strand DNA synthesis present at the 5' end of the plus-strand are indicated. The positions of direct repeat (DR) elements 1 and 2 are illustrated by gray ovals on both plus- and minus-strand DNAs. Only the major transcriptional regulatory elements are indicated. Approximate positions of the four viral promoters (S1, S2, X, core), enhancers I and II, and the common polyadenylation signal (Poly A) are shown. The overlapping open reading frames for the surface proteins X protein, core and precore, and the polymerase are presented. The large (L) surface protein includes all regions of the surface ORF; the middle (M) surface protein includes the preS2 and surface domains but not the preS1 domain. HBeAg, produced through post-translational processing of the precore protein, is not indicated in the figure.

a partially double-stranded DNA molecule with a viral protein covalently attached to the 5' end of the minus-strand of DNA. This protein is the viral polymerase and serves to prime minus-strand DNA synthesis [63–67]. A small RNA is covalently attached to the 5' end of the plus-strand of DNA. This RNA is derived from the 5' end of the viral pregenomic RNA during replication [68–70]. The minus strand DNA also has a short terminal redundancy and is greater than unit length. The plus strand DNA is of variable length but is consistently less than unit length. This differential between the plus- and minus-strands of the hepadnavirus genome results in a single-stranded gap in the genome that varies in length for different types of hepadnaviruse [69, 71–73].

Several regulatory elements are present on the hepadnavirus genome. These include elements involved in transcription of viral mRNA from the viral template generated upon infection, replication of the viral genome, and packaging of the viral genome into the nucleocapsid (Fig. 1). Four promoters direct transcription initiation of HBV pregenomic and mRNA species, specifically the core (precore/polymerase/core/pregenomic), S1, S2, and X promoters. Overlapping the X promoter and the core promoter lie enhancers I and II, respectively. A negative regulatory element is present, as well as a glucocorticoid-responsive element (GRE). Enhanced transcription of HBV mRNA results from steroid stimulation of the GRE [74]. A single polyadenylation signal is present on the genome with recognition dependent on positive (PET) [75] and negative (NET) [76] effectors of transcription. The 11–12-base-long direct repeat (DR) elements 1 and 2 are critical for the initiation of minus- and plus-strand DNA synthesis. The degree of conservation of these elements among all hepadnaviruses varies. For instance, the length and nucleotide sequences of the DR elements are distinct for each hepadnavirus.

The hepadnaviruses display a very efficient use of the small genome. All three frames are used to generate the viral proteins with the open reading frames (ORFs) overlapping both one another and the *cis*-acting regulatory elements present on the genome (Fig. 1). The hepadnaviruses have four overlapping ORFs encoding structural and nonstructural proteins. In some cases, multiple proteins are derived from a single reading frame through the use of different translation initiation sites or through post-translational processing. The HBV structural proteins include core, S, M, and L. The HBeAg shares some sequences with the core protein and is present in some core particles affecting the nucleocapsid structure *in vitro* but is generally a secreted protein. The only nonstructural protein encoded by all avian hepadnaviruses is the viral polymerase. However the classification of this protein solely as nonstructural is subject to debate since the viral polymerase has been reported to be packaged within the viral nucleocapsid and impacts nucleocapsid assembly [77]. Mammalian hepadnaviruses also encode the nonstructural X protein. Recently, a hidden regulatory X-like protein ORF has been described for duck HBV [78] but no role for this putative protein has been found in viral infection and replication studies using 'X'-knockout mutants [79]. At least one additional protein, termed HBV splice-generated protein (HBSP), has been detected arising from a spliced message encoding the N-terminus of the polymerase with other sequences arising from a frame shift. The function of this protein is unclear but it may have a role in viral pathogenesis [80].

The structural proteins

The hepadnaviruses have either two (avian) or three (mammalian) surface ORFs on the genome. In each case, the surface proteins have different promoters and initiate at different codons but share a single stop codon and have the same C-terminal amino acids. Thus, for the orthohepadnaviruses, the largest surface protein (L) contains the entire amino acid sequence of the smaller surface proteins M and S. M contains the entire amino acid sequence of S. The preS1 and preS2 domains represent the N-terminal portions of L and M not present in S (Fig. 1). Each of these proteins is necessary for virion secretion. The preS1 domain of L is critical for both assembly into virions and for receptor interactions [81, 82]. The surface proteins are post-translationally modified by N- and O-linked glycosylation during maturation within the endoplasmic reticulum and Golgi apparatus [83–85]. In addition, the large protein, L, is subject to myristoylation, which seems to be critical for infectivity [86, 87]. In the avihepadnavirus system, the preS domain is phosphorylated [88] but the significance of this is unclear.

Three proteins are derived from the core open reading frame. These are produced from different transcripts and, in the case of HBV, are termed hepatitis B core antigen (HBcAg), precore protein, and HBeAg. HBcAg is the core or capsid protein that dimerizes and assembles to form the viral nucleocapsid that contains the RNA pregenome and the viral polymerase. Ultimately, this nucleocapsid is also present in the infectious virus particle. The core protein is post-translationally modified by specific phosphorylation. Precore protein is an intermediate produced in the generation of HBeAg. In one system, overexpression of precore protein has been shown to result in decreased nucleocapsid detection. Thus it may be that the precore protein has a regulatory role in viral replication [89]. HBeAg is produced from precore protein by a series of post-translational modifications that include removal of N-terminal amino acids by a host signalase with subsequent removal of C-terminal amino acids. HBeAg is generally secreted but assembled nucleocapsids composed of HBeAg have been characterized [90]. HBeAg has toleragenic activity and influences the establishment of persistent infection [91–93], although specific details on the mechanisms of these activities are lacking.

Heterologous HBV structural particles composed of different hepadnavirus proteins are present in the serum of infected individuals. These include large quantities of subviral structures composed of only the surface proteins lacking the viral genome and smaller quantities of complete virus particles. With mammalian hepadnaviruses, both spherical and filamentous surface antigen structures are detected. The spherical particles have a diameter of ~20 nm. The filaments are ~20 nm in width and of variable length. Only spherical surface antigen particles of ~60 nm are detected in avian hepadnavirus infections. Filamentous particles contain all three of the surface glycoproteins: S (small), M (middle, preS2), and L (large, preS1). Only S and M are found in spherical particles. In both particle forms, the S protein is the main constituent. Members of the Hepadnaviridae family share certain characteristics with respect to the virion (or Dane particle) structure. Virus particles are approximately 39–45 nm in diameter and consist of a lipoprotein envelope surrounding the viral nucleocapsid. The nucleocapsid is composed of the viral core protein packaging the viral genome and one to two viral polymerase molecules. All viral surface proteins are detected on the surface of the virion. Host components are apparent in the envelope, which is derived from the endoplasmic reticulum reflecting the virus maturation pathway, and are critical in modifications of the viral proteins.

The nonstructural proteins

The largest gene encodes the viral polymerase, which is critical for both packaging and DNA replication. Functional assays and homology evaluations have elucidated four regions or domains of this protein. At the N-terminus lies the terminal protein which provides a priming function for minus-strand DNA synthesis and through which the protein is attached to the DNA. A spacer region of unknown function separates the terminal protein from the polymerase domain. The RNA- and DNA-dependent DNA polymerase domain is responsible for both reverse transcription of the pregenomic RNA template in the generation of minus-strand DNA and for plus-strand viral DNA synthesis using the minus-strand DNA as a template. The reverse transcriptase domain has been further divided into a number of subdomains involved in nucleotide binding, primer interactions, and genome translocation by homology mapping and modeling with the human immunodeficiency virus (HIV) reverse transcriptase. At the C-terminal region of the protein lie amino acids encoding the RNase H activity [94] that are responsible for degrading the pregenomic RNA in replication. Despite the multifunctional character of this enzyme, the polymerase lacks a proofreading capacity. The annual mutation rate for HBV is estimated to be $\sim 3 \times 10^{-5}$ mutations/position on the genome. Thus, in chronic hepatitis B patients, mutations in the HBV genome occur and variant sequences may emerge under selective pressure (immunological or antiviral).

X is a distinguishing point between the avihepadnaviruses and the orthohepadnaviruses. For the orthohepadnaviruses, X is essential *in vivo* but not critical *in vitro* [95, 96]. While for the avihepadnaviruses, a reading frame encoding an X-like protein is present, there is no obvious effect of this on viral replication [97, 98]. Many intracellular activities have been described for the

X protein in cell culture leading to putative gene-regulatory and immunomodulatory effects. To name only a few, X indirectly transactivates many host and viral promoters [99–101] and affects numerous functions of the cell. X has impact on signal transduction by stimulating Ras-GTP cycling, which activates the Raf mitogen-activated protein kinase (MAP kinase) signaling cascade [102]. In human cells, X interacts with a human homolog of a monkey protein (UV-damaged DN-binding protein) involved in DNA repair mechanisms [103], thus potentially impacting genome replication and host DNA repair. X impacts cytosolic calcium levels and signaling cascades involving Src kinases perhaps by X acting upon mitochondrial voltage-dependent anion channels and affecting the mitochondrial transition pore [104–106]. Thus, X affects not only viral but also host-protein interactions with viral replication and affects the cellular environment in which the virus replicates. For both the ortho- and avihepadnaviruses, the C-terminal half of the X protein is key to suppressing *in vitro* transformation of cultured fibroblasts by c-Ha-ras and c-Myc and inducing programmed cell death, apoptosis through a Bcl-2-sensitive pathway [110, 111]. Thus, X also directly affects the life cycle of the cell. Indirect effects on the infected cell are also apparent for X through the regulation of histocompatibility locus antigen (HLA)-DR, and activation of major histocompatibility complex (MHC) class I genes revealing a role for X in the host immunological responses [107, 108]. In addition, X has been shown to interact with XAPC7 in the proteasome complex [109]. With such a broad spectrum of activities identified for X, it is difficult to pinpoint only one specific role as key in viral replication. Since X is involved in the activation of a large array of host activities as well as affecting viral replication directly, it may be that X impacts the establishment and maintenance of the persistent infection by affecting the intracellular environment optimal for viral replication, influencing cell turnover, modulating the host response and, at least for the orthohepadnaviruses, playing a role in carcinogenesis. Proving which functions of X delineated *in vitro* are substantiated *in vivo* will be key to identifying the true role of X in HBV replication and potentially identifying mechanisms for antiviral opportunities.

The viral life cycle

Many aspects of the viral life cycle (Fig. 2) have been well characterized by both biochemical and genetic studies [112–114] but large gaps exist in our knowledge in some areas. The early events in the HBV life cycle are only beginning to be understood. Experiments on receptor interactions and attachment of the virus to the host cell, viral entry and uncoating, and movement of the genome within the cell are ongoing, addressing the host and virus components in each of these areas. Details on steps involved in the conversion of the relaxed circular genome to the covalently closed circular (ccc) DNA template are being resolved but many steps remain unclear.

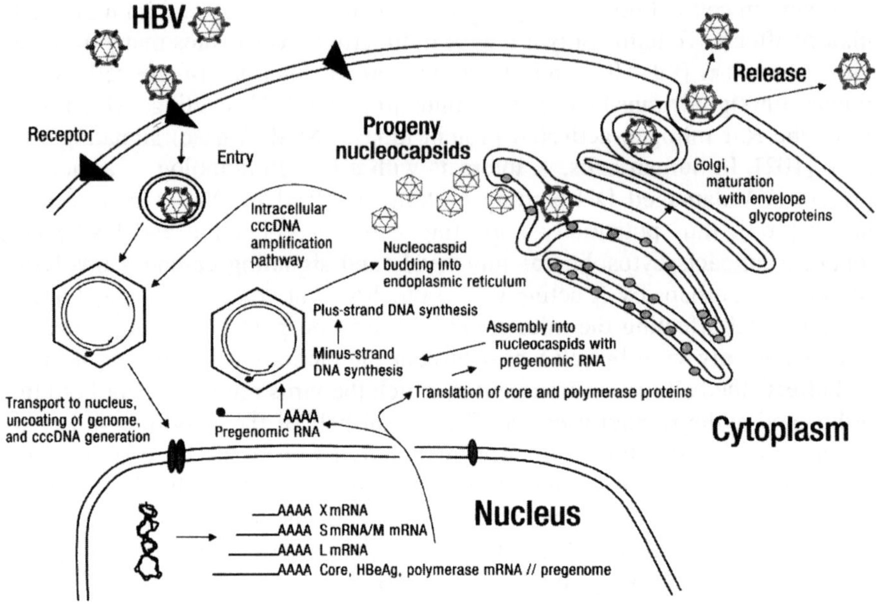

Figure 2. The hepadnavirus life cycle. A general infection and replication pathway for hepadnaviruses is indicated. The virus attaches to a receptor on the cell surface and is internalized and uncoated with transport of the nucleocapsid to the cell nucleus. The viral genome is processed to form the template (cccDNA) for all viral RNAs. The viral RNAs are exported from the nucleus and translated to produce the viral proteins. Not shown on this figure is the translation of the surface glycoproteins at the endoplasmic reticulum. Pregenomic RNA together with the viral polymerase and assorted host proteins is packaged within a nucleocapsid composed of the viral core protein. Viral DNA synthesis occurs within the nucleocapsid in a multistep process involving several intramolecular translocations of the growing DNA strand. Genome maturation may trigger budding of the nucleocapsid into the endoplasmic reticulum to obtain the outer lipid envelope and surface proteins. In this pathway, mature virus is processed through the Golgi apparatus and released from the cell. Alternatively, the nucleocapsid may be directed into a pathway to amplify the viral nuclear cccDNA template and not mature into infectious particles.

Early events in the viral life cycle

The initial step in the viral replication cycle is mediated by the interaction of the viral envelope proteins with the host receptor (see below). There is considerable evidence that the L (preS or preS1 domains) surface antigen is critical for this interaction. Virus neutralizing antibodies exist to the N-terminal portion (preS) of the large surface antigen [115–117] and preS is critical in determining host range in the duck hepadnavirus system [118, 119]. Likewise, this same portion (preS1) of the woolly monkey hepadnavirus L protein is key to determining species specificity in infections of human hepatocyte cultures [9]. Infectivity is also affected by post-translational modifications, with myristoylation of the preS1 domain of the HBV L surface protein required for infection in cultured hepatocytes [86, 87, 120]. In the duck hepadnavirus system, the middle portion of the preS domain of the large surface antigen protein binds to a receptor (see below) for the duck hepadnavirus but not to the chicken or human homologs [119].

The primary receptor for the duck hepadnavirus has been identified as the enzyme carboxypeptidase D [119–129]. This molecule cycles between the Golgi apparatus and the plasma membrane and changes distribution within the cell upon infection of the hepatocyte by the duck hepadnavirus [122, 123]. These changes are mediated by interaction with the L surface protein and may account in part for the decreased efficiency of superinfection seen with the hepadnaviruses [122]. Although carboxypeptidase D homologs exist in mammalian systems [130], the orthohepadnaviruses may not use the same receptors as the avihepadnaviruses. A number of molecules have been described as potential receptors for these viruses (e.g. annexin V [131, 132]) but the data are inconclusive. There is some evidence suggesting that co-receptors are involved in mediating host specificity and infection [120, 125, 127] as documented for other viruses such as the HIV, but none have been defined.

After the virus binds to the receptor, it is generally accepted that the virus is internalized through endocytosis, resulting in the removal of the viral envelope and surface antigens. This may occur through an endosomal pathway. The nucleocapsids containing the virus genome are transported to the nucleus as directed by a nuclear localization signal in the core protein [133]. There has been some debate about where capsid disassembly and genome release occur. An early suggestion derived from work using transgenic mice was that the nucleocapsid is transported to the nucleus with disassembly occurring at the cytoplasmic side of the nuclear membrane and only the viral genome is transported into the nucleus [134, 135]. Even though empty nucleocapsids were detected in the nucleus of these mice, which is consistent with detection of nuclear HBcAg in patients infected with HBV, it was suggested that this represented assembly within the nucleus. More recent work suggests that the nucleocapsids may actually be transported across the nuclear membrane through nuclear pore complexes [133]. Phosphorylation of the core proteins in the nucleocapsid is required for this importin α– and β-mediated interaction with the nuclear pore complex [135, 136] and may involve cellular protein kinases packaged within the nucleocapsid along with the pregenomic RNA and the viral polymerase. Biochemical and kinetic analyses of the dynamics of nucleocapsid assembly and disassembly suggest that dissociation of nucleocapsids may require some triggering event [137]. Such a triggering event could again be the phosphorylation or dephosphorylation of the core protein in the nucleocapsid [136, 138]. In the duck hepadnavirus system, the stability of the nucleocapsid icosahedron is modulated by a Cdc2 (threonine/proline) kinase-like recognition signal forming a junction between the interior of the capsid and the exposed surface [138]. No definitive host or viral components have been identified that are involved in the disassembly of nucleocapsid to release the viral genome into the nucleus for subsequent steps in replication.

Viral template generation

Once released into the nucleus, the infecting, partially double-stranded DNA of the viral genome must be converted into the cccDNA species that serves as the template for subsequent transcription of mRNA and viral replication. Although the specific order and mechanism of this conversion process are unclear, several steps must occur. The viral polymerase covalently attached to the 5' end of the minus-strand DNA of the viral genome must be removed. The minus-strand DNA is reduced to a single genomic length with removal of the terminal redundancy. Viral DNA synthesis to elongate the incomplete plus-strand of the DNA occurs and the RNA primer attached to the 5' end of the plus-strand DNA of the viral genome is removed. The revealed ends of the double-stranded hepadnavirus DNA are ligated together to form the supercoiled template molecule, cccDNA, which will ultimately be found as a heterogeneous population of minichromosomes within infected cells [139]. Both host and viral enzymes are involved in the steps generating these minichromosomes. Inhibitors of the viral polymerase, such as antiviral nucleoside and nucleotide analogs, can prevent cccDNA formation and initiation of infection [7, 140]. Host DNA repair components are likely involved in the removal of the protein primer and perhaps ligation of the free ends of the genome. The conversion of the relaxed circular genomic DNA form to the supercoiled cccDNA form may depend upon host topoisomerases. Topoisomerase I likely plays a role in the integration of HBV sequences into the host chromosomes [141, 142], as occasionally found in some hepatocellular carcinoma specimens. Thus, although only one viral enzyme is defined for the replication of the hepadnavirus genome, many enzymatic activities carried out by this protein are critical for the establishment of cccDNA and the initiation of replication.

Transcription

For replication of the viral genome to continue past generation of cccDNA, transcription by RNA polymerase II must occur. This activity is highly regulated as reviewed in detail in Moolla et al. [112]. Transcription of viral RNA complementary to minus-strand DNA is directed by four viral promoters, is unidirectional, and results in four well-characterized, partially overlapping, capped, unspliced RNA transcripts covering the entire HBV genome. These four mRNA families are approximately 3.5, 2.4, 2.1, and 0.8 kb in size and encode the core/precore/polymerase, L surface antigen (preS1), M (preS2)/S surface antigens, and X, respectively (Fig. 2). The 3.5 kb mRNA is actually composed of a few distinct RNA molecules with slight nucleotide-length variations at the 5' end. The shortest mRNA of this group is the translational template for core and polymerase proteins while the longest mRNA encodes the precore protein, which is post-translationally modified and secreted as HBeAg. The shorter of the 3.5-kb transcripts also serves as the viral pregenome and is the template for viral replication. Similarly, the 2.1-kb mRNA is also a family of molecules with defined but heterogeneous 5' ends.

The promoters with affiliated regulatory elements are distinct for each RNA group and impact levels of viral RNA synthesized. Disruption of transcriptional control can inhibit virion secretion [143]. However, relative abundance of each RNA species is tightly controlled at multiple levels. The transcripts consistently found in the greatest abundance are the 2.1-kb M/S mRNAs. Although this is consistent with the balance seen with the viral surface proteins where S and M are in excess to L surface antigen, there is not always a direct reflection of promoter activity with RNA level or protein level. The X and core promoters are more active in transfection experiments than the L and M/S promoters [112] but X mRNA and X protein are consistently present only at very low levels in cells.

The core promoter directs transcription of all of the 3.5 kb RNA molecules [112, 144–148]. The structure of the core promoter is complex and is composed of a basal core promoter (BCP) element with a minimal length of ~100 bp [147, 148] and an upstream regulatory sequence [146]. The core upstream regulatory sequence (CURS) located immediately 5' of the BCP can activate the BCP 200–2000-fold [146–148]. Although the BCP lacks a TATA box, each mRNA is actually produced from promoter sequences including a TATA-box-like element [144, 147]. A portion of the CURS forms part of the enhancer II [148]. Enhancer II activates transcription from the HBV core promoter with both position and orientation restrictions [112]. The second HBV enhancer element, enhancer I, can activate transcription of pregenomic/precore mRNAs. A negative regulatory element (NRE) located further upstream can suppress core promoter activity [145]. Both ubiquitous and liver-enriched transcription factors interact with the core promoter to control transcription of the 3.5 kb transcripts (reviewed in [112]). Enhancer II activity is moderated by the binding of these transcription factors. The requirement for various liver-enriched transcription factors contributes to both the liver-enhanced replication and the species-specific infection that are characteristic of the hepadnaviruses [12, 149].

Control of surface antigen transcription is also complex. The S1 promoter controls production of the L surface antigen while the S2 promoter controls production of the M and S surface antigens. In each case, the promoters are influenced by both enhancers I and II [112, 150, 151]. The S1 promoter contains the canonical TATA-box sequence [152]. Similar to the situation seen with the core promoter, both ubiquitous and liver-enriched transcription factors positively and negatively impact the level of activity [112], but in this case only two *cis* elements seem to be required; an hepatocyte nuclear factor (HNF)-1 and the TATA-binding protein motif [153]. A NRE for this promoter is located in the S2 promoter [154]. S2 is a TATA-less promoter containing a CCAAT motif as well as other transcription initiation elements [155, 156]. Multiple elements within S2 control activity through interacting with both ubiquitous and liver-enriched transcription factors. S2 promoter is indirectly regulated by the impact of the L surface antigen on transcription factor activation [157].

The X promoter directs transcription of the X mRNAs. This promoter is located just upstream of the 5' end of the X transcript overlapping the 3' end of enhancer I. As with the core and S2 promoters, the traditional TATA box is missing. Both enhancer I and enhancer II can activate transcription of X mRNA [112, 150, 151, 158, 159]. Again, transcription is controlled by the interaction of numerous ubiquitous transcription factors.

In each case, although the transcription of the viral mRNAs originates at different positions on the genome and is driven by different promoters, all four of these transcript families use the same polyadenylation signal resulting in a common 3' end. However, because the hepadnaviruses have circular genomes, transcription of the precore/core/polymerase/pregenome mRNAs must proceed through the polyadenylation signal once without recognizing the signal to generate the terminally redundant RNA molecules. A positive effector of transcription (PET), which promotes lack of recognition of the polyadenylation signal during the first passage, has been mapped between transcription initiation and the polyadenylation signal [75]. A negative effector of transcription (NET) has also been identified as important in promoting recognition of the polyadenylation signal [76]. Recognition of NET resulting in associated polyadenylation is suppressed by PET [76].

Recently, a series of spliced mRNAs have been identified [80, 160, 161]. Doubly and singly spliced mRNAs derived from the pregenomic RNA of ~2.2 kb have been described in cell culture and liver extracts. In one case, the RNA represents a splice between core sequences and the surface domain. Very little is known about the mechanism that drives the generation of these spliced messages when the primary mRNAs of the hepadnaviruses are exported from the nucleus without splicing. Little is understood about the function of these messages or the products encoded by these messages, however, they may have roles in modulating pathogenesis and the production of defective interfering particles.

Translation

Translation of the viral proteins is also highly regulated and key to maintaining balanced levels of the viral proteins. Translation of the core/precore/polymerase mRNA occurs in the cytoplasm of the cell while the surface antigens are translated on the rough endoplasmic reticulum. This differential compartmentalization of the viral proteins is consistent with the maturation pathway of the virus. As described above, one of the 3.5-kb transcripts has three functions serving as the viral pregenome for replication and as mRNA for core and polymerase production. Although core and polymerase are translated from the same mRNA, the ratio of HBcAg to polymerase is approximately 250:1. Core protein is synthesized if translation initiates at the initial AUG in the mRNA [123]. If initiation occurs at the less-favored distal AUG, then polymerase is produced. When polymerase is produced, the polymerase binds to the 5' end of the RNA molecule and the RNA is directed to serve as the pregenome and no longer serves as mRNA for the core or polymerase proteins (reviewed in [114]).

Capsid assembly

This tight regulation promotes the *cis*-specific packaging of the pregenome in the nucleocapsid [162–164]. Assembly and maturation of the virion requires viral DNA synthesis [165, 166]. The polymerase interacts with a stem-loop structure near the 5' end of the viral pregenomic RNA referred to as epsilon [167], and triggers packaging of core protein dimers to form the icosahedral nucleocapsid [163, 167–169]. Although *in vitro* the core dimers assemble spontaneously into capsid structures [170–172] with both $T = 3$ and $T = 4$ symmetry [90, 171, 173], the core protein has a nucleic acid-binding 'protamine' domain that may be involved in the recognition of the viral nucleic acid during assembly into nucleocapsids or in subsequent interactions with the packaged viral nucleic acids [172]. A host kinase may also be packaged within the nucleocapsid [174]. A number of cellular proteins (chaperones) are directly and indirectly involved in assembly of the nucleocapsid and interaction of the polymerase with the RNA and, subsequently, DNA [175–179], but details on these interactions are not yet clear.

Viral genome synthesis

All hepadnaviruses require reverse transcription of the pregenomic RNA for replication. Within the nucleocapsids, the viral polymerase copies the pregenomic RNA into minus-strand DNA and then copies the minus-strand DNA into plus-strand DNA degrading the RNA pregenome to generate the relaxed circular genome present in the mature virion. Thus, as stated before, the viral polymerase is a multifunctional enzyme with RNA-dependent DNA polymerase (reverse transcriptase), DNA-dependent DNA polymerase, and RNase H activities. Replication of the hepadnavirus genome is surprisingly complex with multiple template switches. Using a small portion within epsilon in the RNA pregenome as a template [180–182], the polymerase serves as a self-primer for first-strand DNA synthesis [63–65]. Priming of minus-strand DNA initiates with the formation of a covalent bond (esterification) between a tyrosine of the viral polymerase and the first nucleotide to be added to the minus-strand DNA [66, 67]. A short segment of the minus-strand DNA (3–4 nucleotides) is synthesized at epsilon. The nascent DNA with the attached polymerase then switches template position by translocating to the 3' end of the pregenome at the DR1 locus [183, 184]. Specifics of this translocation are not fully understood. Synthesis of the rest of the minus-strand occurs with concomitant degradation of the RNA pregenome catalyzed by the RNase H activity of the polymerase [185, 186], leaving only a short segment at the 5' end of the pregenome containing an intact copy of the sequence DR1. This small RNA will serve as the primer for plus-strand DNA synthesis.

Plus-strand synthesis begins with the translocation of the RNA primer to an acceptor site on the minus-strand DNA [68, 70]. This site, DR2, is located near the 5' end of the minus-strand. The complementary character between DR1 on the RNA primer and DR2 on the DNA aids in directing and stabilizing this translocation. Synthesis of plus-strand DNA then proceeds to copy the portion of the redundant 5' end of the HBV minus-strand. This portion of the terminally redundant HBV genome enables a second translocation to occur in plus-strand generation [187, 188]. This second template switch is key to the formation of the relaxed circular viral genome. The growing plus-strand DNA shifts in base pairing from the nucleotides in the 5' end of the minus-strand to homologous sequence found in the redundant portion of the 3' end of the minus-strand. Not surprisingly, the GC/AT content of this region impacts the readiness with which this second template switch occurs [189]. Once this switch occurs, the polymerase continues to extend the growing plus-strand DNA that further stabilizes the relaxed circular character of the genome. The role of additional *cis*-acting sequences and structures regulating the generation of the relaxed circular structure is being explored for both avi- and orthohepadnaviruses [188–197].

Intracellular template amplification versus virus maturation

New viral nucleocapsids containing the relaxed circular viral genome follow two pathways within the cell. Nucleocapsids may mature by budding into the endoplasmic reticulum obtaining the viral envelope with subsequent maturation through the Golgi and release (reviewed in [198, 199]). Alternatively, the nucleocapsids may be transported back to the nucleus following the pathway of newly infecting virus [200]. This route results in an amplification of the nuclear cccDNA pool. The levels of cccDNA within the cell are regulated so that much higher levels of intermediate and relaxed circular DNA are detected. The presence and the amount of surface proteins is key to determining which pathway is followed [201]. If surface proteins are absent or the intracellular concentration of these is low, then the nucleocapsid is directed back to the nucleus resulting in the amplification of the cccDNA template. If large amounts of surface proteins are present, then the nucleocapsids are assembled into virions and secreted [201, 202].

Molecular targets for prophylactic and therapeutic opportunities

Approximately 350 million people are chronically infected with HBV [203, 204] and, as such, are persistent carriers of the virus. Of these, <5% will resolve the infection without therapeutic intervention each year [205]. Chronic carriers remain at risk for progressive liver disease including fibrosis, cirrhosis, hepatocellular carcinoma, and liver failure with an increase in morbidity and mortality [206] as well as serve as vectors for the spread of the virus. The global medical and economic burden associated with chronic hepatitis B infection is enormous. As such, tremendous effort has gone into the identification and development of both prophylactic and therapeutic intervention strategies.

Historically, intervention strategies have been more focussed toward host-mediated mechanisms for affecting viral replication. However with advances made in the knowledge of HBV replication, efforts are now targeting specific aspects of the viral life cycle. For the treatment of patients with HBV infection, interferon was not initially considered a direct antiviral but rather a broad-spectrum agent with potential antiviral benefit. Mechanistic studies defining interferon effects on HBV through signal transduction pathways have certainly supported this concept [207]. Therapeutic efforts have focussed on interferon to promote sustained HBeAg and HBV DNA reductions to below detectable limits with hopes for establishment of protective seroconversion. Studies with interferon α have proven successful in <50% of chronic hepatitis B patients. However, largely because of its broad spectrum of activities, therapy with interferon α is often associated with intolerable and numerous adverse effects.

Knowledge of viral antigens and the immune response has been key to the development of the HBV vaccine. The most common prophylactic vaccines contain only recombinant HBsAg. While the immune system in most people responds with the development of protective immune responses, some do not. The proportion of patients failing to develop protective immune responses varies with age and many other factors. As only one example, infants born to mothers with a high level of viremia often develop chronic hepatitis B despite vaccination. In these cases, studies are being conducted combining vaccine strategies with approved therapeutic antivirals to reduce viral levels in the mother and thus reduce transmission while offering immunologic stimulation to the infants. In addition novel vaccines containing combinations of HBV proteins (L, M, and S or S and core) with various adjuvants as well as alternative delivery are being considered.

The success of targeting specific aspects of the viral life cycle is dependent on an understanding of the molecular virology associated with HBV. Using this knowledge, therapeutic efforts can target specific steps in the viral life cycle. In part, this approach has been taken in the hope that side effects may be minimized when specificity is improved. Some investigators are evaluating macromolecular approaches such as small interfering RNA, antisense oligonucleotides and ribozymes [208–210]. However, the most advanced and successful of these strategies involves discovery and development of more traditional small-molecule inhibitors of defined viral enzymatic activities such as polymerase inhibitors.

Two antivirals (lamivudine and adefovir dipivoxil) now have approval in many countries for the treatment of chronic hepatitis B virus infections. Both lamivudine, a nucleoside analog, and adefovir dipivoxil, a nucleotide analog, block viral replication by inhibiting the viral polymerase. Many chronic hepatitis B patients treated for 1 year with these agents exhibit multilog reductions in serum viral DNA levels and show improvements in liver histology [211–217]. However, as with interferon, not all patients respond with sustained reductions in HBeAg and HBV DNA to levels below the limits of detection and with sustainable clinical improvements at the end of treatment. In addition, while many of the adverse effects apparent with interferon are not detected, drug-resistant mutants emerge in a proportion of patients on therapy with these agents [211, 215, 218–222]. The initial clinical benefits of treatment such as the improvements in liver histology may not be maintained after mutants emerge [221, 223]. Consistent with the mode of action of these antiviral agents, these drug-resistant mutants contain amino acid changes in the viral polymerase [218–220, 222]. More polymerase inhibitors are in various stages of development and will be discussed in other chapters. It is likely that some of these will prove more potent in virological suppression with associated clinical benefit. But it is also likely that not all patients will develop sustained responses when antiviral treatment is stopped and that viruses resistant to these agents will emerge. As such, a need exists for additional treatments focussed on different molecular targets in the life cycle of the virus.

Additional approaches evaluating small-molecule inhibitors of viral replication that do not target the viral polymerase are being explored. Recently, non-nucleosidic inhibitors of viral replication have been identified and shown to be potent in both *in vitro* and *in vivo* models [224–230]. Some mediate N-glycan processing of the viral envelope proteins and affect transport within the cell by host chaperones [226–229]. One of the most promising series includes heteroaryldihydropyrimidines, which function by preventing the correct assembly of the viral nucleocapsid and through this mechanism, promote the degradation of intracellular core protein [225]. It remains to be seen if any these approaches will prove efficacious in the clinic. However, now that progress has been made in this area, there is hope that additional novel agents may be discovered that affect various steps of the viral life cycle.

Whether any of these agents can effectively clear the cccDNA template pool remains to be determined. The pool of cccDNA within the cell is remarkably stable. In the duck hepadnavirus system, 5–50 copies of cccDNA are maintained per cell [231]. It is likely that cccDNA levels will also have characteristic levels in the orthohepadnaviruses as well. In the natural infection, two pathways have been suggested for clearing viral template. One pathway involves a noncytolytic cytokine-mediated clearance of the cccDNA pool defined by interferon γ and tumor necrosis factor α (TNF-α) production (reviewed in [232, 233]) and the CD8+ T cell response. A second pathway predicts cccDNA loss as associated with infected cell-turnover [234, 235]. It seems likely that cccDNA turnover involves a combination of these two pathways.

Summary

In summary, as our knowledge of the molecular aspects of the viral life cycle has evolved, so too has the potential for preventing new HBV infections and for treating existing patients with chronic hepatitis B. It may be that the best approach for treating patients with HBV infection involves a combination of antivirals or a combination of immunologic and antiviral treatments. These prospects will be discussed in detail in subsequent chapters.

Acknowledgements
The authors are indebted to Marchelle I. Allen for excellent editorial assistance.

References

1 Summers J, Smolec J, Snyder R (1978) A virus similar to human hepatitis B virus associated with hepatitis and hepatoma in woodchucks. *Proc Natl Acad Sci USA* 75: 4533–4537
2 Marion PL, Oshiro LS, Regnery DC, Scullard GH, Robinson WS (1980) A virus in Beechey ground squirrels that is related to hepatitis B of humans. *Proc Natl Acad Sci USA* 77: 2941–2945
3 Mason WS, Seal G, Summers J (1980) Virus of Pekin ducks with structural and biological relatedness to human hepatitis B virus. *J Virol* 36: 829–836
4 Sprengel R, Kaleta EF, Will H (1988) Isolation and characterization of a hepatitis B virus endemic in herons. *J Virol* 62: 932–937
5 Lanford RE, Chavez D, Brasky KM, Burns III RB, Rico-Hesse R (1998) Isolation of a hepadnavirus from the woolly monkey, a New World primate. *Proc Natl Acad Sci USA* 95: 5757–5761
6 Barker LF, Chisari FV, McGrath PP, Dalgard DW, Kirschstein RL, Almeida JD, Edgington TS, Sharp DG, Peterson MR (1973) Transmission of type B viral hepatitis to chimpanzees. *J Infect Dis* 127: 648–652
7 Aldrich CE, Coates L, Wu TT, Newbold J, Tennant BC, Summers J, Seeger C, Mason WS (1989) *In vitro* infection of woodchuck hepatocytes with woodchuck hepatitis virus and ground squirrel hepatitis virus. *Virology* 172: 247–252
8 Lanford RE, Chavez D, Rico-Hesse R, Mootnick A (2000) Hepadnavirus infection in captive gibbons. *J Virol* 74(6): 2955–2959
9 Chouteau P, Le Seyec J, Cannie I, Nassal M, Guguen-Guillouzo C, Gripon P (2001) A short N-proximal region in the large envelope protein harbors a determinant that contributes to the species specificity of human hepatitis B virus. *J Virol* 75: 11565–11572
10 Kock J, Nassal M, MacNelly S, Baumert TF, Blum HE, von Weizsacker F (2001) Efficient Infection of Primary Tupaia Hepatocytes with Purified Human and Woolly Monkey Hepatitis B virus. *J Virol* 75: 5084–5089
11 Guidotti LG, Matzke B, Schaller H, Chisari FV (1995) High-level hepatitis B virus replication in mice. *J Virol* 69: 6158–6169
12 Tang H, McLachlan A (2001) Transcriptional regulation of hepatitis B virus by nuclear hormone receptors is a critical determinant of viral tropism. *Proc Natl Acad Sci USA* 98: 1841–1846
13 Hu X, Margolis HS, Purcell RH, Ebert J, Robertson BH (2000) Identification of hepatitis B indigenous to chimpanzees. *Proc Natl Acad Sci USA* 97: 1661–1664
14 Lanford RE, Chavez D. Barrera A. Brasky KM (2003) An infectious clone of woolly monkey hepatitis B virus. *J Virol* 77(14): 7814–7819
15 Makuwa M, Souquiere S, Telfer P, Leroy E, Bourry O, Rouquet P, Clifford S, Wickings EJ, Roques P, Simon F (2003) Occurrence of hepatitis viruses in wild-born non-human primates: a 3 year (1998–2001) epidemiological survey in Gabon. *J Med Primatol* 32: 307–314
16 Starkman SE, MacDonald DM, Lewis JCM, Holmes EC, Simmonds P (2003) Geographic and species association of hepatitis B virus genotypes in non-human primates. *Virology* 314: 381–393
17 Takahashi K, Mishiro S, Prince AM (2001) Novel hepatitis B virus strain from a chimpanzee of Central Africa (Pan troglodytes troglodytes) with an unusual antigenicity of the core protein. *Intervirology* 44: 321–326

18 Tennant BC, Mrosovsky N, McLean K, Cote PJ, Korba BE, Engle RE, Gerin JL, Wright J, Michener GR, Uhl E et al. (1991) Hepatocellular carcinoma in Richardson's ground squirrels (Spermophilus richardsonii); evidence for association with hepatitis B-like virus infection. *Hepatology* 13: 1215–1221
19 Testut P, Renard CA, Terradillos O, Vitvtski TL, Tekaia F, Degott C, Blake J, Boyer B, Buendia MA (1996) A new hepadnavirus endemic in arctic ground squirrels in Alaska. *J Virol* 70: 4210–4219
20 Verschoor EJ, Warren KS, Langenhuijzen S, Heriyanto, Swan RA, Heeney JL (2001) Analysis of two genomic variants of orangutan hepadnavirus and their relationship to other primate hepatitis B-like viruses. *J Gen Virol* 82: 893–897
21 Walter E, Keist R, Niederost B, Pult I, Blum HE (1996) Hepatitis B virus infection of tupaia hepatocytes *in vitro* and *in vivo*. *Hepatology* 24: 1–5
22 Prassolov A, Hohenberg H, Kalinina T, Schneider C, Cova L, Krone O, Frolich K, Will H, Sirna H (2003) New hepatitis B virus of cranes that has an unexpected broad host range. *J Virol* 77: 1964–1976
23 Pult I, Netter HJ, Bruns M, Prassolov A, Sirna H, Hohenberg H, Chang SF, Frolich K, Krone O, Kaleta EF, Will H (2001) Identification and analysis of a new hepadnavirus in white storks. *Virology* 289: 114–128
24 Tang H, McLachlan A (2002) Avian and mammalian hepadnaviruses have distinct transcription factor requirements for viral replication. *J Virol* 76: 7468–7472
25 Blum HE, Geballe AP, Stowring L, Figus A, Haase A, Vyas, G (1984) Hepatitis B virus in non-hepatocytes: demonstration of viral DNA in spleen, bile duct epithelium, and vascular elements by *in situ* hybridization. In GN Vyas, JL Dienstag, JH Hoofnagle (eds): *Viral hepatitis and liver disease*. Grune and Stratton, New York, 634
26 Cabrerizo M, Bartolome J, Carreno V (2002) *In vitro* infection of human peripheral blood mononuclear cells by a defective hepatitis B virus with a deletion in the PreS1 region of the viral genome. *J Viral Hep* 9: 265–271
27 Hadchouel M, Pasquinelli C, Fournier JG, Hugon RN, Scotto J, Bernard O, Brechot C (1988) Detection of mononuclear cells expressing hepatitis B virus in peripheral blood from HbsAg positive and negative patients by *in situ* hybridisation. *J Med Virol* 24: 27–32
28 Halpern MS, England JM, Deery DT, Petcu DJ, Mason WS, Molnar-Kimber KL (1983) Viral nucleic acid synthesis and antigen accumulation in pancreas and kidney of Pekin ducks infected with duck hepatitis B virus. *Proc Natl Acad Sci USA* 80: 4865–4869
29 Michalak TI, Mulrooney PM, Coffin CS (2004) Low doses of hepadnavirus induce infection of the lymphatic system that does not engage the liver. *J Virol* 78: 1730–1738
30 Mulrooney PM, Michalak TI (2003) Quantitative detection of hepadnavirus-infected lymphoid cells by *in situ* PCR combined with flow cytometry: Implications for the study of occult virus persistence. *J Virol* 77: 970–979
31 Ogston CW, Schechter EM, Humes CA, Pranikoff MB (1989) Extrahepatic replication of woodchuck hepatitis virus in chronic infection. *Virology* 169: 9–14
32 Pontisso P, Poon MC, Tiollais P, Brechot C (1984) Detection of hepatitis B virus DNA in mononuclear blood cells. *Br Med J Clin Res* 288: 1563–1566
33 Shimoda T, Shikata T, Karasawa T, Tsukagoshi S, Yoshimura M, Sakurai I (1981) Light microscopic localization of hepatitis B virus antigens in human pancreas. *Gastroenterology* 81: 998–1005
34 Yoffe B, Noonan CA, Melnick JL, Hollinger FB (1986) Hepatitis B virus DNA in mononuclear cells and analysis of cell subsets for the presence of replicative intermediates of viral DNA. *J Infect Dis* 153: 471–477
35 Lok ASF, McMahon BJ (2001) Chronic hepatitis B. *Hepatology* 34: 1225–1241
36 Courouce-Pauty AM, Lemaire JM, Roux JF (1978) New hepatitis B surface antigen subtypes inside the ad category. *Vox Sang* 35: 304–308
37 LeBouvier GL, McCollum RW, Hierholzer WJ, Irwin GR, Krugman S, Giles JP (1972) Subtypes of Australia antigen and hepatitis B virus. *JAMA* 222: 928–930
38 Magnius LO, Norder H (1995) Subtypes, genotypes and molecular epidemiology of the hepatitis B virus as reflected by sequence variability of the S-gene. *Intervirology* 38: 24–34
39 Okamoto H, Tsuda F, Sakagawa H, Sastrosoeqignjo RI, Imai M, Miyakawa Y, Mayumi M (1988) Typing hepatitis B virus by homology in nucleotide sequence: comparison of surface antigen subtypes. *J Gen Virol* 69: 2575–2583

40 Raney AK, McLachlan A (1991) The Biology of Hepatitis B Virus. In: A McLachlan (ed): *Molecular Biology of Hepatitis B virus.* CRC Press, Boca Raton, FL, 1–37
41 Arauz-Ruiz P, Norder H, Robertson BH, Magnius LO (2002) Genotype H: a new Amerindian genotype of hepatitis B revealed in Central America. *J Gen Virol* 83: 2059–2073
42 Kato H, Orito E, Gish RG, Sugauchi F, Suzuki S, Ueda R, Miyakawa Y, Mizokami M (2002) Characteristics of hepatitis B virus isolates of genotype G and their phylogenetic differences from the other six genotypes (A-F). *J Virol* 76: 6131–6137
43 Norder H, Courouce AM, Magnius LO (1994) Complete genomes, phylogenetic relatedness and structural proteins of six strains of the hepatitis B virus, four of which represent two new genotypes. *Virology* 198: 489–503
44 Stuyver L, de Gendt S, van Geyt C, Zoulim F, Fried M, Schinazi RF, Rossau R (2000) A new genotype of hepatitis B virus: complete genome and phylogenetic relatedness. *J Gen Virol* 81: 67–74
45 Devesa M, Rodriguez C, Leon G, Liprandi F, Pujol FH (2004) Clade analysis and surface antigen polymorphism of hepatitis B virus American genotypes. *J Med Virol* 72: 377–384
46 Huy TTT, Ushijima H, Quang VX, Win KM Luengrojanakul P, Kikuchi K, Sata T, Abe K (2004) Genotype C of hepatitis B virus can be classified into at least two subgroups. *J Gen Virol* 85: 283–292
47 Norder H, Arauz-Ruiz P, Blitz L, Pujol FH, Echevarria JM, Magnius LO (2003) The T-1858 variant predisposing to the precore stop mutation correlates with one of two major genotype F hepatitis B virus clades. *J Gen Virol* 84: 2083–2087
48 Robertson BH, Margolis HS (2002) Primate hepatitis B viruses – genetic diversity, geography and evolution. *Rev Med Virol* 12: 133–141
49 Lindh M, Andersson AS, Gusdal A (1997) Genotypes, nt 1858 variants, and geographic origin of hepatitis B virus large-scale analysis using a new genotyping method. *J Infect Dis* 175: 1285–1293
50 Norder H, Hammas B, Lee SD, Bile K, Courouce AM, Mushahwar IK, Magnius LO (1993) Genetic relatedness of hepatitis B viral strains of diverse geographical origin and natural variations in the primary structure of the surface antigen. *J Gen Virol* 74: 1341–1348
51 Chu C-J, Hussain M, Lok ASF (2002) Genotype B is associated with earlier HBeAg Seroconversion compared with hepatitis B virus genotype C. *Gastroenterology* 122: 1756–1762
52 Chu C-J, Keefe, EB, Han, S-H, Perrillo RP, Min AD, Soldevila-Pico C, Carey W, Brown Jr, RS, Luketic VA, Terrault N, Lok ASF, the US HBV Epidemiology Study Group (2003) Hepatitis B virus genotypes in the United States: results of a nationwide survey. *Gastroenterology* 125: 444–451
53 Kao JH (2002) Hepatitis B viral genotypes: clinical relevance and molecular characteristics. *J Gastroenterol Hepatol* 17: 643–650
54 Kao JH, Wu NH, Chen PJ, Lai MY, Chen DS (2000) Hepatitis B genotypes and the response to interferon therapy. *J Hepatol* 33: 998–1002
55 Sumi H, Yokosuka O, Seki N, Arai M, Imazeki F, Kurihara T, Kanda T, Fukai K, Kato M, Saisho H (2003) Influence of hepatitis B virus genotypes on the progression of chronic type B liver disease. *Hepatology* 37: 19–26
56 Gripon P, Rumin S, Urban S, Le Seyec J, Glaise D, Cannie I, Guyomard C, Lucas J, Trepo C, Guguen-Guillouzo C (2002) Infection of a human hepatoma cell line by hepatitis B virus. *Proc Natl Acad Sci USA* 99: 15655–15660
57 Galle PR, Hagelstein J, Kommerell B, Volkmann M, Schranz P, Zentgraf H (1994) *In vitro* experimental infection of primary human hepatocytes with hepatitis B virus. *Gastroenterology* 106: 664–673
58 Gripon P, Diot C, Guguen-Guillouzo C (1993) Reproducible high level infection of cultured adult human hepatocytes by hepatitis B virus: effect of polyethylene glycol on adsorption and penetration. *Virology* 192: 534–540
59 Lu XY, Block TM, Gerlich WH (1996) Protease-induced infectivity of hepatitis B virus for a human hepatoblastoma cell line. *J Virol* 70: 2277–2285
60 Lu XY, Hazboun T, Block T (2001) Limited proteolysis induces woodchuck hepatitis virus infectivity for human HepG2 cells. *Virus Res* 73: 27–40
61 Mason WS, Halpern MS, England JM, Seal G, Egan, J, Coates L, Aldrich C, Summers J (1983) Experimental transmission of duck hepatitis B virus. *Virology* 131: 375–384
62 Tuttleman JS, Pugh JC, Summers JW (1986) *In vitro* experimental infection of primary duck hepatocyte cultures with duck hepatitis B virus. *J Virol* 58: 17–25

63 Lanford RE, Notvall L, Beames B (1995) Nucleotide priming and reverse transcriptase activity of hepatitis virus polymerase expressed in insect cells. *J Virol* 69: 4431–4439
64 Tavis JE, Ganem D (1993) Expression of functional hepatitis B virus polymerase in yeast reveals it to be the sole viral protein required for correct initiation of reverse transcription. *Proc Natl Acad Sci USA* 90: 4107–4111
65 Wang GH, Seeger C (1992) The reverse transcriptase of hepatitis B virus acts as a protein primer for viral DNA synthesis. *Cell* 71: 663–670
66 Weber M, Bronsema V, Bartos H, Bosserhoff A, Bartenschlager R, Schaller H (1994) Hepadnavirus P protein utilizes a tyrosine residue in the TP domain to prime reverse transcription. *J Virol* 68: 2994–2999
67 Zoulim F, Seeger C (1994) Reverse transcription in hepatitis B viruses is primed by a tyrosine residue of the polymerase. *J Virol* 68: 6–13
68 Lien JM, Aldrich CE, Mason WS (1986) Evidence that a capped oligoribonucleotide is the primer for duck hepatitis B virus plus-strand DNA synthesis. *J Virol* 57: 229–236
69 Lien JM, Petcu DJ, Aldrich CE, Mason WS (1987) Intitation and termination of duck hepatitis B virus DNA synthesis during virus maturation. *J Virol* 61: 3832–3840
70 Seeger C, Ganem D, Varmus HE (1986) Biochemical and genetic evidence for the hepatitis B virus replication strategy. *Science* 232: 477–484
71 Hruska JF, Clayton DA, Rubenstein JL, Robinson WS (1977) Structure of hepatitis B Dane particle DNA before and after the Dane particle DNA polymerase reaction. *J Virol* 21: 666–672
72 Landers TA, Greenberg HB, Robinson WS (1977) Structure of hepatitis B Dane particle DNA and nature of the endogenous DNA polymerase reaction. *J Virol* 23: 368–376
73 Summers J, O'Connell A, Millman I (1975) Genome of hepatitis B virus: restriction enzyme cleavage and structure of DNA extracted from Dane particles. *Proc Natl Acad Sci USA* 72: 4597–4601
74 Tur-Kaspa R, Shaul Y, Moore DD, Burk RD, Okret S, Poellinger L, Shafritz DA (1988) The glucocorticoid receptor recognizes a specific nucleotide sequence in hepatitis B virus DNA causing increased activity of the HBV enhancer. *Virology* 167: 630–633
75 Huang M, Summers J (1994) Pet, a small sequence distal to the pregenome cap site, is required for expression of the duck hepatitis B virus pregenome. *J Virol* 68: 1564–1572
76 Beckel-Mitchener A, Summers J (1997) A novel transcriptional element in circular DNA monomers of the duck hepatitis B virus. *Virology* 71: 7917–7922
77 Bartenschlager R, Junker-Niepmann M, Schaller H (1990) The P gene product of hepatitis B virus is required as a structural component for genomic RNA encapsidation. *J Virol* 64: 5324–5332
78 Chang SF, Netter HJ, Hildt E, Schuster R, Schaefer S, Hsu YC, Rang A, Will H (2001) Duck hepatitis B virus expresses a regulatory HBx-like protein from a hidden open reading frame. *J Virol* 75: 161–170
79 Meier P, Scougall CA, Will H, Burrell CJ, Jilbert AR (2003) A duck hepatitis B virus strain with a knockout mutation in the putative X ORF shows similar infectivity and *in vivo* growth characteristics to wild-type virus. *Virology* 317: 291–298
80 Soussan P, Tuveri R, Nalpas B, Garreau F, Zavala F, Masson A, Pol S, Brechot C, Kremsdorf D (2003) The expression of hepatitis B spliced protein (HBSP) encoded by a spliced hepatitis B virus RNA is associated with viral replication and liver fibrosis. *J. Hepatology* 38: 343–348
81 Bruss V, Lu X, Thomssen R, Gerlich WH (1994) Post-translational alterations in transmembrane topology of the hepatitis B virus large envelope protein. *EMBO J* 13: 2273–2279
82 Guo J-T, Pugh JC (1997) The topology of the large envelope protein of duck hepatitis B virus suggests a mechanism for membrane translocation during particle morphogenesis. *J Virol* 71: 1107–1114
83 Lu XY, Mehta A, Dwek R, Butters T, Block T (1995) Evidence that N-linked glycosylation is necessary for hepatitis B virus secretion. *Virology* 213: 660–665
84 Mehta A, Lu X, Block TM, Blumberg BS, Dwek RA (1997) Hepatitis B virus (HBV) envelope glycoproteins vary drastically in their sensitivity to glycan processing: Evidence that alteration of a single N-linked glycosylation site can regulate HBV secretion. *Proc Natl Acad Sci USA* 94: 1822–1827
85 Schmitt S, Glebe D, Alving K, Tolle TK, Linder M, Geyer H, Linder D, Peter-Katalinic J, Gerlich WH (1999) Analysis of the pre-S2 N- and O-linked glycans of the M surface protein from human hepatitis B virus. *J Biol Chem* 274: 11945–11957
86 Bruss V, Hagelstein J, Gerhardt E, Galle PR (1996) Myristylation of the large surface protein is

required for hepatitis B virus *in vitro* infectivity. *Virology* 218: 396–409
87 Gripon, P, LeSeyec J, Rumin S, Guguen-Guillouzo C (1995) Myristylation of the hepatitis B virus large surface protein is essential for viral infectivity. *Virology* 213: 292–299
88 Borel C, Sunyach C, Hantz O, Trepo C, Kay A (1998) Phosphorylation of DHBV pre-S: identification of the major site of phosphorylation and effects of mutations on the virus life cycle. *Virology* 242: 90–98
89 Guidotti LG, Matzke B, Pasquinelli C, Shoenberger JM, Rogler CE, Chisari FV (1996) Hepatitis B virus (HBV) precore protein inhibits HBV replication in transgenic mice. *J Virol* 70: 7056–7061
90 Wingfield, PT, Stahl, SJ, Williams, RW, Steven, AC (1995) Hepatitis Core antigen produced in Escherichia coli: subunit composition, conformational analysis, and *in vitro* capsid assembly. *Biochemistry* 34: 4919–4932
91 Chen HS, Kew MC, Hornbuckle WE, Tennant BC, Cote PJ, Gerin JL, Purcell RH, Miller RH (1992) The precore gene of the woodchuck hepatitis virus genome is not essential for viral replication in the natural host. *J Virol* 66: 5682–5684
92 Hadziyannis SJ, Vassilopoulos D (2001) Hepatitis B e antigen negative chronic hepatitis B. *Hepatology* 34: 617–624
93 Milich DR, Jones JE, Hughes JL, Price J, Raney AK, McLachlan A (1990) Is a function of the secreted hepatitis B e antigen to induce immunologic tolerance *in utero*? *Proc Natl Acad Sci USA* 87: 6599–6603
94 Oberhaus SM, Newbold JE (1995) Detection of an RNase H activity associated with hepadnaviruses. *J Virol* 69: 5697–5704
95 Chen HS, Kaneko S, Girones R, Anderson RW, Hornbuckle WE, Tennant BC, Cote PJ, Gerin JL, Purcell RH, Miller RG (1993) The woodchuck hepatitis virus X gene is important for establishment of virus infection in woodchucks. *J Virol* 67: 1218–1226
96 Zoulim F, Saputelli J, Seeger C (1994) Woodchuck hepatitis virus X protein is required for viral infection *in vivo*. *J Virol* 68: 2026–2030
97 Chang SF, Netter HJ, Hildt E, Schuster R, Schaefer S, Hsu YC, Rang A, Will H (2001) Duck hepatitis B virus expresses a regulatory HBx-like protein from a hidden open reading frame. *J Virol* 75: 161–170
98 Meier P, Scougall CA, Will H, Burrell CJ, Jilbert AR (2003) A duck hepatitis B virus strain with a knockout mutation in the putative X ORF shows similar infectivity and *in vivo* growth characteristics to wild-type virus. *Virology* 317: 291–298
99 Balsano C, Avantaggiati ML, Natoli G, De ME, Will H, Perricaudet M, Levrero M (1991) Full-length and truncated versions of the hepatitis B virus (HBV) X protein (pX) transactivate the cmyc protooncogene at the transcriptional level. *Biochem Biophy Res Commun* 176: 985–992
100 Colgrove R, Simon G, Ganem D (1989) Transcriptional activation of homologous and heterologous genes by the hepatitis B virus X gene product in cells permissive for viral replication. *J Virol* 63: 4019–4026
101 Haviv I, Shamay M, Doitsh G, Shaul Y (1998) Hepatitis B virus pX targets TFIIB in transcription coactivation. *Mol Cell Biol* 18: 1562–1569
102 Benn J, Schneider RJ (1994) Hepatitis B virus HBx protein activates Ras-GTP complex formation and establishes a Ras, Raf, MAP kinase signaling cascade. *Proc Natl Acad Sci USA* 91: 10350–10354
103 Lee TH, Elledge SJ, Butel JS (1995) Hepatitis B virus X protein interacts with a probable cellular DNA repair protein. *J Virol* 69: 1107–1114
104 Bouchard MJ, Puro RJ, Wang L, Schneider RJ (2003) Activation and inhibition of cellular calcium and tyrosine kinase signaling pathways identify targets of the HBx protein involved in hepatitis B virus replication. *J Virol* 77: 7713–7719
105 Bouchard MJ, Wang JH, Schneider RJ (2001) Calcium signaling by HBx protein in hepatitis B virus DNA replication. *Science* 294: 2376–2378
106 Melegari M, Scaglioni PP, Wands JR (1998) Cloning and characterization of a novel hepatitis B virus × binding protein that inhibits viral replication. *J Virol* 72: 1737–1743
107 Hu KQ, Vierling JM, Siddiqui A (1990) Trans-activation of HLA-DR gene by hepatitis B virus X gene product. *Proc Natl Acad Sci USA* 87: 7140–7144
108 Zhou DX, Taraboulos A, Ou JH, Yen TS (1990) Activation of class I major histocompatibility complex gene expression by hepatitis B virus. *J Virol* 64: 4925–4928
109 Huang J, Kwong J, Sun E-Y, Liang T (1996) Proteasome complex as a potential cellular target of

hepatitis B virus X protein. *J Virol* 70: 5582–5591

110 Schuster R, Gerlich WH, Schaefer S (2000) Induction of apoptosis by the transactivating domains of the hepatitis B virus X gene leads to suppression of oncogenic transformation of primary rat embryo fibroblasts. *Oncogene* 19: 1173–1180

111 Schuster R, Hildt E, Chang SF, Terradillos O, Pollicino T, Lanford R, Gerlich WH, Will H, Schaefer S (2002) Conserved transactivating and pro-apoptotic functions of hepadnaviral X protein in ortho- and avihepadnaviruses. *Oncogene* 21: 6606–6613

112 Moolla N, Kew M, Arbuthnot P (2002) Regulatory elements of hepatitis B virus transcription. *J Viral Hep* 9: 323–331

113 Seeger C, Mason WS (1996) Replication of the hepatitis virus genome. In: ML DePamphilis (ed): *DNA replication in eukaryotic cells*. Cold Spring Harbor Press, Cold Spring Harbor, NY, 815–831

114 Seeger C, Mason WS (2000) Hepatitis B virus biology. *Microbiol Molec Biol Rev* 64: 51–68

115 Lambert V, Chassot S, Kay A, Trepo C, Cova L (1991) *In vivo* neutralization of duck hepatitis B virus by antibodies specific to the N-terminal portion of pre-S protein. *Virology* 185: 446–450

116 Lambert V, Fernholz D, Sprengel R, Fourel I, Deleage G, Wildner G, Peyret C, Trepo C, Cova L, Will H (1990) Virus-neutralizing monoclonal antibody to a conserved epitope on the duck hepatitis B virus pre-S protein. *J Virol* 64: 1290–1297

117 Yuasa S, Cheung RC, Pham Q, Robinson WS, Marion PL (1991) Peptide mapping of neutralizing and nonneutralizing epitopes of duck hepatitis B virus pre-S polypeptide. *Virology* 181: 14–21

118 Ishikawa T, Ganem D (1995) The pre-S domain of the large viral envelope protein determines host range in avian hepatitis B viruses. *Proc Natl Acad Sci USA* 92: 6259–6263

119 Spangenberg HC, Lee HB, Li JS, Tan FL, Skidgel R, Wands JR, Tong SP (2001) A short sequence within domain C of duck carboxypeptidase D is critical for duck hepatitis B virus binding and determines host specificity. *J Virol* 75: 10630–10642

120 Urban S, Gripon P (2002) Inhibition of duck hepatitis B virus infection by a myristoylated pre-S peptide of the large viral surface protein. *J Virol* 76: 1986–1990

121 Breiner KM, Schaller H, Knolle PA (2001) Endothelial cell-mediated uptake of a hepatitis B virus: a new concept of liver targeting of hepatotropic microorganisms. *Hepatology* 34: 803–808

122 Breiner KM, Urban S, Glass B, Schaller H (2001) Envelope protein-mediated down regulation of hepatitis B virus receptor in infected hepatocytes. *J Virol* 75: 143–150

123 Breiner KM, Urban S, Schaller H (1998) Carboxypeptidase D (gp180), a golgi-resident protein, functions in the attachment and entry of avian hepatitis B viruses. *J Virol* 72: 8098–8104

124 Eng FJ, Novikova EG, Kuroki K, Ganem D, Fricker LD (1998) Gp180, a protein that binds duck hepatitis B virus particles, has metallo carboxypeptidase D-like enzymatic activity. *J Biol Chem* 273: 8382–8388

125 Kuroki K, Cheung R, Marion PL, Ganem D (1994) A cell surface protein that binds avian hepatitis B virus particles. *J Virol* 68: 2091–2096

126 Kuroki K, Eng F, Ishikawa T, Turck C, Harada F, Ganem D (1995) gp 180, a host cell glycoprotein that binds duck hepatitis B virus particles, is encoded by a member of the carboxypeptidase gene family. *J Biol Chem* 270: 15022–15028

127 Li J, Tong SP, Lee HB, Perdigoto AL, Spangenberg HC, Wands JR (2004) Glycine decarboxylase mediates a postbinding step in duck hepatitis B virus infection. *J Virol* 78: 1873–1881

128 Tong SP, Li JS, Wands JR (1999) Carboxypeptidase D is an avian hepatitis B virus receptor. *J Virol* 73: 8696–8702

129 Urban S, Kruse C, Multhaup G (1999) A soluble form of the avian hepatitis B virus receptor – biochemical characterization and functional analysis of the receptor ligand complex. *J Biol Chem* 274: 5707–5715

130 Mcguire GB, Tan FL, Michel B, Rehli M, Skidgel RA (1997) Identification of a membrane-bound carboxypeptidase as the mammalian homolog of duck gp180, a hepatitis B virus-binding protein. *Life Sci* 60: 715–724

131 De Meyer S, Gong ZJ, Depla E, Maertens G, Yap SH (1999) Involvement of phosphatidylserine and nonphospholipid components of the hepatitis B virus envelope in human Annexin V binding and in HBV infection *in vitro*. *J Hepatol* 31: 783–790

132 Gong ZJ, de Meyer S, van Pelt J, Hertogs K, Depla E, Soumillion A, Fevery J, Yap SH (1999) Transfection of a rat hepatoma cell line with a construct expressing human liver annexin V confers susceptibility to hepatitis B virus infection. *Hepatology* 29: 576–584

133 Rabe B, Vlachou A, Pante N, Helenius A, Kann M (2003) Nuclear import of hepatitis B virus capsids and release of the viral genome. *Proc Natl Acad Sci USA* 100: 9849–9854
134 Guidotti LG, Martinez V, Loh YT, Rogler CE, Chisari FV (1994) Hepatitis B virus nucleocapsid particles do not cross the hepatocyte nuclear membrane in transgenic mice. *J Virol* 68: 5469–5475
135 Kann M, Sodeik B, Vlachou A, Gerlich WH, Helenius A (1999) Phosphorylation-dependent binding of hepatitis B virus core particles to the nuclear pore complex. *J Cell Biol* 145: 45–55
136 Kock J, Kann M, Putz G, Blum HE, von Weizsacker F (2003) Central role of a serine phosphorylation site within duck hepatitis B virus core protein for capsid trafficking and genome release. *J Biol Chem* 278: 28123–28129
137 Singh S, Zlotnick A (2003) Observed hysteresis of virus capsid disassembly is implicit in kinetic models of assembly. *J Biol Chem* 278: 18249–18255
138 Barrasa MI, Guo J-T, Saputelli J, Mason WS, Seeger C (2001) Does a cdc2 kinase-like recognition motif on the core protein of hepadnaviruses regulate assembly and disintegration of capsids? *J Virol* 75: 2024–2028
139 Newbold JE, Xin H, Tencza M, Sherman G, Dean J, Bowden S, Locarnini S (1995) The covalently closed duplex form of the hepadnavirus genome exists *in situ* as a heterogeneous population of viral minichromosomes. *J Virol* 69: 3350–3357
140 Moraleda G, Saputelli J, Aldrich CE, Averett D, Condreay L, Mason WS (1997) Lack of effect of antiviral therapy in nondividing hepatocyte cultures on the closed circular DNA of woodchuck hepatitis virus. *J Virol* 71: 9392–9399
141 Pourquier P, Jensen AD, Gong SS, Pommier Y, Rogler CE (1999) Human DNA topoisomerase I mediated cleavage and recombination of duck hepatitis B virus DNA *in vitro*. *Nucleic Acids Res* 27: 1919–1925
142 Schirmacher P, Wang HP, Stahnke G, Will H, Rogler CE (1995) Sequences and structures at hepadnaviral integration – recombination sites implicate topoisomerase I in hepadnaviral DNA rearrangements and integration. *J Hepatol* 22 (S1): 21–33
143 Lu CC, Chen M, Ou JH, Yen TS (1995) Key role of a CCAAT element in regulating hepatitis B virus surface protein expression. *Virology* 206: 1155–1158
144 Chen IH, Huang CJ, Ting LP (1995) Overlapping initiator and TATA box funtions in the basal core promoter of hepatitis B virus. *J Virol* 69: 3647–3657
145 Chen M, Ou JH (1995) Cell type-dependent regulation of the activity of the negative regulatory element of the hepatitis B virus core promoter. *Virology* 214: 198–206
146 Yaginuma K, Koike K (1989) Identification of a promoter region for 3.6-kilobase mRNA of hepatitis B virus and specific cellular binding protein. *J Virol* 63: 2914–2920
147 Yu X, Mertz JE (1996) Promoters for synthesis of the pre-C and pregenomic mRNAs of human hepatitis B virus are genetically distinct and differentially regulated. *J Virol* 70: 8719–8726
148 Yuh CH, Chang YL, Ting LP (1992) Transcriptional regulation of precore and pregenomic RNAs of hepatitis B virus. *J Virol* 66: 4073–4084
149 Tang H, McLachlan A (2002) Avian and mammalian hepadnaviruses have distinct transcription factor requirements for viral replication. *J Virol* 76: 7468–7472
150 Guo WT, Bell KD, Ou JH (1991) Characterization of the hepatitis B virus EnhI enhancer and X promoter complex. *J Virol* 65: 6686–6692
151 Yuh CH, Ting LP (1990) The genome of hepatitis B virus contains a second enhancer: cooperation of two elements within this enhancer is required for its function. *J Virol* 64: 4281–4287
152 Siddiqui A, Jameel S, Mapoles J (1986) Transcriptional control elements of hepatitis B surface antigen gene. *Proc Natl Acad Sci USA* 83: 566–570
153 Raney AK, Easton AJ, McLachlan A (1994) Characterization of the minimal elements of the hepatitis B virus large surface antigen promoter. *J Gen Virol* 75: 2671–2679
154 Bulla GA, Siddiqui A (1989) Negative regulation of the hepatitis B virus pre-S1 promoter by internal DNA sequences. *Virology* 170: 251–260
155 Cattaneo R, Will H, Hernandez N, Schaller H (1983) Signals regulating hepatitis B surface antigen transcription. *Nature* 305: 336–338
156 Zhou DX, Yen TS (1991) The hepatitis B virus S promoter comprises a CCAAT motif and two initiation regions. *J Biol Chem* 266: 23416–23421
157 Xu Z, Jensen G, Yen TS (1997) Activation of hepatitis B virus S promoter by the viral large surface protein via induction of stress in the endoplasmic reticulum. *J Virol* 71: 7387–7392
158 Fukai K, Takada S, Yokosuka O, Saisho H, Omata M, Koike K (1997) Characterization of a spe-

cific region in the hepatitis B virus Enhancer I for the efficient expression of X gene in the hepatic cell. *Virology* 236: 279–287
159 Hu KQ, Siddiqui A (1991) Regulation of the hepatitis B virus gene expression by the enhancer element I. *Virology* 181: 721–726
160 Loeb DD, Mack AA, Tian R (2002) A Secondary structure that contains the 5' and 3' splice sites suppresses splicing of duck hepatitis B virus pregenomic RNA. *J Virol* 76: 10195–10202
161 Obert S, Zachmann-Brand B, Deindl E, Tucker W, Bartenschlager R, Schaller H (1996) A spliced hepadnavirus RNA that is essential for virus replication. *EMBO J* 15: 2565–2574
162 Bartenschlager R, Junker-Niepmann M, Schaller H (1990) The P gene product of hepatitis B virus is required as a structural component for genomic RNA encapsidation. *J Virol* 64: 5324–5332
163 Hirsch RC, Lavine JE, Chang LJ, Varmus HE, Ganem D (1990) Polymerase gene products of hepatitis B viruses are required for genomic RNA packaging as well as for reverse transcription. *Nature* 344: 552–555
164 Junker-Niepmann M, Bartenschlager R, Schaller H (1990) A short cis-acting sequence is required for hepatitis B virus pregenome encapsidation and sufficient for packaging of foreign RNA. *EMBO J* 9: 3389–3396
165 Gerelsaikhan T, Tavis JE, Bruss V (1996) Hepatitis B virus nucleocapsid envelopment does not occur without genomic DNA synthesis. *J Virol* 70: 4269–4274
166 Wei Y, Tavis JE, Ganem D (1996) Relationship between viral DNA synthesis and virion envelopment in hepatitis B viruses. *J Virol* 70: 6455–6458
167 Pollack JR, Ganem D (1993) An RNA stem-loop structure directs hepatitis B virus genomic RNA encapsidation. *J Virol* 67: 3254–3263
168 Bartenschlager R, Schaller H (1992) Hepadnaviral assembly is initiated by polymerase binding to the encapsidation signal in the viral RNA genome. *EMBO J* 11: 3413–3420
169 Wang GH, Zoulim F, Leber EH, Kitson J, Seeger C (1994) Role of RNA in enzymatic activity of the reverse transcriptase of hepatitis B viruses. *J Virol* 68: 8437–8442
170 Birnbaum F, Nassal (1990) Hepatitis B virus nucleocapsid assembly: primary structure requirements in the core protein. *J Virol* 64: 3319–3330
171 Ceres P, Zlotnick A (2002) Weak protein–protein interactions are sufficient to drive assembly of hepatitis B virus capsids. *Biochemistry* 41: 11525–11531
172 Watts NR, Conway JF, Cheng NQ, Stahl SJ, Belnap DM, Steven AC, Wingfield PT (2002) The morphogenic linker peptide of HBV capsid protein forms a mobile array on the interior surface. *EMBO J* 21: 876–884
173 Zlotnick A, Cheng N, Stahl SJ, Conway JF, Steven AC, Wingfield PT (1997) Localization of the C terminus of the assembly domain of hepatitis B virus capsid protein – implications for morphogenesis and organization of encapsidated RNA. *Proc Natl Acad Sci USA* 94: 9556–9561
174 Gerlich WH, Goldmann U, Muller R, Stibbe W, Wolff W (1982) Specificity and localization of the hepatitis B virus-associated protein kinase. *J Virol* 42: 761–766
175 Hu J, Seeger C (1996) Hsp90 is required for the activity of a hepatitis B virus reverse transcriptase. *Proc Natl Acad Sci USA* 93: 1060–1064
176 Hu J, Toft DO, Seeger C (1997) Hepadnavirus assembly and reverse transcription require a multicomponent chaperone complex which is incorporated into nucleocapsids. *EMBO J* 16: 59–68
177 Hu JM, Toft D, Anselmo D, Wang XT (2002) *In vitro* reconstitution of functional hepadnavirus reverse transcriptase with cellular chaperone proteins. *J Virol* 76: 269–279
178 Ostrow KM, Loeb DD (2002) Characterization of the cis-acting contributions to avian hepadnavirus RNA encapsidation. *J Virol* 76: 9087–9095
179 Wang XT, Grammatikakis N, Hu JM (2002) Role of p50/CDC37 in hepadnavirus assembly and replication. *J Biol Chem* 277: 24361–24367
180 Nassal M, Rieger A (1996) A bulged region of the hepatitis B virus RNA encapsidation signal contains the replication origin for discontinuous first-strand DNA synthesis. *J Virol* 70: 2764–2773
181 Pollack JR, Ganem D (1994) Site-specific RNA binding by a hepatitis B virus reverse transcriptase initiates two distinct reactions: RNA packaging and DNA synthesis. *J Virol* 68: 5579–5587
182 Wang GH, Seeger C (1993) Novel mechanism for reverse transcription in hepatitis B viruses. *J Virol* 67: 6507–6512
183 Tavis JE, Ganem D (1995) RNA sequences controlling the initiation and transfer of duck hepatitis B virus minus-strand DNA. *J Virol* 69: 4283–4291

184 Tavis JE, Perri S, Ganem D (1994) Hepadnavirus reverse transcription initiates within the stem-loop of the RNA packaging signal and employs a novel strand transfer. *J Virol* 68: 3536–3543
185 Radziwell G, Tucker W, Schaller H (1990) Mutational analysis of the hepatitis B virus P gene product: domain structure and RNase H activity. *J Virol* 64: 613–620
186 Summers J, Mason WS (1982) Replication of the genome of a hepatitis B-like virus by reverse transcription of an RNA intermediate. *Cell* 29: 403–415
187 Loeb DD, Gulya KJ, Tian R (1997) Sequence identity of the terminal redundancies on the minus-strand DNA template is necessary but not sufficient for the template switch during hepadnavirus plus-strand DNA synthesis. *J Virol* 71: 152–160
188 Loeb DD, Tian R (2001) Mutations that increase *in situ* priming also decrease circularization for duck hepatitis B virus. *J Virol* 75: 6492–6497
189 Habig JW, Loeb DD (2003) Template switches during plus-strand DNA synthesis of duck hepatitis B virus are influenced by the base composition of the minus-strand terminal redundancy. *J Virol* 77: 12412–12420
190 Havert MB, Li J, Loeb DD (2002) Analysis of duck hepatitis B virus reverse transcription indicates a common mechanism for the two template switches during plus-strand DNA synthesis. *J Virol* 76: 2763–2769
191 Havert MB, Loeb DD (1997) cis-acting sequences in addition to donor and acceptor sites are required for template switching during synthesis of plus-strand DNA for duck hepatitis B virus. *J Virol* 71: 5336–5344
192 Liu N, Ji L, Maguire ML, Loeb DD (2004) cis-acting sequences that contribute to the synthesis of relaxed-circular DNA of human hepatitis B virus. *J Virol* 78: 642–649
193 Liu N, Tian R, Loeb DD (2003) Base pairing among three *cis*-acting sequences contributes to template switching during hepadnavirus reverse transcription. *Proc Natl Acad Sci USA* 100: 1984–1989
194 Mueller-Hill K, Loeb DD (2002) cis-acting sequences 5E, M, and 3E interact to contribute to primer translocation and circularization during reverse transcription of avian hepadnavirus DNA. *J Virol* 76: 4260–4266
195 Perri S, Ganem D (1996) A host factor that binds near the termini of hepatitis B virus pregenomic RNA. *J Virol* 70: 6803–6809
196 Perri S, Ganem D (1997) Effects of mutations within and adjacent to the terminal repears of hepatitis B virus pregenomic RNA on viral DNA synthesis. *J Virol* 71: 8448–8455
197 Tang H, McLachlan A (2002) A pregenomic RNA sequence adjacent to DR1 and complementary to epsilon influences hepatitis B virus replication efficiency. *Virology* 303: 199–210
198 Ganem D, Varmus HE (1987) The molecular biology of the hepatitis B viruses. *Annu Rev Biochem* 56: 651–693
199 Huovila A-PJ, Eder AM, Fuller SD (1992) Hepatitis B surface antigen assembles in a post-ER, pre-Golgi compartment. *J Cell Biol* 118: 1305–1320
200 Tuttleman JS, Pourcel C, Summers J (1986) Formation of the pool of covalently closed circular viral DNA in hepadnavirus-infected cells. *Cell* 47: 451–460
201 Summers J, Smith PM, Horwich AL (1990) Hepadnavirus envelope proteins regulate covalently closed circular DNA amplification. *J Virol* 64: 2819–2824
202 Summers J, Smith PM, Huang MJ, Yu MS (1991) Morphogenetic and regulatory effects of mutations in the envelope proteins of an avian hepadnavirus. *J Virol* 65: 1310–1317
203 Maynard JE (1990) Hepatitis B: global importance and need for control. *Vaccine* 8: S18–S20
204 Zuckerman AJ (1996) Progress towards the comprehensive control of hepatitis B – introduction. *Gut* 38 (suppl): S1
205 Sampliner RE, Hamilton FA, Iseri OA, Tabor E, Boitnott J (1979) The liver histology and frequency of clearance of the hepatitis B surface antigen (HbsAg) in chronic carriers. *Am J Med Sci* 277: 17–22
206 Lee WM (1997) Hepatitis B virus infection. *N Engl J Med* 337: 1733–1745
207 Robek MD, Boyd BS, Wieland SF, Chisari FV (2004) Signal transduction pathways that inhibit hepatitis B virus replication. *Proc Natl Acad Sci USA* 101: 1743–1747
208 Bartholomew RM, Carmichael EP, Findeis MA, Wu CH, Wu GY (1995) Targeted delivery of antisense DNA in woodchuck hepatitis virus-infected woodchucks. *J Viral Hepatol* 2: 273–278
209 Klein C, Bock CT, Wedemeyer H, Wustefeld T, Locarnini S, Dienes HP, Kubicka S, Manns MP, Trautwein C (2003) Inhibition of hepatitis B virus replication *in vivo* by nucleoside analogues and siRNA. *Gastroenterology* 125: 9–18

210 Welch PJ, Tritz R, Yei S, Barber J, Yu M (1997) Intracellular application of hairpin ribozyme genes against hepatitis B virus. *Gene Ther* 4: 736–743

211 Dienstag JL, Schiff ER, Mitchell M, Casey DE, Gitlin N, Lissoos T, Gelb LD, Condreay L, Crowther L, Rubin M, Brown N (1999) Extended lamivudine retreatment for chronic hepatitis B: maintenance of viral suppression after discontinuation of therapy. *Hepatology* 30: 1082–1087

212 Dienstag JL, Schiff ER, Wright TL, Perrillo RP, Hann HWL, Goodman Z, Crowther L, Condreay LD, Woessner M, Rubin M, Brown NA (1999) Lamivudine as intital treatment for chronic hepatitis B in the United States. *N Eng J Med* 341: 1256–1263

213 Hadziyannis SJ, Tassopoulos NC, Heathcote EJ, Chang TT, Kitis G, Rizzetto M, Marcellin P, Lim SG, Goodman Z, Wulfsohn MS et al. (2003) Adefovir dipivoxil for the treatment of hepatitis B e antigen-negative chronic hepatitis B. *N Engl J Med* 348: 800–807

214 Marcellin P, Chang T, Lim SG, Tong MJ, Sievert W, Shiffman ML, Jeffers L, Goodman Z, Wulfson MS, Xiong S, Fry J, Brosgart CL (2003) Adefovir dipivoxil for the treatment of hepatitis B e antigen-positive chronic hepatitis B. *N Engl J Med* 348: 808–816

215 Perrillo RP, Wright T, Rakela J, Levy G, Schiff E, Gish R, Martin P, Dienstag J, Adams P, Dickson R et al. (2001) A multicenter United States-Candadian trial to assess lamivudine monotherapy before and after liver transplantation for chronic hepatitis. *Hepatology* 33: 424–432

216 Schiff ER, Dienstag JL, Karayalcin S, Grimm IS, Perrillo RP, Husa P, de Man RA, Goodman Z, Condreay LD, Crowther LM et al. (2003) Lamivudine and 24 weeks of lamivudine/interferon combination therapy for hepatitis B e antigen-positive chronic hepatitis B in interferon nonresponders. *J Hepatol* 38: 818–826

217 Tassopoulos NC, Volpes R, Pastore G, Heathcote J, Buti M, Goldin RD, Hawley S, Barber J, Condreay L, Gray DF (1999) Efficacy of lamivudine in patients with heptatitis B e antigen-negative hepatitis B virus DNA-positive (precore mutant) chronic hepatitis B. *Hepatology* 29: 889–896

218 Allen MI, Deslauriers M, Andrews CW, Tipples GA, Walters KA, Tyrrell DLJ, Brown N, Condreay LD (1998) Identification and characterization of mutations in hepatitis B virus resistant to lamivudine. *Hepatology* 27: 1670–1677

219 Bartholomew MM, Jansen RW, Jefferes LJ, Reddy KR, Johnson LC, Bunzehdahl H, Condreay LD, Tzakis AG, Schiff ER, Brown NA (1997) Hepatitis B virus resistance to lamivudine given for recurrent infection after orthotopic liver transplantation. *Lancet* 349: 20–22

220 Ling R, Mutimer D, Ahmed M, Boxall EH, Elias E, Dusheiko GM, Harrison TJ (1996) Selection of mutations in the hepatitis B virus polymerase during therapy of transplant recipients with lamivudine. *Hepatology* 24: 711–713

221 Wong VWS, Chan HLY, Wong ML, Tam JSL, Leung NWY (2004) Clinical course after stopping lamivudine in chronic hepatitis B patients with lamivudine-resistant mutants. *Alimentary. Pharmacol Ther* 19: 323–329

222 Angus P, Vaughan R, Xiong S, Yang HL, Delaney W, Gibbs C, Brosgart C, Colledge D, Edwards R, Ayres A et al. (2003) Resistance to adefovir dipivoxil therapy associated with the selection of a novel mutation in the HBV polymerase. *Gastroenterology* 125: 292–297

223 Leung NWY, Lai, C-L, Chang T-T, Guan R, Lee C-M, Ng K-Y, Lim S-G, Wu P-C, Dent JC, Edmundson S et al. (2001) Extended lamivudine treatment in patients with chronic hepatitis B enhances hepatitis e antigen seroconversion rates: results after 3 years of therapy. *Hepatology* 33: 1527–1532

224 Delaney WE, Edwards R, Colledge D, Shaw T, Furman P, Painter G, Locarnini S (2002) Phenylpropenamide derivatives AT-61 and AT-130 inhibit replication of wild-type and lamivudine-resistant strains of hepatitis B virus *in vitro*. *Antimicrob Agents Chemother* 46: 3057–3060

225 Deres K, Schroder CH, Paessens A, Goldmann S, Hacker HJ, Weber O, Kramer T, Niewohner U, Pleiss U, Stoltefuss J et al. (2003) Inhibition of hepatitis B virus replication by drug-induced depletion of nucleocapsids. *Science* 299: 893–896

226 Lu XY, Lu YF, Geschwindt R, Dwek RA, Block TM (2001) Hepatitis B virus MHBs antigen is selectively sensitive to glucosidase-mediated processing in the endoplasmic reticulum. *DNA Cell Biol* 20: 647–656

227 Lu XY, Tran T, Simsek E, Block TM (2003) The alkylated imino sugar, n-(n-nonyl)-deoxygalactonojirimycin, reduces the amount of hepatitis B virus nucleocapsid in tissue culture. *J Virol* 77: 11933–11940

228 Mehta A, Convers B, Tyrrell DLJ, Walters KA, Tipples GA, Dwek RA, Block TM (2002) Structure-activity relationship of a new class of anti-hepatitis B virus agents. *Antimicrob Agents*

Chemother 46: 4004–4008
229 Mehta A, Zitzmann N, Rudd PM, Block TM, Dwek RA (1998) Alpha-glucosidase inhibitors as potential broad based anti-viral agents. *FEBS Lett* 430 (SI): 17–22
230 Weber O, Schlemmer K-H, Hartmann E, Hagelschuer I, Paessens A, Graef E, Deres K, Goldmann S, Niewoehner U, Stoltefuss J et al. (2002) Inhibition of human hepatitis B virus (HBV) by a novel non-nucleosidic compound in a transgenic mouse model. *Antivir Res* 54: 69–78
231 Jilbert AR, Wu TT, England JM, Hall PM, Carp NZ, O'Connell AP, Mason WS (1992) Rapid resolution of duck hepatitis B virus infection occurs after massive hepatocellular involvement. *J Virol* 66: 1377–1388
232 Guidotti LG, Chisari FV (2001) Noncytolytic control of viral infections by the innate and adaptive immune response. *Annu Rev Immunol* 19: 65–91
233 Wieland SF, Spangenberg HC, Thimme R, Purcell RH, Chisari FV (2004) Expansion and contraction of the hepatitis B virus transcriptional template in infected chimpanzees. *Proc Natl Acad Sci USA* 101: 2129–2134
234 Summers J, Jilbert AR, Yang WG, Aldrich CE, Saputelli J, Litwin S, Toll E, Mason WS (2003) Hepatocyte turnover during resolution of a transient hepadnaviral infection. *Proc Natl Acad Sci USA* 100: 11652–11659
235 Summers J, Mason WS (2004) Residual integrated viral DNA after hepadnavirus clearance by nucleoside analog therapy. *Proc Natl Acad Sci USA* 101: 638–640

Hepatitis Prevention and Treatment
Edited by J.M. Colacino and B.A. Heinz

Molecular virology of hepatitis C virus

Guangxiang (George) Luo

Department of Microbiology, Immunology and Molecular Genetics, University of Kentucky College of Medicine, 800 Rose Street, MN475, UKMC, Lexington, KY 40536, USA

Virus genome

Hepatitis C virus (HCV) is a small enveloped RNA virus containing a single-stranded and positive-sense RNA molecule of approximately 9.6 kb in length [1]. The viral RNA genome consists of the 5' untranslated region (5'UTR) of 341 nucleotides, a single long open reading frame (ORF) encoding a large viral polypeptide of 3010–3040 amino acids, and the 3' untranslated region (3'UTR) of variable length (Fig. 1A) [1, 2]. The viral proteins are translated as a single polyprotein, which is proteolytically processed into mature structural and nonstructural viral proteins. The genes for the structural proteins C (core), E1 and E2 (envelope), and possibly p7, are located in the 5' portion of the genome, while the nonstructural proteins (NS2, NS3, NS4A, NS4B, NS5A, and NS5B) are encoded by the remaining 3' portion of the ORF (Fig. 1A). The features of HCV genome organization are similar to those of other flaviviruses and pestiviruses in the family Flaviviridae [3, 4]. Accordingly, HCV was classified as a separate genus, Hepacivirus, of the Flaviviridae family [5]. Sequence analysis and comparison studies revealed that both the 5'UTR and 3'UTR of the HCV genome are highly conserved [6–8]. In a sharp contrast, sequences of the ORF exhibit a remarkable variation (quasispecies) among various HCV isolates [9]. Based on the nucleotide sequence diversity, HCV was further divided into six major genotypes and more than 90 subtypes.

The 5'UTR of the HCV RNA genome harbors an internal ribosome entry site (IRES), an RNA element mediating Cap-independent internal initiation of translation [10]. The HCV 5'UTR folds into a highly ordered complex structure with multiple stem-loops and a pseudoknot structure (Fig. 1B). The distinct structural features of the 5'UTR allow the highly folded secondary and tertiary RNA elements to be grouped into four major structural domains, designated I, II, III, and IV (Fig. 1B) [11]. Domains II and III, which consist of many stem-loops and bulge loops, are further divided into different subdomains (IIa, IIb, IIIa, IIIb, IIIc, IIIe, and IIIf) (Fig. 1B). Domain I contains a short 5'-proximal stem-loop structure that is well conserved (Fig. 1B) [12]. It enhances the efficiency of the IRES-mediated translation [12, 13]. Additionally, domain I is also essential for HCV RNA replication [12, 13].

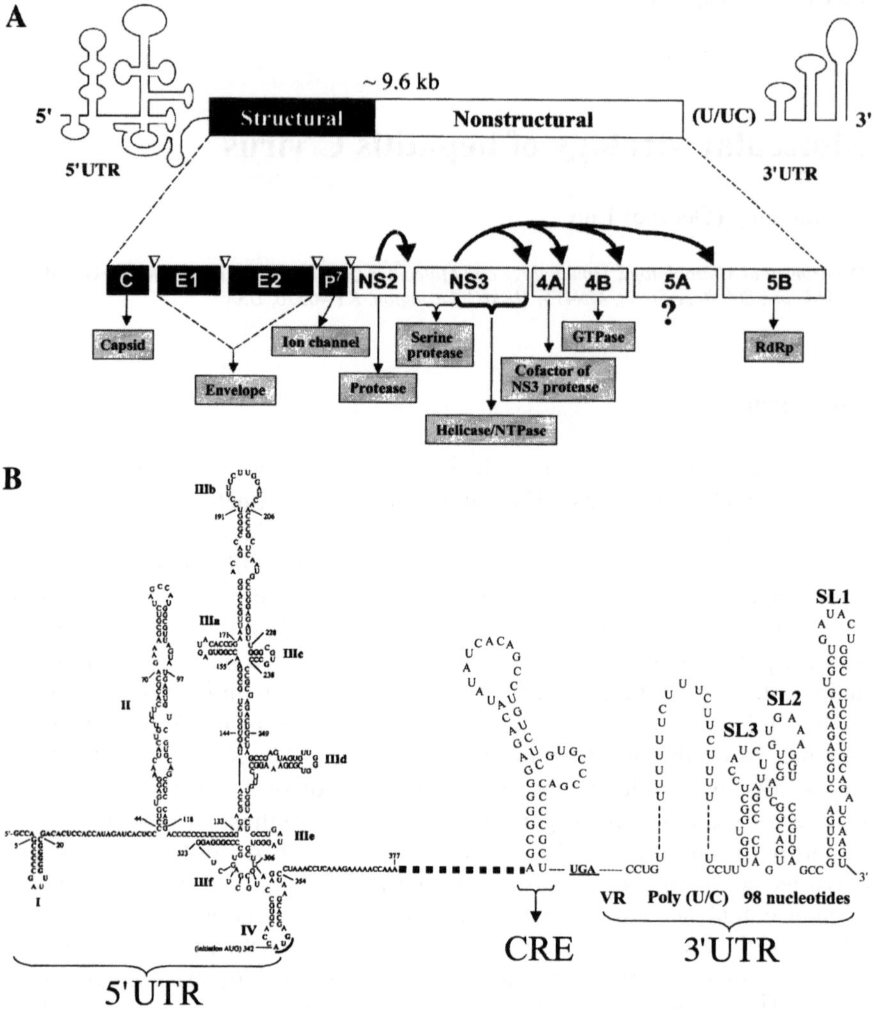

Figure 1. (A) HCV genome organization and viral proteins. The HCV RNA genome is shown schematically at the top of the figure. The ORF is flanked by the 5'UTR and 3'UTR. The viral proteins and their major functions are indicated at the bottom. The structural proteins are shaded in black, and nonstructural proteins are shown as open boxes. Open triangles indicate cleavage sites by cellular signal peptidases, and arrows indicate cleavage sites by either viral NS2/3 protease (at the NS2/NS3 junction) or NS3 serine protease (downstream of NS3). (B) Secondary structures of the 5' and 3'UTRs and a *cis*-acting replication element (CRE) of the HCV RNA genome. The RNA sequences are derived from a subgenomic HCV RNA replicon [67]. Sequences shown in the 5'UTR also include 36 nucleotides immediately downstream of the initiation (AUG) codon. Structural domains of the 5'UTR are labeled I, II, IIIa, IIIb, IIIc, IIId, IIIe, and IIIf according to Rijnbrand and Lemon [11]. A pseudoknot structure is formed by base-pairs between nucleotides 306–311 and 325–330 (domain IIIf). The 3'UTR consists of three distinct regions: a variable region (VR), a poly(U/C) tract, and a highly conserved 98 nucleotides. Sequences of the conserved 98-nucleotide fold to form three stem-loop (SL) structures designated SL1, SL2, and SL3 according to Ito and Lai [21]. A *cis*-acting replication element (CRE) is located in a region juxtaposed to the encoding sequences for the C-terminal 21 amino acid hydrophobic domain of NS5B [20]. The initiation (AUG) and stop (UGA) codons are underlined.

Deletion or disruption of the 5'-proximal stem-loop structure by nucleotide substitutions resulted in a complete abrogation of HCV RNA replication in the cell. However, restoration of the 5'-proximal stem-loop structure by compensatory mutations with different nucleotides rescued the ability of HCV RNA to replicate in the cell [12]. Domains II, III, and IV constitute the IRES element responsible for initiating the translation of HCV polyprotein [11]. These highly structured RNA domains serve as a platform for recruitment of the host translation machinery. The 40 S ribosomal subunit and eukaryotic initiation factor 3 (eIF3) were found to interact directly with the HCV IRES [14, 15]. Domain II is also required for HCV RNA replication [13]. Thus, domain II contains overlapping *cis*-acting RNA elements required for both viral RNA translation and replication. It appeared that domains III and IV might also contribute to optimal replication of the HCV RNA in the cell [13]. What viral and cellular proteins recognize the *cis*-acting RNA elements of the 5'UTR is still poorly understood. The biological relevance of a number of cellular proteins such as polypyrimidine tract-binding protein (PTB), heterogeneous nuclear ribonucleoprotein L (hnRNP L), La autoantigen, and poly(C)-binding protein (PCBP) that were found to interact with the 5'UTR has not yet been determined [12].

The conserved RNA sequences and structures in the ORF also play important roles in the control of HCV RNA translation and replication. The core (C) protein-coding sequence was found to modulate the HCV IRES function [16, 17]. The sequences encoding the N-terminal portion of C contain signal sequences for enhancement of the IRES-mediated translation and ribosome frame shifting from C to Frameshift protein (F) ORF [17, 18]. In contrast, the nucleotides 428–442 of the C-coding region are involved in inhibition of the IRES function via an RNA-RNA interaction with nucleotides 24–38 of the 5'UTR [19]. Like other positive-stranded RNA viruses, HCV RNA also contains a *cis*-acting replication element (CRE) in the sequence encoding the C-terminal portion of NS5B. Mutational analysis demonstrated that the primary sequence as well as the RNA structure of the CRE is required for efficient HCV RNA replication [20].

The 3'UTR consists of three distinct regions: a variable region (VR) immediately downstream of the ORF, a poly(U/C) tract of variable length, and a highly conserved 98-nucleotide sequence at the 3' end (Fig. 1B) [7, 8]. The 98 nucleotides at the 3' end of the RNA can fold into a secondary structure of three stem-loops (Fig. 1B) [7, 21]. It has been shown that both the poly(U/C) tract and the 3'-end 98 nucleotides are essential for HCV RNA replication in a chimpanzee model of HCV infection as well as for HCV replicon replication [22, 23]. The VR sequence appears to be genotype-specific and is required for efficient replication of HCV replicon RNAs in Huh7 cells [24, 25]. Mutagenesis analysis of the 3'UTR revealed that both sequence and structure of the stem-loop I are critical for HCV RNA replication in the cell [26]. Similar to the 5'UTR, the 3'UTR was also found to interact with a number of cellular proteins, including the PTB and glyceraldehyde-3-phosphate dehydrogenase (GAPDH) [27–29]. How binding of cellular proteins to the 3'UTR regulates HCV RNA replication has not yet been determined. There is circumstantial evidence that the 3'-end 98 nucleotides could enhance the IRES-dependent translation due partially to PTB binding [30]. Future studies are needed to define the roles of HCV *cis*-acting RNA elements in the control of HCV RNA translation and replication.

Virus proteins

HCV proteins are translated as a single large polyprotein precursor, which is co- or post-translationally processed by cellular and viral proteases into at least 10 structural and nonstructural viral proteins in the order of NH_2-C-E1-E2-p7-NS2-NS3-NS4A-NS4B-NS5A-NS5B-COOH (Fig. 1A) [2, 3]. A new protein designated F (frameshift) or ARFP (alternate reading frame protein) was recently identified which is synthesized by ribosomal frameshift from the C ORF [18, 31]. The properties and role of the newly identified F/ARFP protein are still not clear. Given that the anti-F antibody can be detected in patients with HCV infection [18, 31], the F/ARFP protein is likely produced during infection and thereby may be physiologically relevant to HCV replication and/or pathogenesis. However, the actual evidence for its role in HCV infection is still lacking. Recently, other new alternative forms of C protein were also reported [32, 33]. Although antibodies reacting with synthetic peptides derived from the alternative forms of C protein were detected in chronic HCV patients, the significance of these proteins in HCV replication and/or pathogenesis remains unclear [33].

Structural proteins

The HCV structural proteins include C and envelope proteins E1 and E2. It is unclear whether p7 is present in the virus particles. Expression of C, E1, and E2 is sufficient to form virus-like particle (VLP) within cells [34]. However, these recombinant VLPs lack the intrinsic property of the infectious HCV virion as they fail to bud from the cell [35].

Core (C)

The C protein is cleaved from the N-terminus of the viral polyprotein by the cellular endoplasmic reticulum (ER) signal peptidase. It exhibits two major forms, consisting of 173 (19 kDa) and 191 (23 kDa) amino acids [36]. The 191-amino acid C protein contains the signal sequence of the E1 glycoprotein upon cleavage at the N-terminal end of E1 by the ER signal peptidase. It is anchored on the ER membrane by the signal sequence, which is subsequently removed by the membranous signal peptide peptidase, resulting in the smaller form of C (173-amino acids). The smaller form (19 kDa) is believed to be the mature C protein found in the virion. Although several other smaller C proteins were also observed [36], their roles in HCV replication and pathogenesis are not clear.

C is a multifunctional protein that has been reported to possess a variety of diverse biological properties. The N-terminal two-thirds of C are hydrophilic, while the C-terminal one-third is hydrophobic. It is phosphorylated mainly at serine residues 53, 93, 96, and 116, most likely by host-cell protein kinases A and C [37]. The putative nuclear localization signals present in C might be responsible for the nuclear localization of the truncated forms of C [36]. Homotypic oligomers of C primarily function as building blocks to form the viral nucleocapsid [38]. C also interacts with the envelope glycoprotein E1 to mediate virion assembly. In addition, C likely has other diverse functions in the HCV life cycle and pathogenesis via interactions with intracellular membranes, nucleic acids (RNA and DNA), and numerous host cell proteins [36]. Binding of C to cellular DNA and RNA may affect normal cellular functions [39]. Interaction of C with the 5'UTR has been implicated in the inhibition of the HCV IRES-mediated translation [40]. However, a role for the binding of C to HCV RNA in the regulation of HCV replication has not been experimentally examined. An increasing number of cellular proteins have been found to interact with C [36, 39]. The protein–protein interactions between different forms of C and various cellular proteins may disrupt normal cellular pathways that control many important biological events in the cell. These disruptions would conceivably contribute to the clinical manifestation and disease progression of hepatitis C. It was found that overexpression of C can interfere with lipid metabolism, resulting in an accumulation of lipid droplets in the cell (steatosis) [41, 42]. Also, C was found to modulate apoptosis and gene expression of some cellular and viral genes and to transform rat primary embryo fibroblasts in cooperation with the *ras* oncogene in cell culture [36, 39]. In transgenic mice, overexpression of C resulted in liver steatosis and hepatocellular carcinoma [42, 43].

E1 and E2

E1 and E2 are envelope glycoproteins that form a noncovalently linked heterodimeric complex, which is believed to be the functional subunit mediating virus attachment and entry [44]. Both E1 and E2 are type I integral transmembrane proteins anchored on the ER membrane. The C-terminal transmembrane domains of E1 and E2 contain sequences responsible for the noncovalent heterodimerization of E1 and E2 (E1E2) as well as retention of the envelope glycoproteins in the ER compartment [41]. However, recent studies found that, contrary to the situation in which E1E2 is retained in the ER, unmodified E1E2 heterodimers were expressed on the cell surface and were also incorporated into pseudotyped retroviral particles [45–47]. Retroviruses pseudotyped with both E1 and E2, but not individual E1 or E2, were able to mediate pH-dependent cell entry [48]. Furthermore, the infectivity of pseudotyped retroviral particles was neutralized by E2-specific monoclonal antibodies [45] as well as patient sera [44]. These new developments have furthered our understanding of the molecular mechanisms underlying virus–cell membrane fusion and will facilitate the identification of HCV receptors and/or coreceptors expressed on the surface of susceptible cells.

E2 is believed to be responsible for virus attachment by binding the virion to HCV receptor, while E1 is thought to play a role in virus uncoating by mediating virus–cell membrane fusion. These viral envelope glycoproteins were found to interact specifically with a variety of cell-surface molecules such as human CD81 [49], the low-density-lipoprotein receptor (LDLR) [50], scavenger receptor class B type 1 (SR-B1) [51], dendritic cell-specific intercellular adhesion molecule 3 grabbing nonintegrin (DC-SIGN), liver/lymph node-specific SIGN (L-SIGN) [52], and heparan sulfate proteoglycans (HSPG) [53]. Whether any of these molecules serve as a receptor or coreceptor for HCV infection has not been experimentally confirmed. Cell-surface expression of CD81, LDLR, or SR-B1 individually or combined, in nonpermissive cell types was not sufficient for the infectivity of retroviral pseudotypes, suggesting a requirement for other unidentified molecule(s) in the entry of retroviruses pseudotyped with HCV E1 and E2.

In addition to mediating virus attachment, unglycosylated E2 was found to act as a specific inhibitor of the interferon (IFN)-inducible and double-stranded RNA-activated protein kinase (PKR) [54, 55]. PKR is thought to be a major protein responsible for mediating the IFN-induced intracellular antiviral defense. Inhibition of PKR by E2 was postulated as a mechanism by which HCV circumvents the intracellular antiviral defense [55].

p7

HCV p7 is a small hydrophobic protein of 63 amino acids (7 kDa). It contains two membrane-spanning domains connected by a small hydrophilic loop, with both its N- and C-termini oriented toward the lumen of the ER [56]. *In vitro* cross-linking analysis of recombinant p7 suggested that p7 might exist as hexamers in the cell [57]. The hexameric form of p7 could function as an ion channel when examined in artificial lipid membranes *in vitro* [57, 58]. The ion-channel activity of p7 could be blocked by amantadine and various long-alkyl-chain aminosugar derivatives [57, 58]. *In vivo*, p7 is essential for infectivity of HCV as determined in a chimpanzee model of HCV infection [59]. In addition, findings derived from mutational analysis of p7 for infectivity in chimpanzees indicate that the sequence for the function of p7 is genotype-specific [59]. In analogy to the influenza virus M2, it is conceivable that the ion-channel activity of p7 would prevent the viral envelope glycoproteins from undergoing a premature conformational change resulting in viral inactivation during transport through the acidic environment of the trans-Golgi network [60]. Conversely, the p7 ion channel may play an essential role in the initiation of HCV infection by acidifying the virus-attached endosome, in which the viral envelope glycoproteins undergo a low pH-triggered conformational rearrangement, a prerequisite for virus-cell membrane fusion and virus uncoating [61].

Nonstructural proteins

NS2

NS2 is a hydrophobic protein of approximately 23 kDa. It is a type III integral membrane protein with multiple internal transmembrane domains [62]. Both the N- and C-termini of NS2 are oriented in the lumen of the ER [62]. Little is known about the function of NS2 other than that it is an essential component of the NS2/3 protease responsible for the *cis*-cleavage between the NS2 and NS3 proteins. Although NS2 itself lacks enzymatic activity, the NS2/3 protease is a distinct enzyme requiring both NS2 and the N-terminal protease domain of NS3 [63]. Also, the NS3 protease activity is dispensable for the function of the NS2/3 protease since the active site mutation of NS3 completely inactivated the NS3 protease activity but did not have any effect on the NS2/3 activity [63]. The catalytic mechanism of the NS2/3 protease is unknown; however, it is thought to be either a metalloprotease or cysteine protease [64]. Furthermore, the activity of the NS2/3 protease requires the cellular chaperone protein heat-shock protein (Hsp)90, which directly interacts with the NS2/3 protein to form a complex that is presumably required for maturation of the active NS2/3 protease [65]. It is known that NS2 is not required for HCV RNA replication although a regulatory role has not been ruled out [66, 67]. NS2 may also play a role in viral pathogenesis by interfering with cellular functions.

NS3 and NS4A

NS3 is the largest HCV nonstructural protein (70 kDa) possessing multiple enzymatic activities that are essential for HCV replication [2, 22]. The N-terminal one-third of NS3 is part of the aforementioned NS2/3 metalloprotease [63, 64]. This region of NS3 also functions as a serine protease responsible for all cleavages of the polyprotein downstream of NS3 with the help of NS4A as a cofactor. The remaining two-thirds of the molecule has helicase and NTPase activities, both of which are essential for HCV RNA replication *in vivo* [68, 69]. The enzymatic activities of NS3 have been extensively characterized, and the atomic structures of the NS3 protease and helicase/NTPase have all been determined [68, 69]. These advances contributed to the rational design and discovery of the peptidomimetic inhibitors of the NS3/4A serine protease, which are currently under clinical development [70, 71].

Besides possessing enzymatic activities, NS3 plays a pivotal role in viral pathogenesis by interfering with cellular functions. A recent breakthrough study demonstrated that the NS3/4A serine protease can prevent the IFN regulatory factor-3 (IRF-3) from becoming phosphorylated and mediating subsequent biological effects [72]. As a result, HCV can efficiently replicate in the cell by evading the intracellular antiviral response mediated through the IRF-3 pathway [72]. NS3 has also been shown to interact with a number of cellular proteins, including protein kinases A and C, p53, histones, Sm-D1, and protein arginine N-methyltransferase (PRMT) [39]. However, the biological significance of these protein–protein interactions remains unclear. NS3 is methylated, probably by the action of PRMT via protein–protein interaction [73]. As mentioned above, NS4A is known to be a cofactor of the NS3 serine protease [2]. Additionally, NS4A interacts with NS4B and NS5A, suggesting that it plays a role in the assembly of the HCV replication complex [74].

NS4B

NS4B is a hydrophobic transmembrane protein with a molecular mass of approximately 27 kDa. NS4B has four membrane-spanning domains through which it is associated with intracellular membranes [75]. Biochemical and cell biological studies revealed that NS4B is a component of the membrane-bound HCV replication complex although the function of NS4B in HCV replication is unknown [76]. Interestingly, recent findings demonstrate that NS4B contains a GTP-binding pocket and possesses a GTPase activity (Einav S, Elazar M, Danieli T, and Glenn JS, unpublished observations). Mutations of the GTP-binding site of NS4B resulted in the inactivation of the GTPase and consequently the abrogation of a subgenomic HCV replicon replication in Huh7 cells, demonstrating that the GTPase activity of NS4B is essential for HCV RNA replication in the cell. It has also been reported that NS4B can transform NIH 3T3 cells in cooperation with Ha-*ras* oncogene [39]. Whether NS4B is associated with the development of hepatocellular carcinoma is not yet clear.

NS5A

NS5A is a phosphorylated protein that is usually detected as two major electrophoretic forms with estimated molecular masses of 56 and 58 kDa. The smaller form (p56) represents the basal phosphorylated NS5A, while p58 is hyperphosphrylated primarily on multiple serine residues by a cellular kinase(s) [77]. Hyperphosphorylation of NS5A occurs when it is expressed in a polyprotein with the other nonstructural proteins NS3, NS4A, and NS4B [78]. Given that NS5A interacts with a variety of cellular proteins [39], it may possess diverse functions in the HCV life cycle as a result of differential phosphorylation.

NS5A was found to colocalize with the replicating HCV RNA as well as other nonstructural proteins NS3, NS4A, NS4B, and NS5B in the membrane-bound HCV replication complex, suggesting a role in RNA replication [76]. NS5A is anchored on the intracellular membranes via the N-terminal amphipathic helix [79]. Mutations that disrupt the amphipathic helix of NS5A abolished replication of a subgenomic HCV RNA, consistent with an important role for the membrane association of NS5A [79]. Additionally, NS5A interacts with a human VAMP-associated protein of 33 kDa (hVAp-33) [80], but the exact role of hVAP-A in HCV replication is unknown. Furthermore, NS5A has been found to interact with a growing body of cellular proteins. Surprisingly, purified recombinant NS5A inhibited the *in vitro* RNA-dependent RNA polymerase (RdRp) activity of NS5B [81]. Although the biological relevance of the protein–protein interactions between NS5A and cellular proteins has not yet been determined, NS5A likely plays a role in the modulation of apoptosis, cell growth, and proliferation, and intracellular signaling pathways [39]. Interaction of NS5A with PKR is postulated as a mechanism to antagonize the IFN-induced intracellular antiviral response [82]. Like the C protein [42], NS5A has been implicated in the interference of intracellular lipid and lipoprotein metabolism that may lead to the development of steatosis, a clinically characteristic feature of HCV infection [41].

NS5B

NS5B is the RdRp consisting of 591 amino acids (65 kDa) [83]. It contains functional motifs characteristic of all known RNA polymerases. As shown by mutagenesis studies, the conserved functional motifs are important for the RdRp activity [84]. The C-terminal domain of 21 amino acids is highly hydrophobic and deletion of this domain does not affect the *in vitro* RdRp activity but does ablate RNA replication in the cell ([85]; Ahn B and Luo G, unpublished observations). The C-terminal hydrophobic domain of NS5B functions to anchor it to the membrane of the ER [86]. The NS5B protein is phosphorylated although the role of phosphorylation for RdRp activity and HCV RNA replication has not been determined [87]. Evidence derived from biochemical studies suggests that purified recombinant NS5B forms a dimmer *in vitro*, which is critical for the *in vitro* RdRp activity [88, 89]. It will be interesting to determine whether the oligomerization of NS5B is physiologically important for the formation and function of the HCV replication complex in the cell.

The X-ray crystallographic structure of NS5B has been determined to 1.9-Å resolution [90]. It resembles the canonical structure of other polymerases with the characteristic fingers, palm, and thumb subdomains except that there is extensive interactions between the fingers- and thumb domains. The polymerase active site is situated in a 15-Å wide and 18-Å deep cavity at the center of the molecule with the palm as the base. The palm domain contains the conserved DXXXXD (motif A) and GDD (motif C) signature residues responsible for the nucleotidyl-transfer reaction [90]. The thumb subdomain is predominantly α-helical, leaving open the possibility for a role in interaction with other viral/cellular proteins. The structure of NS5B in complex with rNTP has also been determined and a low-affinity GTP-specific binding site was identified on the surface of the enzyme about 30 Å away from the active site [91]. Mutagenesis analysis has revealed that the GTP-specific binding site is not required for the *in vitro* RdRp activity but is important for HCV RNA replication in the cell (Cai Z and Luo G, unpublished observations).

NS5B is the central component of the HCV replication complex responsible for RNA synthesis during RNA replication (see section on viral replication, below). It also interacts with other nonstructural proteins NS3, NS4B, and NS5A as well as cellular proteins [39, 80]. The HCV proteins NS3, NS4B, and NS5A were found to modulate the *in vitro* RdRp activity of NS5B [92].

However, it is not known what the roles of the NS5B-associated cellular proteins are in HCV RNA replication and viral pathogenesis.

Precursor proteins

Unlike many other positive-stranded RNA viruses, very little is known about the nature and roles of the HCV precursor proteins in HCV replication and pathogenesis. Although the precursor proteins 4AB, 4AB5A, and 4B5A can readily be detected in Huh7 cells that harbor a subgenomic HCV replicon [93], the biological roles of these precursor proteins in the HCV life cycle have not been determined. One recent study found that NS4AB, but not 4A or 4B, singly or in combination, inhibited ER-to-Golgi traffic when expressed in the cell, suggesting that the NS4AB precursor protein probably plays a critical role in viral pathogenesis by disturbing the host-cell protein secretion and/or transport [94]. This observation opens up the possibility that other putative HCV precursor proteins may be physiologically relevant to HCV replication and pathogenesis *in vivo*. This is an area of HCV research that has not been adequately explored and should be addressed in future investigations.

Virus replication strategy

cDNA-derived HCV RNAs were infectious in chimpanzees upon intrahepatic inoculation [95, 96]. Genetic mutagenesis analysis of the infectious HCV RNAs has further demonstrated that all HCV enzymes (NS2/3 protease, NS3/4A serine protease, NS3 helicase/NTPase, and NS5B RdRp) and the conserved 3'UTR sequences are required for productive infections of HCV in chimpanzees [22]. These studies also provide a confirmation that all virally encoded enzymes play pivotal roles in the HCV life cycle and therefore are viable targets for antiviral intervention. For unknown reasons, however, HCV RNA transcripts that were infectious in chimpanzees were unable to produce infectious virus in cultured cells. The lack of a reliable tissue-culture system for HCV propagation and the lack of a small-animal model of HCV infection have significantly hampered the understanding of the molecular aspects of the HCV life cycle and molecular viral pathogenesis. Current knowledge about HCV replication strategy is derived mainly from studies using *in vitro* recombinant systems involving the expression of HCV proteins as well as the recently developed cell-based system for subgenomic HCV RNA (replicon) replication.

In general, it is thought that HCV follows a replication strategy similar to that of other positive-stranded RNA viruses (Fig. 2) [3, 97]. HCV enters the target-cell probably via receptor-mediated endocytosis. A number of cell-surface proteins have been postulated to be HCV receptors or coreceptors (see above), but none of them have yet been confirmed. Accumulating evidence points to a role for CD81 in HCV entry. For instance, a recent study found that

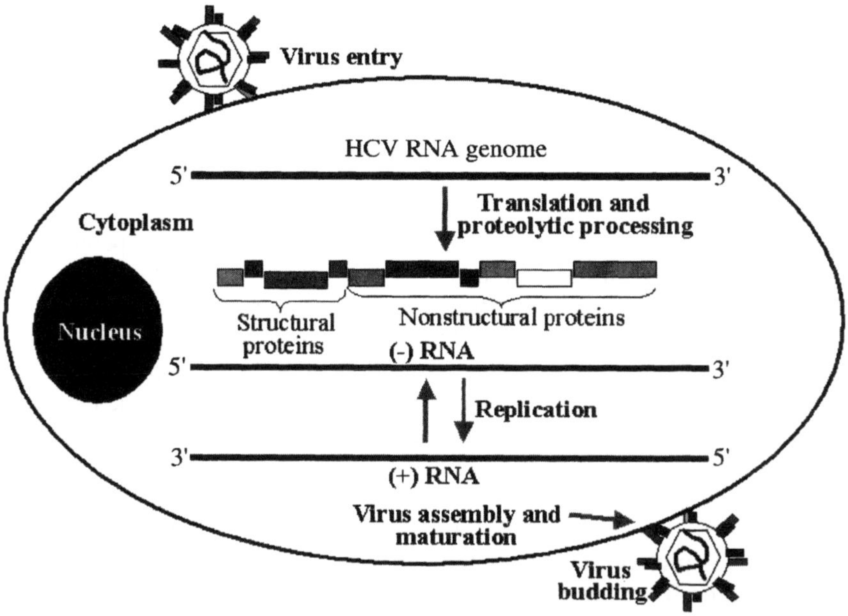

Figure 2. Schematic of the HCV life cycle. Each step of the HCV life cycle is shown schematically. Virus attachment to and entry into the target-cell are mediated by the HCV envelope glycoproteins E1 and E2 and HCV receptors on the cell surface. Upon uncoating, HCV RNA genome initially serves as an mRNA for translation of viral polyprotein precursor, which is subsequently cleaved by cellular and viral proteases into mature viral proteins. RNA replication is then initiated by RNA synthesis from positive-strand to complementary negative-strand RNA, which in turn acts as a template for synthesis of more genomic RNA. The nascent HCV RNA genome and structural proteins are assembled into virions that egress from the infected cell.

CD81 was required but not sufficient for cell entry of retroviruses pseudotyped with the HCV envelope glycoproteins E1 and E2, suggesting that CD81 is either a receptor or coreceptor for virus entry [46, 47]. Upon cell entry, HCV undergoes uncoating, a process that is likely mediated by the viral envelope glycoproteins and results in fusion between viral and cellular membranes. Recent studies with retroviral pseudotypes demonstrated that the HCV envelope glycoproteins E1 and E2 are capable of mediating both cell entry and membrane fusion of recombinant retroviruses [45–47]. Following virus entry and uncoating, HCV RNA is released and initially serves as an mRNA for translation of the viral polyprotein. Subsequently, the HCV polyprotein is proteolytically processed into individual mature viral proteins. HCV RNA replication is then initiated by synthesis of complementary negative-strand RNA, which in turn acts as a template for replication of more positive-strand RNA genome. Finally, nascent HCV RNA genome and viral structural proteins are assembled to form virions that egress from the infected cells (Fig. 2).

Despite the fact that HCV is refractory to genetic analysis and manipulation in tissue culture, significant progress in determining the molecular mechanisms of HCV RNA replication has recently been made through utilization of recombinant systems and HCV replicon RNAs [12, 13, 20, 25, 26, 66, 67, 76, 83–85, 98–100]. The virally encoded RdRp is responsible for catalyzing RNA synthesis during RNA replication. The RdRp activity of HCV NS5B catalyzes *in vitro* RNA synthesis on both HCV-specific and nonviral RNA templates, implying that NS5B itself lacks template specificity [83–85, 100, 101]. It has also been demonstrated that purified NS5B is capable of synthesizing RNA *in vitro* by primer-dependent (elongation) as well as primer-independent (*de novo*) initiation mechanisms. In the absence of a primer, NS5B either extends the 3' end of the RNA template itself (self-priming or copy-back) or initiates RNA synthesis *de novo* [85, 100–102]. Initiation of RNA synthesis *de novo* is the mechanism commonly used for RNA replication by most positive-, negative-, and double-strand RNA viruses. Thus, *de novo* initiation is the most likely mechanism used for HCV RNA replication *in vivo*. This is supported by biochemical studies that have revealed that purified recombinant HCV RdRp is able to initiate RNA synthesis *de novo* using purine (A and G), but not pyrimidine (C and U), nucleotides [85, 100, 102]. In the cell, replication of positive-strand RNA is preferentially initiated with ATP and GTP, while replication of negative-strand RNA favors an ATP as the initiation nucleotide [103]. In contrast, CTP and UTP were also found at the 5' ends of the HCV RNAs extracted from serum samples of chimpanzees, indicating that replication of positive-strand RNA in chimpanzees may also be initiated with CTP and UTP (Cai et al., unpublished observations). Whether HCV RNA replication is selectively controlled by different initiation nucleotides remains to be determined. If so, it will also be interesting to determine the molecular basis underlying the selective initiation of HCV RNA replication *in vivo*.

Despite of many reports claiming virus propagation in cultured cells, none has proven to be efficient and reliable. In the absence of an efficient HCV infection system, a cell-based replicon replication system has become the method of choice for the study of HCV RNA replication in the cell [66, 67, 98]. Both subgenomic and full-length genomic HCV RNAs have been demonstrated to replicate efficiently in a human hepatoma cell line, Huh7 [66, 67, 98, 104–107]. Additionally, subgenomic HCV RNAs were reported to replicate in non-hepatic human cell types such as Hela and HEK-293 cells [108–110]. As such, HCV RNA replication is not restricted by hepatocyte-specific factors. In addition, HCV RNA replicates in mouse hepatoma cells [109]. These findings indicate that cell tropism of HCV is not strictly determined at the level of RNA replication. However, replication of subgenomic HCV RNAs in nonhepatic and mouse hepatoma cells is limited to a very small fraction of cells [109, 110], suggesting that specific cellular factors that are present in some but not all cells are required for HCV RNA replication. Identification of cellular factors important for HCV replication will provide insights to the underlying mechanism of HCV replication as well as novel targets for anti-HCV intervention. Replication of the subgenomic HCV RNA encoding the nonstructural proteins NS3 to NS5B also suggests that the nonstructural viral proteins

NS3, NS4A, NS4B, NS5A, and NS5B are sufficient for HCV RNA replication in the cell [66, 67]. Whether these nonstructural proteins are all required for RNA replication has yet to be defined. Deletion analysis of the subgenomic HCV replicon indicated that NS4B and NS5A are essential for HCV RNA replication (Luo G et al., unpublished observations). As described above, the GTPase activity of NS4B is required for replication of an HCV replicon RNA (Einav S, Elazar M, Danieli T, and Glenn JS, unpublished observations). The NS3 helicase/NTPase activity is also believed to play a role in RNA replication. Accordingly, the nonstructural proteins NS3, NS4A, NS4B, NS5A, and NS5B may represent the minimal set of viral proteins required for HCV RNA replication in the cell.

Replication of RNA viruses is a complex process requiring multiple viral and cellular proteins acting through protein–RNA and protein–protein interactions. Like other positive-stranded RNA viruses, HCV RNA replication occurs in a membrane-bound viral replication complex (RC) consisting of viral and cellular proteins and viral RNA. The nonstructural HCV proteins NS3, NS4A, NS4B, NS5A, and NS5B were all found to colocalize with each other to the perinuclear ER membranes in Huh7 cells harboring a replicating subgenomic HCV RNA [76]. In addition, each of these nonstructural viral proteins was predominantly colocalized with the newly synthesized HCV RNA in the perinuclear ER membranes as punctate spots [76, 111]. Through electron-microscopic analysis, the intracellular site of HCV RNA replication appears as a membranous web, apparently an alteration of the ER membrane induced by HCV protein(s) and/or RNA replication [94, 112]. Additional evidence for membrane association of HCV RNA replication comes from subcellular fractionation studies and Western-blot analyses. These studies demonstrated that the nonstructural HCV proteins were colocalized with HCV RNA mainly in the membrane fractions [76, 111]. Conversely, disruption of the membrane by treatment with detergents resulted in disassociation of the HCV RC. HCV RNA present in the membrane-bound RC was resistant to RNase treatment, whereas it became sensitive to RNases once the membranes were disrupted by treatment with detergent [76]. Taken together, these findings clearly demonstrate that the HCV RC is membrane-associated, is assembled within the membrane structures, and requires cellular membranes. Furthermore, the HCV RC likely contains cellular proteins that are necessary for HCV RNA replication and/or regulation. That cellular proteins play a role in HCV RNA replication was initially suggested by findings derived from biochemical studies. The HCV proteins required for RNA replication, as discussed above, were found to interact with various cellular proteins. In addition, a number of cellular proteins were identified which bind to the 5' and 3'UTRs of the HCV RNA genome. For example, knockdown expression of PTB by small interfering RNA (siRNA) resulted in a reduction in HCV RNA replication (Chang KS and Luo G, unpublished observations). Evidence for the role of cellular proteins in HCV RNA replication also comes from a recent finding that inhibition of geranylgeranylation of cellular protein(s) by lovastatin resulted in a block of HCV

RNA replication [113]. However, the cellular proteins involved in HCV RNA replication have not been identified. Their identification and characterization await future investigation. Other questions regarding HCV replication include whether HCV structural proteins play any regulatory role in the control of HCV RNA replication and how HCV translation is switched to replication. Why is HCV RNA, which is infectious in chimpanzees, not infectious in the cultured cells? What are the determinants of virion assembly, maturation, and egression? Answers to these questions are critically important for a more thorough understanding of the HCV replication strategy. Moreover, determination of the exact roles of viral and cellular proteins and their structure–function relationship in HCV replication will inevitably propel the discovery and development of specific and efficacious therapies for the successful treatment of HCV infection.

Acknowledgment
I thank Dr Jeffrey S. Glenn (Stanford University) for sharing information prior to publication and Kathy Grzech (Sponsored Program Development, University of Kentucky) for editing. Research in the author's laboratory is supported by grants from the National Institutes of Health (NIH) and in part by a grant from the Kentucky Science and Engineering Foundation (KSEF-148-502-02-18).

References

1 Choo QL, Kuo G, Weiner AJ, Overby LR, Bradley DW, Houghton M (1989) Isolation of a cDNA clone derived from a blood-borne non-A, non-B viral hepatitis genome. *Science* 244: 359–362
2 Reed KE, Rice CM (2000) Overview of hepatitis C virus genome structure, polyprotein processing, and protein properties. *Curr Top Microbiol Immunol* 242: 55–84
3 Rice CM (1996) Flaviviridae: the viruses and their replication. Lippincott-Raven, Philadelphia, PA. 931–956
4 Miller RH, Purcell RH (1990) Hepatitis C virus shares amino acid sequence similarity with pestiviruses and flaviviruses as well as members of two plant virus supergroups. *Proc Natl Acad Sci USA* 87: 2057–2061
5 Robertson B, Myers G, Howard C, Brettin T, Bukh J, Gaschen B, Gojobori T, Maertens G, Mizokami M, Nainan O et al. (1998) Classification, nomenclature, and database development for hepatitis C virus (HCV) and related viruses: proposals for standardization. International Committee on Virus Taxonomy. *Arch Virol* 143: 2493–2503
6 Bukh J, Purcell RH, Miller RH (1992) Sequence analysis of the 5' noncoding region of hepatitis C virus. *Proc Natl Acad Sci USA* 89: 4942–4946
7 Kolykhalov AA, Feinstone SM, Rice CM (1996) Identification of a highly conserved sequence element at the 3' terminus of hepatitis C virus genome RNA. *J Virol* 70: 3363–3371
8 Tanaka T, Kato N, Cho MJ, Sugiyama K, Shimotohno K (1996) Structure of the 3' terminus of the hepatitis C virus genome. *J Virol* 70: 3307–3312
9 Bukh J, Miller RH, Purcell RH (1995) Genetic heterogeneity of hepatitis C virus: quasispecies and genotypes. *Semin Liver Dis* 15: 41–63
10 Tsukiyama-Kohara K, Iizuka N, Kohara M, Nomoto A (1992) Internal ribosome entry site within hepatitis C virus RNA. *J Virol* 66: 1476–1483
11 Rijnbrand RC, Lemon SM (2000) Internal ribosome entry site-mediated translation in hepatitis C virus replication. *Curr Top Microbiol Immunol* 242: 85–116
12 Luo G, Xin S, Cai Z (2003) Role of the 5'-proximal stem-loop structure of the 5' untranslated region in replication and translation of hepatitis C virus RNA. *J Virol* 77: 3312–3318
13 Friebe P, Lohmann V, Krieger N, Bartenschlager R (2001) Sequences in the 5' nontranslated region of hepatitis C virus required for RNA replication. *J Virol* 75: 12047–12057

14 Hellen CU, Pestova TV (1999) Translation of hepatitis C virus RNA. *J Viral Hepat* 6: 79–87
15 Lukavsky PJ, Otto GA, Lancaster AM, Sarnow P, Puglisi JD (2000) Structures of two RNA domains essential for hepatitis C virus internal ribosome entry site function [In Process Citation]. *Nat Struct Biol* 7: 1105–1110
16 Wang TH, Rijnbrand RC, Lemon SM (2000) Core protein-coding sequence, but not core protein, modulates the efficiency of cap-independent translation directed by the internal ribosome entry site of hepatitis C virus. *J Virol* 74: 11347–11358
17 Reynolds JE, Kaminski A, Kettinen HJ, Grace K, Clarke BE, Carroll AR, Rowlands DJ, Jackson RJ (1995) Unique features of internal initiation of hepatitis C virus RNA translation. *EMBO J* 14: 6010–6020
18 Xu Z, Choi J, Yen TS, Lu W, Strohecker A, Govindarajan S, Chien D, Selby MJ, Ou J (2001) Synthesis of a novel hepatitis C virus protein by ribosomal frameshift. *EMBO J* 20: 3840–3848
19 Kim YK, Lee SH, Kim CS, Seol SK, Jang SK (2003) Long-range RNA-RNA interaction between the 5' nontranslated region and the core-coding sequences of hepatitis C virus modulates the IRES-dependent translation. *RNA* 9: 599–606
20 You S, Stump D, Branch AD, Rice CM (2004) A cis-acting replication element in the sequence encoding the NS5B RNA-dependent RNA polymerase is required for hepatitis C virus RNA replication. *J Virol* 78: 1352–1366
21 Ito T, Lai MM (1997) Determination of the secondary structure of and cellular protein binding to the 3'-untranslated region of the hepatitis C virus RNA genome. *J Virol* 71: 8698–8706
22 Kolykhalov AA, Mihalik K, Feinstone SM, Rice CM (2000) Hepatitis C virus-encoded enzymatic activities and conserved RNA elements in the 3' nontranslated region are essential for virus replication *in vivo*. *J Virol* 74: 2046–2051
23 Yanagi M, St Claire M, Emerson SU, Purcell RH, Bukh J (1999) *In vivo* analysis of the 3' untranslated region of the hepatitis C virus after *in vitro* mutagenesis of an infectious cDNA clone. *Proc Natl Acad Sci USA* 96: 2291–2295
24 Friebe P, Bartenschlager R (2002) Genetic analysis of sequences in the 3' nontranslated region of hepatitis C virus that are important for RNA replication. *J Virol* 76: 5326–5338
25 Yi M, Lemon SM (2003) 3' nontranslated RNA signals required for replication of hepatitis C virus RNA. *J Virol* 77: 3557–3568
26 Yi M, Lemon SM (2003) Structure-function analysis of the 3' stem-loop of hepatitis C virus genomic RNA and its role in viral RNA replication. *RNA* 9: 331–345
27 Ito T, Lai MM (1999) An internal polypyrimidine-tract-binding protein-binding site in the hepatitis C virus RNA attenuates translation, which is relieved by the 3'-untranslated sequence. *Virology* 254: 288–296
28 Luo G (1999) Cellular proteins bind to the poly(U) tract of the 3' untranslated region of hepatitis C virus RNA genome. *Virology* 256: 105–118
29 Petrik J, Parker H, Alexander GJ (1999) Human hepatic glyceraldehyde-3-phosphate dehydrogenase binds to the poly(U) tract of the 3' non-coding region of hepatitis C virus genomic RNA. *J Gen Virol* 80: 3109–3113
30 Ito T, Tahara SM, Lai MM (1998) The 3'-untranslated region of hepatitis C virus RNA enhances translation from an internal ribosomal entry site. *J Virol* 72: 8789–8796
31 Walewski JL, Keller TR, Stump DD, Branch AD (2001) Evidence for a new hepatitis C virus antigen encoded in an overlapping reading frame. *RNA* 7: 710–721
32 Vassilaki N, Mavromara P (2003) Two alternative translation mechanisms are responsible for the expression of the HCV ARFP/F/core+1 coding open reading frame. *J Biol Chem* 278: 40503–40513
33 Boulant S, Becchi M, Penin F, Lavergne JP (2003) Unusual multiple recoding events leading to alternative forms of hepatitis C virus core protein from genotype 1b. *J Biol Chem* 278: 45785–45792
34 Baumert TF, Ito S, Wong DT, Liang TJ (1998) Hepatitis C virus structural proteins assemble into viruslike particles in insect cells. *J Virol* 72: 3827–3836
35 Blanchard E, Brand D, Trassard S, Goudeau A, Roingeard P (2002) Hepatitis C virus-like particle morphogenesis. *J Virol* 76: 4073–4079
36 Lai MM, Ware CF (2000) Hepatitis C virus core protein: possible roles in viral pathogenesis. *Curr Top Microbiol Immunol* 242: 117–134
37 Lu W, Ou J (2002) Phosphorylation of hepatitis C virus core protein by protein kinase A and protein kinase C. *Virology* 300: 20–30

38 Kunkel M, Lorinczi M, Rijnbrand R, Lemon SM, Watowich SJ (2001) Self-assembly of nucleocapsid-like particles from recombinant hepatitis C virus core protein. *J Virol* 75: 2119–2129
39 Tellinghuisen TL, Rice CM (2002) Interaction between hepatitis C virus proteins and host cell factors. *Curr Opin Microbiol* 5: 419–427
40 Zhang J, Yamada O, Yoshida H, Iwai T, Araki H (2002) Autogenous translational inhibition of core protein: implication for switch from translation to RNA replication in hepatitis C virus. *Virology* 293: 141–150
41 Shi ST, Polyak SJ, Tu H, Taylor DR, Gretch DR, Lai MM (2002) Hepatitis C virus NS5A colocalizes with the core protein on lipid droplets and interacts with apolipoproteins. *Virology* 292: 198–210
42 Moriya K, Yotsuyanagi H, Shintani Y, Fujie H, Ishibashi K, Matsuura Y, Miyamura T, Koike K (1997) Hepatitis C virus core protein induces hepatic steatosis in transgenic mice. *J Gen Virol* 78: 1527–1531
43 Moriya K, Fujie H, Shintani Y, Yotsuyanagi H, Tsutsumi T, Ishibashi K, Matsuura Y, Kimura S, Miyamura T, Koike K (1998) The core protein of hepatitis C virus induces hepatocellular carcinoma in transgenic mice. *Nat Med* 4: 1065–1067
44 Dubuisson J (2000) Folding, assembly and subcellular localization of hepatitis C virus glycoproteins. *Curr Top Microbiol Immunol* 242: 135–148
45 Bartosch B, Dubuisson J, Cosset FL (2003) Infectious hepatitis C virus pseudo-particles containing functional E1-E2 envelope protein complexes. *J Exp Med* 197: 633–642
46 Bartosch B, Vitelli A, Granier C, Goujon C, Dubuisson J, Pascale S, Scarselli E, Cortese R, Nicosia A, Cosset FL (2003) Cell entry of hepatitis C virus requires a set of co-receptors that include the CD81 tetraspanin and the SR-B1 scavenger receptor. *J Biol Chem* 278: 41624–41630
47 Zhang J, Randall G, Higginbottom A, Monk P, Rice CM, McKeating JA (2004) CD81 is required for hepatitis C virus glycoprotein-mediated viral infection. *J Virol* 78: 1448–1455
48 Hsu M, Zhang J, Flint M, Logvinoff C, Cheng-Mayer C, Rice CM, McKeating JA (2003) Hepatitis C virus glycoproteins mediate pH-dependent cell entry of pseudotyped retroviral particles. *Proc Natl Acad Sci USA* 100: 7271–7276
49 Pileri P, Uematsu Y, Campagnoli S, Galli G, Falugi F, Petracca R, Weiner AJ, Houghton M, Rosa D, Grandi G, Abrignani S (1998) Binding of hepatitis C virus to CD81. *Science* 282: 938–941
50 Agnello V, Abel G, Elfahal M, Knight GB, Zhang QX (1999) Hepatitis C virus and other flaviviridae viruses enter cells via low density lipoprotein receptor. *Proc Natl Acad Sci USA* 96: 12766–12771
51 Scarselli E, Ansuini H, Cerino R, Roccasecca RM, Acali S, Filocamo G, Traboni C, Nicosia A, Cortese R, Vitelli A (2002) The human scavenger receptor class B type I is a novel candidate receptor for the hepatitis C virus. *EMBO J* 21: 5017–5025
52 Pohlmann S, Zhang J, Baribaud F, Chen Z, Leslie GJ, Lin G, Granelli-Piperno A, Doms RW, Rice CM, McKeating JA (2003) Hepatitis C virus glycoproteins interact with DC-SIGN and DC-SIGNR. *J Virol* 77: 4070–4080
53 Barth H, Schafer C, Adah MI, Zhang F, Linhardt RJ, Toyoda H, Kinoshita-Toyoda A, Toida T, Van Kuppevelt TH, Depla E et al. (2003) Cellular binding of hepatitis C virus envelope glycoprotein E2 requires cell surface heparan sulfate. *J Biol Chem* 278: 41003–41012
54 Nicole P, Taylor DR, Lai MMC (2002) Detection of a novel unglycosylated form of hepatitis C virus E2 envelope protein that is located in the cytosol and interacts with PKR. *J Virol* 76: 1265–1272
55 Taylor DR, Shi ST, Romano PR, Barber GN, Lai MM (1999) Inhibition of the interferon-inducible protein kinase PKR by HCV E2 protein [see comments]. *Science* 285: 107–110
56 Carrere-Kremer S, Montpellier-Pala C, Cocquerel L, Wychowski C, Penin F, Dubuisson J (2002) Subcellular localization and topology of the p7 polypeptide of hepatitis C virus. *J Virol* 76: 3720–3730
57 Griffin SD, Beales LP, Clarke DS, Worsfold O, Evans SD, Jaeger J, Harris MP, Rowlands DJ (2003) The p7 protein of hepatitis C virus forms an ion channel that is blocked by the antiviral drug, Amantadine. *FEBS Lett* 535: 34–38
58 Pavlovic D, Neville DC, Argaud O, Blumberg B, Dwek RA, Fischer WB, Zitzmann N (2003) The hepatitis C virus p7 protein forms an ion channel that is inhibited by long-alkyl-chain iminosugar derivatives. *Proc Natl Acad Sci USA* 100: 6104–6108
59 Sakai A, Claire MS, Faulk K, Govindarajan S, Emerson SU, Purcell RH, Bukh J (2003) The p7 polypeptide of hepatitis C virus is critical for infectivity and contains functionally important geno-

type-specific sequences. *Proc Natl Acad Sci USA* 100: 11646–11651
60 Ohuchi M, Cramer A, Vey M, Ohuchi R, Garten W, Klenk HD (1994) Rescue of vector-expressed fowl plague virus hemagglutinin in biologically active form by acidotropic agents and coexpressed M2 protein. *J Virol* 68: 920–926
61 Pinto LH, Holsinger LJ, Lamb RA (1992) Influenza virus M2 protein has ion channel activity. *Cell* 69: 517–528
62 Yamaga AK, Ou JH (2002) Membrane topology of the hepatitis C virus NS2 protein. *J Biol Chem* 277: 33228–33234
63 Reed KE, Grakoui A, Rice CM (1995) Hepatitis C virus-encoded NS2-3 protease: cleavage-site mutagenesis and requirements for bimolecular cleavage. *J Virol* 69: 4127–4136
64 Wu Z, Yao N, Le HV, Weber PC (1998) Mechanism of autoproteolysis at the NS2-NS3 junction of the hepatitis C virus polyprotein. *Trends Biochem Sci* 23: 92–94
65 Waxman L, Whitney M, Pollok BA, Kuo LC, Darke PL (2001) Host cell factor requirement for hepatitis C virus enzyme maturation. *Proc Natl Acad Sci USA* 98: 13931–13935
66 Blight KJ, Kolykhalov AA, Rice CM (2000) Efficient initiation of HCV RNA replication in cell culture [In Process Citation]. *Science* 290: 1972–1975
67 Lohmann V, Korner F, Koch J, Herian U, Theilmann L, Bartenschlager R (1999) Replication of subgenomic hepatitis C virus RNAs in a hepatoma cell line [see comments]. *Science* 285: 110–113
68 De Francesco R, Steinkuhler C (2000) Structure and function of the hepatitis C virus NS3-NS4A serine proteinase. *Curr Top Microbiol Immunol* 242: 149–169
69 Kwong AD, Kim JL, Lin C (2000) Structure and function of hepatitis C virus NS3 helicase. *Curr Top Microbiol Immunol* 242: 171–196
70 Lamarre D, Anderson PC, Bailey M, Beaulieu P, Bolger G, Bonneau P, Bos M, Cameron DR, Cartier M, Cordingley MG et al. (2003) An NS3 protease inhibitor with antiviral effects in humans infected with hepatitis C virus. *Nature* 426: 186–189
71 Lin C, Lin K, Luong YP, Rao BG, Wei Y, Brennan DL, Fulghum JR, Hsiao HM, Ma S, Maxwell JP et al. (2004) *In vitro* resistance studies of hepatitis C virus serine protease inhibitors, VX-950 and BILN 2061: Structural analysis indicates different resistance mechanisms. *J Biol Chem* 279: 17508–17514
72 Foy E, Li K, Wang C, Sumpter R Jr, Ikeda M, Lemon SM, Gale M Jr (2003) Regulation of interferon regulatory factor-3 by the hepatitis C virus serine protease. *Science* 300: 1145–1148
73 Rho J, Choi S, Seong YR, Choi J, Im DS (2001) The arginine-1493 residue in QRRGRTGR1493G motif IV of the hepatitis C virus NS3 helicase domain is essential for NS3 protein methylation by the protein arginine methyltransferase 1. *J Virol* 75: 8031–8044
74 Lin C, Wu JW, Hsiao K, Su MS (1997) The hepatitis C virus NS4A protein: interactions with the NS4B and NS5A proteins. *J Virol* 71: 6465–6471
75 Lundin M, Monne M, Widell A, Von Heijne G, Persson MA (2003) Topology of the membrane-associated hepatitis C virus protein NS4B. *J Virol* 77: 5428–5438
76 El-Hage N, Luo G (2003) Replication of hepatitis C virus RNA occurs in a membrane-bound replication complex containing nonstructural viral proteins and RNA. *J Gen Virol* 84: 2761–2769
77 Reed KE, Rice CM (1999) Identification of the major phosphorylation site of the hepatitis C virus H strain NS5A protein as serine 2321. *J Biol Chem* 274: 28011–28018
78 Koch JO, Bartenschlager R (1999) Modulation of hepatitis C virus NS5A hyperphosphorylation by nonstructural proteins NS3, NS4A, and NS4B. *J Virol* 73: 7138–7146
79 Elazar M, Cheong KH, Liu P, Greenberg HB, Rice CM, Glenn JS (2003) Amphipathic helix-dependent localization of NS5A mediates hepatitis C virus RNA replication. *J Virol* 77: 6055–6061
80 Tu H, Gao L, Shi ST, Taylor DR, Yang T, Mircheff AK, Wen Y, Gorbalenya AE, Hwang SB, Lai MM (1999) Hepatitis C virus RNA polymerase and NS5A complex with a SNARE-like protein. *Virology* 263: 30–41
81 Shirota Y, Luo H, Qin W, Kaneko S, Yamashita T, Kobayashi K, Murakami S (2002) Hepatitis C virus (HCV) NS5A binds RNA-dependent RNA polymerase (RdRP) NS5B and modulates RNA-dependent RNA polymerase activity. *J Biol Chem* 277: 11149–11155
82 Gale MJ Jr, Korth MJ, Katze MG (1998) Repression of the PKR protein kinase by the hepatitis C virus NS5A protein: a potential mechanism of interferon resistance. *Clin Diagn Virol* 10: 157–162
83 Behrens SE, Tomei L, De Francesco R (1996) Identification and properties of the RNA-dependent RNA polymerase of hepatitis C virus. *EMBO J* 15: 12–22

84 Lohmann V, Korner F, Herian U, Bartenschlager R (1997) Biochemical properties of hepatitis C virus NS5B RNA-dependent RNA polymerase and identification of amino acid sequence motifs essential for enzymatic activity. *J Virol* 71: 8416–8428
85 Luo G, Hamatake RK, Mathis DM, Racela J, Rigat KL, Lemm J, Colonno RJ (2000) *De novo* initiation of RNA synthesis by the RNA-dependent RNA polymerase (NS5B) of hepatitis C virus. *J Virol* 74: 851–863
86 Schmidt-Mende J, Bieck E, Hugle T, Penin F, Rice CM, Blum HE, Moradpour D (2001) Determinants for membrane association of the hepatitis C virus RNA-dependent RNA polymerase. *J Biol Chem* 276: 44052–44063
87 Hwang SB, Park KJ, Kim YS, Sung YC, Lai MM (1997) Hepatitis C virus NS5B protein is a membrane-associated phosphoprotein with a predominantly perinuclear localization. *Virology* 227: 439–446
88 Wang QM, Hockman MA, Staschke K, Johnson RB, Case KA, Lu J, Parsons S, Zhang F, Rathnachalam R, Kirkegaard K, Colacino JM (2002) Oligomerization and cooperative RNA synthesis activity of hepatitis C virus RNA-dependent RNA polymerase. *J Virol* 76: 3865–3872
89 Qin W, Luo H, Nomura T, Hayashi N, Yamashita T, Murakami S (2002) Oligomeric interaction of hepatitis C virus NS5B is critical for catalytic activity of RNA-dependent RNA polymerase. *J Biol Chem* 277: 2132–2137
90 Lesburg CA, Cable MB, Ferrari E, Hong Z, Mannarino AF, Weber PC (1999) Crystal structure of the RNA-dependent RNA polymerase from hepatitis C virus reveals a fully encircled active site. *Nat Struct Biol* 6: 937–943
91 Bressanelli S, Tomei L, Rey FA, De Francesco R (2002) Structural analysis of the hepatitis C virus RNA polymerase in complex with ribonucleotides. *J Virol* 76: 3482–3492
92 Piccininni S, Varaklioti A, Nardelli M, Dave B, Raney KD, McCarthy JE (2002) Modulation of the hepatitis C virus RNA-dependent RNA polymerase activity by the non-structural (NS) 3 helicase and the NS4B membrane protein. *J Biol Chem* 277: 45670–45679
93 Pietschmann T, Lohmann V, Rutter G, Kurpanek K, Bartenschlager R (2001) Characterization of cell lines carrying self-replicating hepatitis C virus RNAs. *J Virol* 75: 1252–1264
94 Konan KV, Giddings TH Jr, Ikeda M, Li K, Lemon SM, Kirkegaard K (2003) Nonstructural protein precursor NS4A/B from hepatitis C virus alters function and ultrastructure of host secretory apparatus. *J Virol* 77: 7843–7855
95 Yanagi M, Purcell RH, Emerson SU, Bukh J (1997) Transcripts from a single full-length cDNA clone of hepatitis C virus are infectious when directly transfected into the liver of a chimpanzee. *Proc Natl Acad Sci USA* 94: 8738–8743
96 Kolykhalov AA, Agapov EV, Blight KJ, Mihalik K, Feinstone SM, Rice CM (1997) Transmission of hepatitis C by intrahepatic inoculation with transcribed RNA. *Science* 277: 570–574
97 Pogue GP, Huntley CC, Hall TC (1994) Common replication strategies emerging from the study of diverse groups of positive-strand RNA viruses. *Arch Virol Suppl* 9: 181–194
98 Kato T, Date T, Miyamoto M, Furusaka A, Tokushige K, Mizokami M, Wakita T (2003) Efficient replication of the genotype 2a hepatitis C virus subgenomic replicon. *Gastroenterology* 125: 1808–1817
99 Oh JW, Sheu GT, Lai MM (2000) Template requirement and initiation site selection by hepatitis C virus polymerase on a minimal viral RNA template. *J Biol Chem* 275: 17710–17717
100 Ranjith-Kumar CT, Kim YC, Gutshall L, Silverman C, Khandekar S, Sarisky RT, Kao CC (2002) Mechanism of *De Novo* Initiation by the Hepatitis C Virus RNA-Dependent RNA Polymerase: Role of Divalent Metals. *J Virol* 76: 12513–12525
101 Oh JW, Ito T, Lai MM (1999) A recombinant hepatitis C virus RNA-dependent RNA polymerase capable of copying the full-length viral RNA. *J Virol* 73: 7694–7702
102 Zhong W, Uss AS, Ferrari E, Lau JY, Hong Z (2000) *De novo* initiation of RNA synthesis by hepatitis C virus nonstructural protein 5B polymerase. *J Virol* 74: 2017–2022
103 Cai ZH, Liang J, Luo G (2004) Effects of mutations of the initiation nucleotides on hepatitis C virus RNA replication in the cell. *J Virol* 78: 3633–3643
104 Blight KJ, McKeating JA, Marcotrigiano J, Rice CM (2003) Efficient replication of hepatitis C virus genotype 1a RNAs in cell culture. *J Virol* 77: 3181–3190
105 Blight KJ, McKeating JA, Rice CM (2002) Highly permissive cell lines for subgenomic and genomic hepatitis C virus RNA replication. *J Virol* 76: 13001–13014
106 Guo JT, Bichko VV, Seeger C (2001) Effect of alpha interferon on the hepatitis C virus replicon. *J Virol* 75: 8516–8523

107 Ikeda M, Yi M, Li K, Lemon SM (2002) Selectable subgenomic and genome-length dicistronic RNAs derived from an infectious molecular clone of the HCV-N strain of hepatitis C virus replicate efficiently in cultured Huh7 cells. *J Virol* 76: 2997–3006
108 Bartenschlager R, Kaul A, Sparacio S (2003) Replication of the hepatitis C virus in cell culture. *Antivir Res* 60: 91–102
109 Zhu Q, Guo JT, Seeger C (2003) Replication of hepatitis C virus subgenomes in nonhepatic epithelial and mouse hepatoma cells. *J Virol* 77: 9204–9210
110 Ali S, Pellerin C, Lamarre D, Kukolj G (2004) Hepatitis C virus subgenomic replicons in the human embryonic kidney 293 cell line. *J Virol* 78: 491–501
111 Shi ST, Lee KJ, Aizaki H, Hwang SB, Lai MM (2003) Hepatitis C virus RNA replication occurs on a detergent-resistant membrane that cofractionates with caveolin-2. *J Virol* 77: 4160–4168
112 Gosert R, Egger D, Lohmann V, Bartenschlager R, Blum HE, Bienz K, Moradpour D (2003) Identification of the hepatitis C virus RNA replication complex in Huh-7 cells harboring subgenomic replicons. *J Virol* 77: 5487–5492
113 Ye J, Wang C, Sumpter R Jr, Brown MS, Goldstein JL, Gale M Jr (2003) Disruption of hepatitis C virus RNA replication through inhibition of host protein geranylgeranylation. *Proc Natl Acad Sci USA* 100: 15865–15870

Hepatitis Prevention and Treatment
Edited by J.M. Colacino and B.A. Heinz

Immunology of hepatitis B virus

Tomasz I. Michalak

Molecular Virology and Hepatology Research, Faculty of Medicine, Health Sciences Centre, Memorial University, St. John's, Newfoundland, Canada A1B 3V6

Introduction

Hepatitis B virus (HBV) is perceived as a principally hepatotropic virus that causes liver disease ranging from an apparent absence of symptoms [i.e. serum HBV surface antigen (HBsAg)-reactive, subclinical infection], through symptomatic acute and chronic hepatitis and 'healthy' HBsAg-positive chronic carrier state, to infrequent but highly lethal fulminant hepatitis and life-threatening liver cirrhosis and hepatocellular carcinoma (HCC) [1, 2]. The virus can also induce diseases of kidneys and blood vessels due to deposition of virus antigen–antibody immunocomplexes [3–5]. In addition to these well-defined and, for the most part, readily diagnosable conditions, recent studies indicate that HBV also commonly establishes an asymptomatic, serologically silent (i.e. serum HBsAg-negative) infection which persists for decades after resolution of acute hepatitis B. This occult virus carriage is accompanied by low levels of HBV genomes detectable by nested polymerase chain reaction (PCR) in serum, peripheral blood mononuclear cells (PBMCs) and the liver [6–10]. The data accumulated in recent years also indicate that HBV replicates within the lymphatic system [6, 11, 12], although much less vigorously than in the liver, suggesting that lymphotropism is the virus' intrinsic propensity. In this regard, studies in the woodchuck model of hepatitis B, which is a validated paradigm of molecular and pathological events occurring in HBV-infected humans [13, 14], showed that cells of the lymphatic system are an invariable site of lifelong replication of infectious virus independent of whether the infection is symptomatic or occult [15–20].

Considering the vast majority of symptomatically infected patients, the different courses and outcomes of hepatitis B are not distinctively linked to defined virus subtypes or to a specific pathobiological property of the invading virus. However, this issue remains under investigation since the classification of HBV based on the virus total sequence has been introduced. Thus, it has been suggested that patients with HBV genotype C may have a more severe liver disease [21], whereas genotype B might be associated with a higher tendency to develop HCC in young non-cirrhotic patients in Taiwan [22]. Also, an HBV e antigen (HBeAg)-negative phenotype, which is predominantly caused by a stop-codon mutation in codon 28 of the HBV precore region [23], has been inconsistently linked with fulminant hepatitis or with more severe progression of chronic hepatitis [24]. Nevertheless, it is generally accepted that the outcome of HBV infection is overall determined by the vigor and breadth of the host's immune response. The essentially noncytopathic nature of HBV, as well as other viruses belonging to the hepadnaviral family, argues for believing that hepatitis B is *de facto* an immunological disease which is induced and sustained by the yield of viral products having immunogenic and immunoregulatory properties. The role of the immune system in hepatitis B is exemplified by the rate of development of HBsAg-positive chronic hepatitis in different age groups, which appears to be related to the maturity of the system at the time of virus invasion. Thus, while chronic hepatitis B develops in more than 90% of newborns, it occurs in 5–10% of immunocompetent adults [25].

The focus of this chapter is on the interactions between HBV and the host's immune system and on the contribution of both adaptive (specific) and innate (non-specific) immune responses to hepadnavirus clearance and development of virus-induced liver disease. The evidence gained from both human and animal models of hepatitis B will be considered. The efficacy and safety of the current HBV vaccines and a need for antiviral therapy is also discussed.

Immune response

Humoral immune response

The antibody response to HBV envelope (surface) glycoprotein is a T cell-dependent process. It is generally accepted that these antibodies play a critical role in virus neutralization and clearance. The loss of serum HBsAg and the rise of antibodies to HBsAg (anti-HBs) is the hallmark of resolution of hepatitis in acutely infected patients [2] (Fig. 1), yet the antibodies are not observed in individuals with chronic hepatitis B unless assays detecting immunoglobulins complexed to circulating antigens are applied [26]. The importance of the antibodies has been demonstrated by the universal success of HBsAg-based HBV vaccine. It is thought that anti-HBs antibodies limit virus spread by complexing circulating virions facilitating their removal.

All glycoproteins of HBV envelope, i.e. large (preS1), middle (preS2), and major (S), elicit a specific antibody response [27]. While antibodies to major S protein, commonly referred to as anti-HBs, are identifiable by commercial assays and their kinetics are well defined, anti-preS1 and anti-preS2 can be examined by research assays and their clinical relevance is less clear. It has been shown, however, that anti-preS1 are elicited very early, during the incubation period of hepatitis [27, 28], and that they are the first antibody markers of acute infection that precedes the appearance of antibodies to HBV core protein (anti-HBc) of IgM class [27] (Fig. 1). Similar kinetics have been seen for

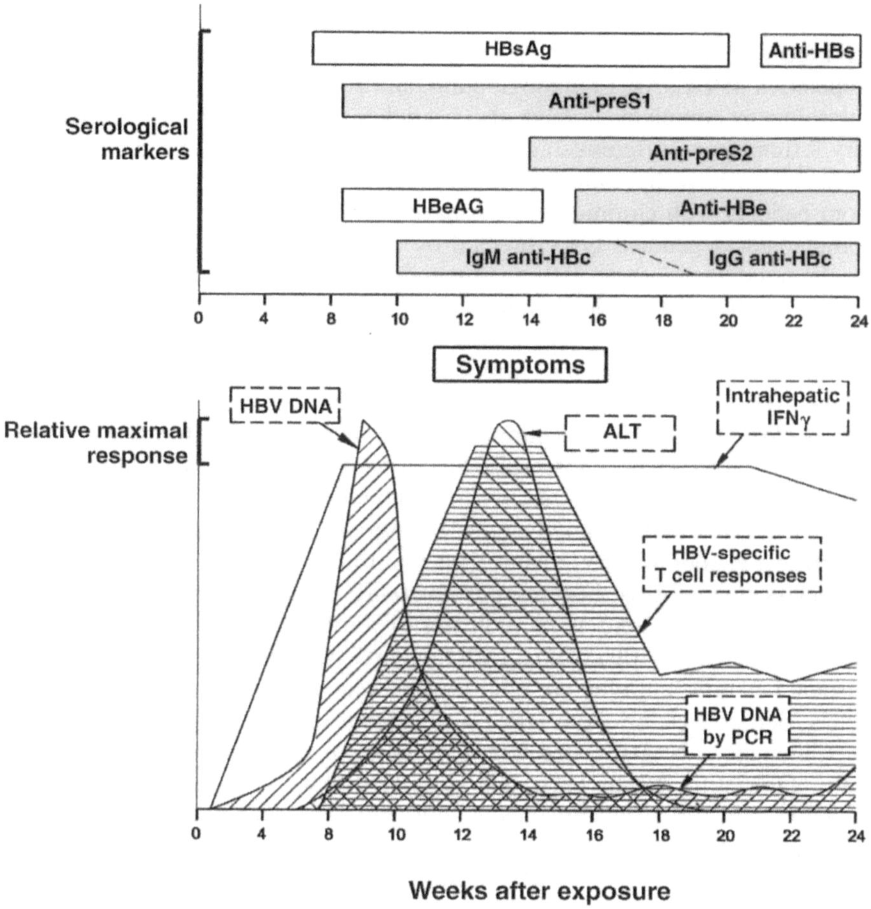

Figure 1. Schematic presentation of the occurrence of HBV infection markers in sera and the profiles of innate, represented by intrahepatic IFNγ expression, and virus-specific peripheral T cell responses in relation to clinical symptoms and liver injury measured by elevation in alanine aminotransferase (ALT) level in acute hepatitis B followed by recovery (extrapolated from references [1, 2, 7, 27, 47, 48]).

antibodies directed against the preS1 immunodominant epitope of woodchuck hepatitis virus (WHV) envelope in acute hepatitis progressing to recovery [29]. In contrast to anti-preS1, anti-preS2 appear in late acute phase prior to resolution of hepatitis when serum HBsAg is still detectable [27, 30]. Both anti-preS1 and anti-preS2 can circulate simultaneously with HBsAg and independently of anti-HBs (Fig. 1).

The humoral response to the HBV envelope proteins may exert a detrimental effect by inducing tissue injury. Extrahepatic diseases caused by HBsAg–antibody immune complexes, such as membranous and membranoproliferative glomerulonephritis [4, 31], necrotizing vasculitis (periarteritis or polyarteritis nodosa) [3, 5, 32], arthritis [33], and other disorders [31], were reported with various frequencies in patients chronically infected with HBV. Deposition of HBsAg, immunoglobulin, and complement, capable of binding heterologous complement, have also been detected in hepatic necroinflammatory lesions from different stages of hepatitis B, suggesting that the complexes may also have liver-damaging potential [34]. In this regard, hepatocytes from patients with chronic hepatitis have been shown to be prone *in vitro* to complement-dependent lysis directed by anti-HBV antibodies [35]. Overall, the humoral response to HBsAg not only remains active in chronic serum HBsAg-positive infection, but it may also occasionally cause pathological effects.

The biological roles of the antibodies to HBV core antigen (HBcAg) and e antigen (HBeAg) are less clear. These antibodies are induced by two structurally closely related but functionally and immunologically different products of the HBV core gene. Thus, while HBcAg acts as both a T cell-independent and a T cell-dependent antigen, response elicited to HBeAg is strictly T cell-dependent [24, 36]. Overall, the non-HBeAg cross-reactive antibody response to HBcAg is significantly greater than that to HBeAg [36]. Anti-HBc of the IgM class emerge early, beginning from 10–12 weeks post-infection, and they are utilized in the clinic as an indicator of a recent onset of HBV infection. Anti-HBc and anti-HBe co-exist with free virus particles during both acute and chronic hepatitis and they are not effective against viremia even at very high levels. Seroconversion from HBeAg-positive to anti-HBe-positive status has been thought to be associated with less infective and perhaps, a more benign state of hepatitis. It has been also suggested that the decline of virus titer, which is observed after seroconversion to anti-HBe, could be the result of the elimination of infected hepatocytes *via* an antibody recognition of HBe protein displayed on the surface of infected hepatocytes [37]. In experimental hepatitis in woodchucks, antibodies (anti-WHe) to WHV e antigen were found bound to the hepatocyte surface only in animals which cleared serum WHeAg and had low histological activity of acute inflammation [38]. These observations raised a possibility that elimination of infected hepatocytes may involve an antibody-mediated process. The antibodies against HBeAg and HBcAg, similarly to that against HBsAg, may cause undesirable tissue injury. Deposition of HBeAg and HBcAg-immune complexes has been associated with the pathogenesis of membranous glomerulonephritis, particularly in children with chronic HBV infection [39, 40].

The antibodies to the HBV polymerase and X proteins are not monitored in the clinical setting, although they have been detected [41, 42]. It has been reported that antibody to the C-terminus of the HBV polymerase may be an early marker of infection and may also reflect active HBV replication [43]. Antibodies to HBxAg have been reported to show quantitative and qualitative heterogeneity. They have been found most frequently and at the highest levels in patients with chronic hepatitis, cirrhosis, or HCC and at lower levels in acutely infected patients and asymptomatic carriers of HBV [44].

Cellular immune response

Antiviral specific T cell response, particularly that mediated by class I major histocompatibility complex (MHC)-restricted (CD8+) cytotoxic T lymphocytes (CTL), is thought to be of primary importance in the control of HBV infection due to the cell capacity to identify and eliminate virus-infected cells [45]. Concomitantly, the same cytotoxic activity as well as pro-inflammatory cytokines released are main contributors to liver damage manifesting as hepatitis [45]. The activated CTLs also play a role in non-cytopathic clearance of hepadnavirus through production of antiviral cytokines, such as interferon γ (IFNγ) and tumor necrosis factor α (TNFα), and by a feedback stimulation of other immune cells, such as natural killer (NK) and NK-T cells and macrophages, as shown in transgenic mouse and chimpanzee models (e.g. [46, 47]). It is believed that the mounting of the effective class II MHC-restricted (CD4+) T helper 1 (Th1) and CTL responses, supplemented by the effectual production of anti-HBs antibodies, is required to clear both intracellular and extracellular virus and resolve hepatitis B (Fig. 2). Surprisingly, the virus is seemingly ignored by the CD4+ and CD8+ T cells for about 2 months after primary infection, even if the infection is finally resolved [48]. Although this needs to be interpreted with caution due to methodological differences, this interval is exceptionally long considering virus-specific T cell kinetics in infections with other viruses where the responses can be detected as early as 2

Virus dose	Innate immunity	Adaptive immunity	Clinical outcome	Long-term virological outcome
Low ($\leq 10^3$ virions)	?	?	Asymptomatic infection	Primary occult persistence
Moderate or high ($\geq 10^4$ virions)	Prompt and strong	B cells: effective (anti-HBs+) Helper T cells: prevailing type 1, broad virus epitope range, high frequency Cytotoxic T cells: polyclonal, multi-epitope specific, high frequency	Recovery	Secondary occult persistence
	Delayed and insufficient	B cells: impaired (anti-HBs−) Helper T cells: prevailing type 2, transient Cytotoxic T cells: narrow virus epitope range, very low frequency	Chronic disease	Serologically evident persistence

Figure 2. Main characteristics of the immunological response associated with recovery or progression to chronic liver disease and with establishment of different forms of long-term viral persistence in HBV infection. Consequences of infection with a low virus dose are inferred from studies in the woodchuck model of hepatitis B [15–18, 20] and from yet sporadic observations in humans (for references see [17] and [20]).

weeks post-infection [49, 50]. This delay is accompanied by a low frequency of HBV-specific CTL, which usually does not exceed 1% of circulating CD8+ T cells during acute phase of self-limited infection [51]. This contrasts with other viral infections in which up to 10% of the peripheral CD8+ T cells might consist of virus-specific CD8+ T lymphocytes.

Studies of acute hepatitis B have shown vigorous, polyclonal, and multi-specific CTL and Th1 responses to HBV antigens in patients who ultimately recover from hepatitis [51–53]. Quantification of HBV-specific CTLs using sensitive class I MHC-peptide tetramer staining, in which biotinylated MHC molecules are loaded with viral peptide and are linked in tetramers to streptavidin, revealed that the highest frequencies of CTLs occur at the peak of hepatocellular damage, as measured by the level of serum alanine aminotransferase (ALT), and that they descend shortly after the enzyme values normalize [51]. Moreover, the quantification of virus DNA in human [48], chimpanzees [47], and woodchucks [54] showed that the maximal reduction in virus load occurred before the peak in virus-specific CD4+ and CD8+ T cell responses and the onset of clinical disease (Fig. 2). These findings challenge the concept that HBV control is predominantly dependent upon CTL-mediated viral clearance. Analysis of serial liver biopsies obtained from experimentally infected chimpanzees [47] or from woodchucks with self-limited acute hepatitis [54–56] showed that recovery from acute inflammation is preceded by enhanced intrahepatic production of IFNγ and TNFα, which are likely produced by activated non-T cells, such as NK and NK-T cells (Fig. 2).

HBV-specific T cell reactivity persists for years, if not for life, after recovery from acute hepatitis B [7, 8]. This occurs in the context of the likewise long presence of low levels of transcriptionally active HBV genomes (e.g. [6, 8]). Physicochemical analyses of HBV DNA-reactive particles circulating years after recovery suggest that trace amounts of virions could be persistently assembled [6]. Studies in the WHV woodchuck model clearly demonstrated that very small quantities of infectious virus are produced for life after apparently complete resolution of acute WHV hepatitis [15]. In adult animals convalescing from hepatitis, the persisting occult infection is accompanied by intermittent minimal to moderate liver inflammation [15, 20, 56] and by an increased risk of developing HCC [15, 57]. These alterations are accompanied by T cell influx, as evidenced by elevated intrahepatic transcription of CD3, and by enhanced IFNγ and TNFα expression [56]. The recent findings of moderate liver inflammation and fibrosis in individuals with occult HBV persistence a decade after the onset of self-limited acute hepatitis [10] are consistent with observations in woodchucks with resolved infection. Development of HCC has also been described in individuals with occult infection due to past exposure to HBV [58, 59]. Overall, it is evident that despite clinical and serological resolution of hepatitis, hepadnavirus persists indefinitely in the asymptomatic hosts. It is proposed that the virus is kept under confinement through continuous immune pressure that perhaps is able to eliminate the existing small number of hepatocytes infected, giving a histological picture of remnant liver inflammation, but is not sufficient to eradicate virus entirely [15, 56]. The data from the woodchuck model also suggest that intrahepatic IFNγ, and to a lesser extent TNFα, plays a central role in the local control of this occult hepadnavirus persistence [56].

In contrast to acute infection progressing to recovery, the establishment of chronic hepatitis is associated with a lower number of circulating and intrahepatic HBV-specific CD8+ and CD4+ T cells. Studies in both neonatal and adult woodchucks infected with WHV revealed that measurements of the hepatic virus load, histological severity of liver injury, the extent of T cell influx, and hepatic expression of IFNγ and TNFα allow for prediction of whether acute hepatitis finally will resolve or progress to chronic disease [55, 56]. It was documented that the development of chronic disease, in contrast to hepatitis followed by recovery, is preceded by an initially greater virus load, a lower level of inflammation, and a decreased transcription of IFNγ and TNFα within the liver.

It is assumed that skewing of the T cell response into a predominant T helper 2 (Th2) pathway early in the infection is one of the main reasons for the development of chronic hepatitis. This is supported by findings in transgenic mice where HBeAg preferentially induces non-inflammatory Th2 response through peripheral deletion of Th1 cells [60]. Since Th1 cells are essential for effective CTL function, this impairment can downgrade (tolerize) the CTL response and, consequently, support the persistence of virus infection. Another piece of evidence that supports the idea that the prevailing Th2 response may predispose to the development of chronic infection can be drawn from the woodchuck model. It was noticed that the occurrence prior to WHV infection of antibodies against hepatic asialoglycoprotein receptor (anti-ASGPR), which can be hepatocyte-pathogenic [35, 61], was connected to a higher rate of progression to chronic hepatitis when antibodies were naturally occurring [62] or induced by immunization with ASGPR [63]. It is possible that a pre-existing antibody response, which is Th2 cell-dependent, may predispose to an antiviral Th2 response that is not sufficient to control virus and terminate disease.

Patients with progressing serum HBsAg-positive chronic hepatitis demonstrate a much less vigorous T cell response than those with acute infection. The reasons behind this are not well understood and the combined effect of different mechanisms could be responsible. Among others, this may include: (1) prolonged exposure of T cells to viral antigens causing tolerance or abrogation of virus-specific T [64] and B [65] cells, (2) interference of viral gene products with pathways responsible for recognition and elimination of infected cells by effector immune cells [66], and/or (3) an impairment of dendritic cell (DC) accessory function [67, 68]. In addition to a potential HBeAg immunomodulatory effect mentioned above, the explicit feature of chronic hepadnaviral infection is persistent production of abundant quantities of viral envelope proteins that circulate freely as subviral particles. In chronically infected woodchucks, this state is accompanied by a massive and irreversible incorporation of the same proteins into the hepatocyte outer membrane [69, 70]. This is a unique feature of chronic hepatitis, which together with circulating excess of the same antigenic material, may provide the means to protect infected hepatocytes against immunoelimination [66, 70]. Importantly, this abundant insertion of virus envelope proteins into the hepatocyte surface co-exists with impairment of class I MHC presentation on the same cells [66]. Although a precise mechanism of this impairment is not yet elucidated, the available data clearly indicate that this is a consequence of a virus-dependent post-transcriptional process. This event could be critical for evasion of the CTL antiviral surveillance, perpetuation of liver disease, and virus persistence.

Innate immune response

The first line of defense against invading pathogens is the innate immune response, and one of the first innate mechanisms activated by virus invasion is production of antiviral cytokines; IFNs and TNFα. As data from other infections indicate, the swiftness and strength of the host's initial response can be critical in determining infection outcome, i.e. recovery *versus* progression to chronic disease [56, 71]. At this stage, the data available do not allow for the delineation of a contribution of individual cell types of the innate immunity, i.e. NK cells, NK-T cells, macrophages, and granulocytes, to the establishment and development of hepadnaviral infection. This is further made confusing by the unique and not yet fully recognized immunological functions of the liver itself. The status of this local milieu might be paramount in determining the progression and outcome of HBV infection. Among others, the liver is characterized by *in situ* production of NK and NK-T cells from stem cells and extrathymic T cells [72], a naturally high prevalence of NK-T cells [71, 73], and by the ability to retain and eliminate activated CD8+ CTL cells [74]. It is postulated that NK-T cells residing in the liver and the NK cells actively recruited to this organ are the principal sources of intrahepatic IFNγ in the pre-acute phase of HBV infection [75]. In this context, an increased number of circulating NK cells was observed in patients during the incubation phase of acute hepatitis 2–4 weeks prior to the appearance of HBV-specific CD8+ cells [48]. Also, an enhanced capacity of peripheral lymphoid cells to elicit perforin-mediated cell killing was found in woodchucks acutely infected with WHV, in contrast to healthy animals or those with chronic WHV hepatitis [76]. Since NK cells kill their cell targets through the perforin pathway and they are one of the main sources of IFNγ, this may imply a role for NK cells in the control of early primary hepadnavirus infection. This is consistent with detection of upregulated transcription of IFNγ in the liver in early acute WHV hepatitis progressing to recovery [47, 55, 56]. Further, it was demonstrated that activation of NK-T cells in the transgenic mouse model of HBV infection by α-galactoceramide, a glycoprotein that activates NK-T cells, forcing them into rapid IFNγ production and subsequent activation of NK cells to secrete IFNγ, IFNα/β, and TNFα, inhibits virus replication [77, 78]. The most convincing evidence supporting the involvement of NK and NK-T cells in the early production of IFNγ, which is capable of inhibiting HBV replication, was provided by studies of acutely infected chimpanzees [47]. In this animal model, a fall in HBV replication was accompanied by an increased hepatic NK and NK-T cell presence and the enhanced production of intrahepatic IFNγ and TNFα which preceded the recruitment of virus-specific CD4+ and CD8+ T cells to the liver.

A role for DCs in HBV infection is not yet fully understood, although the presence of HBV DNA and RNA [67, 79] as well as WHV DNA and its replicative form, covalently closed circular DNA (cccDNA) (Mulrooney PM and Michalak TI; unpublished observations), have been detected in monocyte-derived DCs from chronic hepatitis. DCs constitute a main link between innate and adaptive immunity and play the principal role in the antigen-specific priming of naive CD8+ and CD4+ T cells [80, 81]. These cells secrete chemokines and pro-inflammatory cytokines, such as interleukin (IL)-12 and IL-18, that can induce the release of IFNγ as well as other cytokines, including IL-1, IL-6, and TNFα. Therefore, it could be expected that defects in their maturation, antigen presentation, or production of immunostimulators, which are in a major part dependent on signals delivered by the innate immune system, contribute to a delayed appearance or sub-optimal performance of virus-specific CTL and T helper cells. Currently, there are no data on function of DCs in the early phases of hepadnaviral infection. However, analysis of DCs from chronically infected patients imply an impaired T cell stimulatory capacity and lowered secretion of IL-12, a key cytokine responsible for anti-viral responses, compared to normal controls [67]. In *in vitro* experiments, DC exposed to HBV also showed decreased production of IL-12, IFNγ, TNFα, and IL-2 [68]. The presence of the nucleotide analogue lamivudine (3TC), an inhibitor of HBV replication, restored allostimulatory function of the HBV-exposed DCs and upregulated class II MHC presentation on these cells [68], suggesting that the compromised function of these cells can be reinstated by anti-viral therapy.

Virus response

Targets recognized by CTLs and antibody

Peripheral CTL response against HBV is typically vigorous and directed toward multiple epitopes of both structural (core, envelope) and nonstructural (HBeAg, polymerase) viral proteins in individuals who successfully clear the acute infection [48, 51–53] (Fig. 2). In contrast, CTL reactivity against HBV in chronically infected patients is weak or undetectable and narrowly focused [82]. Similarly, specific CTLs occur at a low frequency in the liver of chronically infected individuals [83, 84]. Despite a high mutational rate of the HBV genome, estimated at 1.75×10^{-5}–7.62×10^{-5} misincorporations/base per year

[85], the emergence of variants capable of escaping CTL response is rare in HBV infection. However, HBV variants carrying mutations in cytotoxic T cell epitopes have been detected and when the CTL response occasionally is strong and narrow, mutational inactivation of CTL epitopes can occur in chronic hepatitis [86].

A number of CTL epitopes have been empirically identified within the HBV sequence in patients with different histocompatibility locus antigen (HLA) backgrounds. Nevertheless, most of the properties of the CTL response in hepatitis B were delineated by analyzing CD8+ T cells from HLA-A-2-positive patients. The HBV epitope first identified as an HLA-A-2 CTL target and, since then, most frequently used to assess the status of HBV-specific CTL response, was the epitope of virus core protein located between amino acids 18 and 27 [87]. Other epitopes relatively frequently tested as CTL targets, using either a classical chromium release assay or novel class I MHC/HBV peptide tetramers, are envelope 335–343 and polymerase 455–463 and 575–583 epitopes (e.g. [48, 88]).

HBV epitopes that are recognized by B cells and targeted by relevant antibodies are multiple and located within all HBV proteins. However, as a result of the use of HBsAg for a population-wide prophylactic vaccination and hepatitis B immunoglobulin (HBIg) to prevent HBV reinfection after liver transplantation, the mutations in the 'a' determinant encoded by the S gene have the greatest practical significance. Up to 99% of antibodies induced by HBV vaccination are directed toward the 'a' immunodominant determinant, located between residues 137 and 149 of the major envelope protein, which is conserved across all HBV subtypes [89]. Breakthrough infections with HBV mutations in the 'a' determinant enabling infection of vaccinated individuals and giving false-negative results in anti-HBs screening tests, have been detected [89–91]. The sequence variation most frequently encountered has been mapped to amino acid 145 (glycine-to-alanine substitution) [92]. A similar mutation may occur when mothers vertically transmit HBV infection and newborns are vaccinated [93]. Most of these individuals become chronic carriers of the variant virus and are indistinguishable from those infected with wild-type HBV. In these carriers there is no reversion to the wild-type sequence, suggesting fitness of this virus even after years of carriage [90, 94].

Virus response to innate immune system

The mechanisms utilized by HBV to escape from the innate immunity are not yet delineated. Indirect evidence from the woodchuck model suggests that an excess of WHV envelope proteins in chronic infection may potentially impair the function of NK cells. This presumption is deduced from a number of related observations. First, a severe downregulation of class I MHC molecules has been documented on hepatocytes in chronically infected woodchucks [66] that coexists with abundant incorporation of WHV envelope proteins into the outer membranes of the same hepatocytes [66, 70]. Second, since virally infected cells with reduced class I MHC antigen display are inherently susceptible to attack from NK cells [95], this implies that non-class I MHC-dependent, cell-mediated cytotoxicity could be involved in virus control and induction of hepatocellular injury in chronic infection. However, circulating lymphoid cells in chronic hepatitis showed that the level of perforin-mediated killing, which is the main pathway of cell elimination by NK cells [96], is significantly lower than that in acute infection and essentially the same as in healthy woodchucks [76]. This may suggest that NK cell activity is in fact inhibited in chronic infection, despite suppressed hepatocyte class I MHC expression. Third, the previous observations have led to the postulation that HBsAg depresses NK cell cytotoxicity *in vitro*, presumably by interfering with their binding to target-cells [97]. If so, the presence of massive quantities of virus envelope material at the hepatocyte surface and in the circulation linked with the reduction of class I MHC presentation on hepatocytes may constitute a strategy that the virus employs to escape elimination by both CTL and NK cells. In addition, a severely reduced presentation of class I MHC molecules was detected not only on hepatocytes but also on organ (splenic) lymphoid cells of chronically infected woodchucks [66]. Since class I molecules are involved in the elaborate network of interactions between cells of the immune system, inhibition of their presentation could deregulate a variety of immune functions the effects of which, although not directly apparent, may profoundly diminish overall effectiveness of antiviral responses. Among others, it has been shown that interruption in class I MHC presentation on lymphoid cells is sufficient to induce autoimmune reactions [98], which are commonly encountered in HBV and WHV infections [62, 99–101]. This mechanism could impair both the innate immunity and the cross-talk between innate and virus-specific responses.

Viral resistance to IFN

IFNα has been licensed for treating chronic hepatitis B since 1992. It is effective in only one-third of treated patients [102, 103]. It induces sustained clearance of HBV DNA (evaluated by assays with sensitivity limits above 10^3 copies/ml) and HBeAg from serum in approximately 35% and HBsAg in 10% of patients [104]. The mechanisms responsible for failure of IFNα therapy remain unclear. It has been observed that patients with HBV genotypes C and D have a lower response rate to IFNα therapy compared with those infected with genotypes A and B [105]. It has been suggested that high levels of HBV replication in chronic carriers inhibit immunomodulatory properties of IFNα. In this regard, it was shown that patients with high levels of serum HBV DNA might be poorer responders to IFNα treatment [106]. The rise of HBV variants not responsive to IFNα with mutations in the core sequence has also been described [107, 108]. Moreover, HBV has been shown to induce cell resistance to IFNs *in vitro* [109] and is able to suppress the expression of human IFNβ gene [110]. Abolishing these inhibitory effects of HBV replication by applying a combination therapy may increase the effectiveness of IFNα treatment. The use of new IFNs, such as PEGylated IFN or consensus IFN, might be advantageous by enhancing immune responsiveness to high loads of the virus.

HBV vaccine

The first hepatitis B vaccine was licensed for general use in 1981 [111]. This plasma-derived product was the first commercially available subunit viral vaccine and the first vaccine that could be applied as a preventive measure against virus-induced cancer in humans. At present, this HBsAg vaccine has been largely replaced by second-generation recombinant yeast-derived vaccines carrying as immunogen the virus S protein [112]. Recently, a third-generation recombinant hepatitis B vaccine containing preS1, pres2 and S antigenic components of viral surface antigen subtypes *adw* and *ayw* has been developed and clinically tested [113]. It is expected that preS domains can enhance protective antibody response and stimulate T cell immune reactivity allowing bypass of inadequate responsiveness or genetic nonresponsiveness to the S antigen that is now occasionally encountered.

Efficacy

The overall protective efficacy of the conventional plasma-derived and recombinant vaccines is 90–95% [114]. In adults, prophylactic efficacy in immune responders is near 100% [111]. The seroconversion rates are lower in the elderly, smokers, and immunocompromised persons including hemodialysis patients, giving overall efficacy of about 95%. The effectiveness of vaccine exceeds 75% in newborn infants born to HBeAg-positive mothers and it prevents development of the serologically evident HBV chronic carrier state in successfully immunized babies [111]. In infants born to non-carrier mothers, the protective efficacy is near 100%. The protection after vaccination is long lasting and the immunological memory seems to endure for at least 13 years in immunocompetent individuals [115]. Currently, there is no universal recommendation for routine booster vaccination of immunocompetent persons who have responded to primary vaccination; however, some governments have such a policy in regard to certain risk groups [116]. A breakthrough infection occurs in 5–19% of children immunized in endemic countries [91] and, as already indicated, they are mainly caused by mutations in the immunodominant epitope of the 'a' determinant in the S protein.

Safety

In recent years, concerns have been raised that HBV vaccination could increase the risk of relapse in patients compromised by multiple sclerosis or induce the onset of the disease in previously healthy individuals. However, several large epidemiological studies have documented that the vaccination is associated neither with the risk of exacerbation of the existing disease nor with the development of multiple sclerosis in healthy people [117–119]. There was also no causal relation found between immunization against HBV and other demyelinating diseases [120, 121]. Similarly, there is no link between hepatitis B vaccination and type I diabetes [122]. Adverse events, such as anaphylactic shock, occur with an estimated frequency of one per 600 000 vaccine doses administered. However, no severe or fatal anaphylactic reactions have been reported [123].

Why the need for anti-viral therapy?

If the global vaccination programs against hepatitis B are implemented as planned [112, 122], a substantial decrease in the number of newly infected cases can be expected in a predictable time frame. Nevertheless, during the last 20 years since hepatitis B vaccine became available, the chronic carrier rate worldwide remains unchanged, although some unquestionable regional successes in control of HBV transmission have been achieved. It has been shown that universal immunization in high endemic areas can reduce the occurrence of long-term pathological consequences of HBV infection, as in the case of HCC in Taiwanese children [123]. The immense pool of serum HBsAg-positive chronically infected persons (estimated at 350–400 million people worldwide), who are potentially infectious to non-immune individuals, requires development of more effective therapies able to produce sustained inhibition of virus synthesis, including elimination of viral cccDNA in chronic carriers. The experience gained so far implies that a combination therapy applying more than one anti-viral agent or anti-viral agent and an immunomodulator may be required. One such possibility is the use of an anti-viral drug paired with a hepatitis B vaccine to reduce the virus load and to break humoral and cell-mediated immune tolerance characterizing chronic HBV infection. It was demonstrated that therapeutic vaccination alone, or administered either before or after treatment with IFNα, can induce transient anti-HBs antibody response and enhance virus-specific T cell reactivity accompanied by moderate reduction in HBV viremia in some chronically infected patients [124–126]. It is expected that combining therapeutic vaccination with a potent anti-viral agent might allow for the induction of a profile of immune responses similar to that in patients recovering from self-limited acute hepatitis. Although this approach may not eliminate the virus entirely or inhibit progression of liver disease completely, it may limit these processes to the levels that would not be imminently harmful to the infected host.

References

1 Chisari FV, Ferrari C (1996) Viral hepatitis. In: R Ahmed, KV Holmes, F Gonzalez-Scarano, FA Murphy, DE Griffin, HL Robinson (eds): *Viral pathogenesis.* Lippincott-Raven Publishers, Philadelphia, 745–778

2 Hollinger B (1990) Hepatitis B virus. In: BN Fields, DM Knipe, RM Chanock, MS Hirsch, JL Melnick, TP Monath, B Roizman (eds): *Virology (2nd edition).* Raven Press, New York, 2171–2236

3 Gocke DJ, Morgan C, Lockshin M, Hsu H, Bombardieri S, Christian CL (1970) Association between polyarteritis nodosa and Australia antigen. *Lancet* 2: 1149–1153

4 Combes B, Stastny P, Shorey J, Eigenbrodt EH, Barrera A, Hull AR, Carter NW (1971) Glomerulonephritis with deposition of Australia antigen-antibody complexes in glomerular basement membrane. *Lancet* 2: 234–237

5 Michalak T (1978) Immune complexes of hepatitis B surface antigen in the pathogenesis of periarteritis nodosa: a study of seven necropsy cases. *Am J Pathol* 90: 619–632

6 Michalak TI, Pasquinelli C, Guilhot S, Chisari FV (1994) Hepatitis B virus persistence after recovery from acute viral hepatitis. *J Clin Invest* 93: 230–239

7 Penna A, Artini M, Caballi A, Levrero M, Bertoletti A, Pilli M, Chisari FV, Rehermann B, Del Prete G, Fiaccadori F, Ferrari C (1996) Long-lasting memory T cell responses following self-limited acute hepatitis B. *J Clin Invest* 98: 1185–1194

8 Rehermann B, Ferrari C, Pasquinelli C, Chisari FV (1996) The hepatitis B virus persists for decades after patients' recovery from acute viral hepatitis despite active maintenance of a cytotoxic T-lymphocyte response. *Nat Med* 2: 1104–1108

9 Yotsuyanagi H, Yasuda K, Iino S, Moriya K, Shintani Y, Fujie H, Tsutsumi T, Kimura S, Koike K (1998) Persistent viremia after recovery from self-limited acute hepatitis B. *Hepatology* 27: 1377–1382

10 Yuki N, Nagaoka T, Yamashiro M, Mochizuki K, Kaneko A, Yamamoto K, Omura M, Hikiji K, Kato M (2003) Long-term histologic and virologic outcomes of acute self-limited hepatitis B. *Hepatology* 37: 1172–1179

11 Stoll-Becker S, Repp R, Glebe D, Schaefer S, Kreuder J, Kann M, Lampert F, Gerlich WH (1997) Transcription of hepatitis B virus in peripheral blood mononuclear cells from persistently infected patients. *J Virol* 71: 5399–5407

12 Laskus T, Radkowski M, Wang L-F, Nowicki M, Rakela J (1999) Detection and sequence analysis of hepatitis B virus integration in peripheral blood mononuclear cells. *J Virol* 73: 1235–1238

13 Tennant BC, Gerin JL (1994) The woodchuck model of hepatitis B virus infection. In: IM Arias, JL Boyer, N Fausto, WB Jakoby, DA Schachter, DA Shafritz (eds): *The liver: biology and pathobiology.* Raven Press, New York, 1455–1466

14 Michalak TI (1998) The woodchuck animal model of hepatitis B. *Viral Hepatitis Rev* 4: 139–165

15 Michalak TI, Pardoe IU, Coffin CS, Churchill ND, Freake DS, Smith P, Trelegan CL (1999) Occult life-long persistence of infectious hepadnavirus and residual liver inflammation in woodchucks convalescent from acute viral hepatitis. *Hepatology* 29: 928–938

16 Coffin CS, Michalak TI (1999) Persistence of infectious virus in the offspring of woodchuck mothers recovered from viral hepatitis. *J Clin Invest* 104: 203–212

17 Michalak TI (2000) Occult persistence and lymphotropism of hepadnaviral infection: insights from the woodchuck viral hepatitis model. *Immunol Rev* 174: 98–111

18 Lew Y-Y, Michalak TI (2001) *In vitro* and *in vivo* infectivity and pathogenicity of the lymphoid cell-derived woodchuck hepatitis virus. *J Virol* 75: 1770–1782

19 Mulrooney PM, Michalak TI (2003) Quantitative detection of hepadnavirus-infected lymphoid cells by *in situ* PCR combined with flow cytometry: implications for the study of occult virus persistence. *J Virol* 77: 970–979

20 Michalak TI, Mulrooney PM, Coffin CS (2004) Low doses of hepadnavirus induce infection of the lymphatic system that does not engage the liver. *J Virol* 78: 1730–1738

21 Kao JH, Chen PJ, Lai MY, Chen DS (2000) Hepatitis B genotypes correlate with clinical outcomes in patients with chronic hepatitis B. *Gastroeneterology* 118: 554–559
22 Fujie H, Moriya K, Shintani Y, Yotsuyanagi H, Iino S, Koike K (2001) Hepatitis B virus genotypes and hepatocellular carcinoma in Japan. *Gastroeneterology* 120: 1564–1565
23 Baumet TF, Liang TJ (1996) Precore mutants revised. *Hepatology* 23: 184–186
24 Milich D, Liang TJ (2003) Exploring the biological basis of hepatitis e antigen in hepatitis B virus infection. *Hepatology* 38: 1075–1086
25 Peters M, Vierling J, Gershwin ME, Milich D, Chisari FV, Hoofnagle JH (1991) Immunology and the liver. *Hepatology* 13: 977–994
26 Maruyama T, McLachlan A, Lino S, Koike K, Kurokawa K, Milich DR (1993) The serology of chronic B infection revised. *J Clin Invest* 91: 2586–2595
27 Alberti A, Pontissio P, Fraiese A, Milanesi C, Stenico D, Gianni M, Chemello L, Realdi G (1987) The preS/anti-preS systems in hepatitis B virus infection. In: W Robinson, K Koike, H Will (eds): *Hepadna viruses.* Alan R. Liss, New York, 397–410
28 Klinkert MQ, Theilmann L, Pfaff E, Schaller H (1986) Pre-S1 antigens and antibodies early in the course of acute hepatitis B infection. *J Virol* 58: 522–525
29 Jin Y-M, Churchill ND, Michalak TI (1996) Protease-activated lymphoid cell and hepatocyte recognition site in the preS1 domain of the large woodchuck hepatitis virus envelope protein. *J Gen Virol* 77: 1837–1846
30 Budkowska A, Dubreuil, Capel F, Pillot J (1986) Hepatitis B virus preS gene-encoded antigenic specificity and anti-pre-S antibody: relationship between anti-pre-S response and recovery. *Hepatology* 6: 360–368
31 Johnson RJ, Couster WG (1990) Hepatitis B virus infection and renal disease: clinical, immunopathogenic and therapeutical considerations. *Kidney Int* 37: 663–676
32 Michalak T, Krawczynski K (1981) Vascular lesions in hepatitis virus infection. In: E Bartoli, L Chaindussi, S Sherlock (eds): *Systemic effects of HBsAg immune complexes.* Piccini Medical Books, Padova, 36–50
33 Wands JR, Alpert E, Isselbacher KJ (1975) Arthritis associated with chronic active hepatitis: complement activation and characterization of circulating immune complexes. *Gastroenterology* 69: 1286–1291
34 Nowoslawski A, Krawczynski K, Brzosko WJ, Madalinski K (1972) Tissue localization of Australia antigen immune complexes in acute and chronic hepatitis and liver cirrhosis. *Am J Pathol* 68: 31–55
35 Michalak TI, Lau JYN, McFarlane BM, Alexander GJM, Eddleston ALWF, Williams R (1995) Antibody-directed complement-mediated cytotoxicity to hepatocytes from patients with chronic hepatitis B. *Clin Exp Immunol* 100: 227–232
36 Milich DR, McLachlan A (1986) The nucleocapsid of hepatitis B virus in both a T-cell-independent and a T-cell-dependent antigen. *Science* 234: 1398–1401
37 Schlicht H-J, Von Brunn A, Theilmann L (1991) Antibodies in anti-HBe-positive patient sera bind to an HBe protein expressed on the surface of human hepatoma cells: implications for virus clearance. *Hepatology* 13: 57–61
38 Michalak TI, Lin B, Churchill ND, Dzwonkowski P, Desousa JRB (1990) Hepadna virus nucleocapsid and surface antigens and the antigen-specific antibodies associated with hepatocyte plasma membranes in experimental woodchuck acute hepatitis. *Lab Invest* 62: 680–689
39 Slusarczyk J, Michalak T, Nazarewicz-de Mezer T, Krawczynski K, Nowoslawski A (1980) Membranous glomerulopathy associated with hepatitis B core antigen immune complexes in children. *Am J Pathol* 98: 29–43
40 Yoshikawa N, Ito H, Yamada Y, Hashimoto H, Katayama Y, Matsuyama S, Hasegawa O, Okada S, Hajikano H, Yoshiwara H et al. (1985) Membranous glomerulonephritis associated with hepatitis B antigen in children: a comparison with idiopathic membranous glomerulonephritis. *Clin Nephrol* 23: 28–34
41 Pfaff E, Salfeld J, Gmelin K, Schaller H, Theilmann L (1987) Synthesis of the X-protein of hepatitis B virus *in vitro* and detection of anti-X-antibodies in human sera. *Virology* 158: 456–460
42 Chang L-J, Dienstag J, Ganem D, Varmus H (1989) Detection of antibodies against hepatitis B virus polymerase antigen in hepatitis B-virus infected patients. *Hepatology* 10: 332–335
43 Weimer T, Schodel F, Jung MC, Pape GR, Alberti A, Fattovich G, Beljaars H, van Eerd PM, Will H (1990) Antibodies to the RNAase H domain of the hepatitis B virus P protein are associated with ongoing viral infection. *J Virol* 64: 5665–5668

44 Stemler M, Weimer T, Tu ZX, Wan DF, Levrero M, Jung C, Pape GR, Will H (1990) Mapping of B-cell epitopes of the human hepatitis B virus X protein. *J Virol* 64: 2802–2809
45 Chisari FV (2000) Viruses, immunity, and cancer: lessons from hepatitis B. *Am J Pathol* 156: 1118–1132
46 Guidotti LG, Ishikwa T, Hobbs MV, Matzke B, Schreiber R, Chisari FV (1996) Intracellular inactivation of the hepatitis B virus by cytotoxic T lymphocytes. *Immunity* 4: 25–36
47 Guidotti LG, Rochford R, Chung J, Shapiro M, Purcell R, Chisari FV (1999) Viral clearance without destruction of infected cells during acute HBV infection. *Science* 284: 825–829
48 Webster GJM, Reignat S, Maini MK, Whalley SA, Ogg GS, King A, Brown D, Amlot PL, Williams R, Vergani D et al. (2000) Incubation phase of acute hepatitis B in man: dynamic of cellular immune mechanisms. *Hepatology* 32: 1117–1124
49 Koup RA, Safrit JT, Cao Y, Andrews CA, McLeod G, Borkowsky W, Farthing C, Ho DD (1994) Temporal association of cellular immune responses with the initial control of viremia in primary human immunodeficiency virus type 1 syndrome. *J Virol* 68: 4650–4655
50 Rentenar RJ, Gamdia LE, van DerHock N, van Diepen FN, Boom R, Weel JF, Wertheim-van Dillen PM, van Lier RA, ten Berge IJ (2000) Development of virus-specific CD4(+) T cells during primary cytomegalovirus infection. *J Clin Invest* 105: 541–548
51 Maini MK, Boni C, Ogg GS, King AS, Reignat S, Lee CK, Larrubia JR, Webster GJ, McMichael AJ, Ferrari C et al. (1999) Direct *ex vivo* analysis of hepatitis B virus-specific CD8+ T cells associated with the control of infection. *Gastroenterology* 117: 1386–1396
52 Missale G, Redeker A, Person J, Fowler P, Guilhot S, Schjicht H, Ferrari C, Chisari FV (1993) HLA-A31 and Aw68 restricted cytotoxic T cell responses to a single hepatitis B virus nucleocapsid epitope during acute viral hepatitis. *J Exp Med* 177: 751–762
53 Rehermann B, Flower P, Sidney J, Person P, Redeker A, Brown M, Moss B, Sette A, Chisari FV (1995) The cytotoxic T cell response to multiple hepatitis B virus polymerase epitopes during and after acute viral hepatitis. *J Exp Med* 181: 1047–1058
54 Guo J-T, Zhou H, Liu C, Aldrich C, Saputelli J, Whitaker T, Barrasa MI, Mason WS, Seeger C (2000) Apoptosis and regeneration of hepatocytes during recovery from transient hepadnavirus infections. *J Virol* 74: 1495–1505
55 Nakamura I, Nupp JT, Cowlen M, Hall WC, Tennant BC, Casey JL, Gerin JL, Cote PJ (2001) Pathogenesis of experimental neonatal woodchuck hepatitis virus infection: chronicity as an outcome of infection is associated with a diminished acute hepatitis that is temporally deficient for the expression of interferon gamma and tumor necrosis factor-alpha messenger RNAs. *Hepatology* 33: 439–447
56 Hodgson PD, Michalak TI (2001) Augmented hepatitis interferon gamma expression and T-cell influx characterize acute hepatitis progressing to recovery and residual lifelong virus persistence in experimental adult woodchuck hepatitis virus infection. *Hepatology* 34: 1049–1059
57 Korba BE, Wells F, Baldwin B. Cote PJ, Tennant BC, Popper H, Gerin JL (1989) Hepatocellular carcinoma in woodchuck hepatitis virus-infected woodchucks: presence of viral DNA in tumor tissue from chronic carriers and animals serologically recovered from acute infection. *Hepatology* 9: 461–470
58 Liang TJ, Blum HE, Wands JR (1990) Characterization and biological properties of a hepatitis B virus isolated from a patient without hepatitis B virus serologic markers. *Hepatology* 15: 32–36
59 Pollicino T, Squadrito G, Cerenzia G, Cacciola I, Raffa G, Crax A, Farinati F, Missale G, Smedile A, Tiribelli C et al. (2004) Hepatitis B virus maintains its pro-oncogenic properties in the case of occult HBV infection. *Gastroenterology* 126: 102–110
60 Milich DR, Chen M, Hughes JL, Jones J (1998) The secreted hepatitis B precore antigen can modulate the immune response to the nucleocapsid: a mechanism for persistence. *J Immunol* 160: 7009–7013
61 Diao J, Churchill ND, Michalak TI (1998) Complement-mediated cytotoxicity and inhibition of ligand binding to hepatocytes by woodchuck hepatitis virus-induced autoantibodies to asialoglycoprotein receptor. *Hepatology* 27: 1623–1631
62 Diao J, Michalak TI (1997) Virus-induced anti-asialoglycoprotein receptor autoimmunity in experimental hepadnaviral hepatitis. *Hepatology* 25: 689–696
63 Diao J, Slaney DM, Michalak TI (2003) Modulation of the outcome and severity of hepadnaviral hepatitis in woodchucks by antibodies to hepatic asiloglycoprotein receptor. *Hepatology* 38: 629–638
64 Reignat S, Webster GJ, Brown D, Ogg GS, King A, Seneviratne SL, Dusheiko G, Williams R,

Maini MK, Bertoletti A (2002) Escaping high viral load exhaustion: CD8 cells with altered tetramer binding in chronic hepatitis B virus infection. *J Exp Med* 195: 1089–1101
65 Barnaba V, Franco A, Alberti A, Benvenuto R, Balsano F (1990) Selective killing of hepatitis B virus envelope antigen-specific B cells by class I-restricted, exogenous antigen-specific T lymphocytes. *Nature* 345: 258–260
66 Michalak TI, Hodgson PD, Churchill ND (2000) Posttranscriptional inhibition of class I major histocompatibility complex presentation on hepatocytes and lymphoid cells in chronic woodchuck hepatitis virus infection. *J Virol* 74: 4483–4494
67 Arima S, Akbar SMF, Michitaka K, Horiike N, Nuriya H, Kohara M, Onji M (2003) Impaired function of antigen-presenting dendritic cells in patients with chronic hepatitis B: localization of HBV DNA and HBV RNA in blood DC by *in situ* hybridization. *Int J. Mol Med* 11: 164–174
68 Beckebaum S, Cicinnati VR, Zhang X, Ferencik S, Frilling A, Grosse-Wilde H, Broelsch CE, Gerken G (2003) Hepatitis B virus-induced defect of monocyte-derived dendritic cells leads to impaired T helper type 1 response *in vitro*: mechanisms for viral immune escape. *Immunology* 109: 487–495
69 Michalak TI, Churchill ND (1988) Interaction of woodchuck hepatitis surface antigen with hepatocyte plasma membrane in woodchuck chronic hepatitis. *Hepatology* 8: 499–506
70 Michalak TI, Lin B (1994) Molecular species of hepadnavirus core and envelope polypeptides in hepatocyte plasma membrane of woodchucks with acute and chronic viral hepatitis. *Hepatology* 20: 275–286
71 Brion CA, Nguyen KB, Pien GC, Cousens LP, Salazar-Mather TP (1999) Natural killer cells in antiviral defense: function and regulation by innate cytokines. *Annu Rev Immunol* 17: 189–220
72 Seki S, Habu Y, Kawamura T, Takeda K, Dobashi H, Ohkawa T, Hiraide H (2000) The liver as a crucial organ in the first line of host defense: the roles of Kupffer cells, natural killer (NK) cells and NK1.1 Ag^+ T cells in T helper 1 immune responses. *Immunol Rev* 174: 35–46
73 Doherty DG, Norris S, Madrigal-Estebas L, McEntee G, Traynor O, Hegarty JE, O'Farrelly C (1999) The human liver contains multiple populations of NK cells, T cells, and CD3+CD56+ natural T cells with distinct cytotoxic activities and Th1, Th2, and Th0 cytokine secretion patterns. *J Immunol* 163: 2314–2321
74 Dao T, Klugewitz K, Mehal WZ, Mrtz DP (2000) The liver as a site of T-cell apoptosis: graveyard, or killing field? *Immunol Rev* 174: 47–62
75 Guidotti LG, Chisari FV (1999) Cytokine-induced viral purging-role in viral pathogenesis. *Curr Opin Microbiol* 2: 388–391
76 Hodgson PD, Grant MD, Michalak TI (1999) Perforin and Fas/Fas ligand-mediated cytotoxicity in acute and chronic woodchuck viral hepatitis. *Clin Exp Immunol* 118: 63–70
77 Kakimi K, Guidotti LG, Koezuka Y, Chisari FV (2000) Natural killer T cell activation inhibits hepatitis B virus replication *in vivo*. *J Exp Med* 199: 921–930
78 Kakimi K, Lane TE, Chisari FV, Guidotti LG (2001) Cutting edge: inhibition of hepatitis B virus replication by activated NK T cells does not require inflammatory cell recruitment to the liver. *J Immunol* 167: 6701–6705
79 Beckebaum S, Cicinnati VR, Dworacki G, Muller-Berghaus J, Stolz D, Harnaha J, Whiteside TL, Thomson AW, Lu L, Fung JJ, Bonham A (2002) Reduction of the circulating pDC1/pDC2 ratio and impaired function of *ex vivo*-generated DC1 in chronic hepatitis B infection. *Clin Immunol* 104: 138–150
80 Guermonprez P, Valladeau J, Zitvogel L, Thery C, Amigorena S (2002) Antigen presentation and T cell stimulation by dendritic cells. *Annu Rev Immunol* 20: 621–667
81 Becker Y (2003) Immunological and regulatory functions of uninfected and virus infected immature and mature subtypes of dendritic cells – a review. *Virus Genes* 26: 119–130
82 Bertoletti A, Costanzo A, Chisari FV, Levrero M, Arini M, Sette A, Penna A, Giuberti T, Fiaccadori F, Ferrari C (1994) Cytotoxic T lymphocyte response to a wild-type hepatitis B virus epitope in patients chronically infected by variant viruses carrying substitutions with the epitope. *J Exp Med* 180: 933–943
83 Ferrari C, Penna A, Giuberti T, Tong M, Ribera E, Fiaccadori F, Chisari FV (1987) Intrahepatic nucleocapsid antigen-specific T cells in chronic active hepatitis B. *J Immunol* 139: 2050–2058
84 Barnaba V, Franco A, Balsano A, Alberti C, Benvenuto R, Balsano F (1989) Recognition of hepatitis B envelope proteins by liver-infiltrating T lymphocytes in chronic HBV infection. *J Immunol* 143: 2650–2655
85 Orito E, Mizokami M, Ina Y, Moriyama EN, Kameshima N, Yamamoto M, Gojobori T (1995)

Host-independent evolution and a genetic classification of the hepadnavirus family based on nucleotide sequences. *Proc Natl Acad Sci USA* 86: 7059–7062
86 Bertoletti A, Sette A, Chisari FV, Penna A, Levrero M, DeCarli M, Fiaccadori F, Ferrari C (1994) Natural variants of cytotoxic epitopes are T cell receptor antagonists for antiviral cytotoxic T cells. *Nature* 369: 407–410
87 Penna A, Chisari FV, Bertoletti A, Missale G, Flower P, Giuberti T, Fiaccadori F, Ferrari C (1991) Cytotoxic T lymphocytes recognize an HLA-A2-restricted epitope within the hepatitis B virus nucleocapsid antigen. *J Exp Med* 174: 1565–1570
88 Rehermann B, Lau D, Hoofnagle JH, Chisari FV (1996) Cytotoxic T lymphocyte responsiveness after resolution of chronic hepatitis B virus infection. *J Clin Invest* 97: 1655–1665
89 Howard CR, Allison LM (1995) Hepatitis B surface antigen variation and protective immunity. *Intervirology* 38: 35–40
90 Carman WF, Thomas HC (1992) Genetic variation in hepatitis B virus. *Gastroenterology* 102: 711–719
91 Whittle HC, Inskip H, Hall AJ, Mendy M, Doenes R, Hoare S (1991) Vaccination against hepatitis B and protection against chronic viral carriage in The Gambia. *Lancet* 337: 747–750
92 Lee PI, Chang LY, Lee CY. Huang LM, Chang MH (1997) Detection of hepatitis B surface gene mutation in carrier children with or without immunoprophylaxis at birth. *J Infect Dis* 176: 427–430
93 Ngui SL, O'Connell S, Eglin RP, Heptonstall J, Teo CG (1997) Low detection rate and maternal provenance of hepatitis B virus S gene mutants in cases of failed postnatal immunoprophylaxis in England and Wales. *J Infect Dis* 176: 1360–1365
94 Hsu HY, Chang MH, Ni YH, Lin HH, Wang SM, Chen DS (1997) Surface gene mutants of hepatitis B virus in infants who develop acute or chronic infections despite immunoprophylaxis. *Hepatology* 26: 786–791
95 Brutkiewicz RR, Welsh RM (1995) Major histocompatibility complex class I antigens and the control of viral infections by natural killer cells. *J Virol* 69: 3967–3971
96 Kagi D, Ledermann B, Burki K, Seiler P, Odermatt B, Olsen KJ, Podack ER, Zinkernagel RM, Hengartner H (1994) Cytotoxicity mediated by T cells and natural killer cells is greatly impaired in perforin-deficient mice. *Nature* 369: 31–37
97 Azzari C, Rossi ME, Resti M, Caldini AL, Carbonella R, Ciappi S, Vierucci A (1992) VIP restores natural killer cell activity depressed by hepatitis B surface antigen. *Viral Immunol.* 5: 195–200
98 Fu Y, Nathan DM, Li F,. Li X, Faustman DL (1993) Defective major histocompatibility complex class I expression on lymphoid cells in autoimmunity. *J Clin Invest* 91: 2301–2307
99 Farrow LJ, Holborow EJ, Johnson GD, Lamb SG, Stewart JS, Taylor PE, Zuckerman AJ (1970) Autoantibodies and the hepatitis-associated antigen in acute infective hepatitis. *Br Med J* 2: 693–695
100 Michalak T, Nazarewicz T, Nowoslawski A (1976) Autoantibodies in chronic hepatitis and chronic glomerulonephritis in children infected with hepatitis type B virus. *Pediat. Pol.* 51: 1–7
101 Dzwonkowski P, Michalak TI (1990) Autoantibody pattern in a woodchuck model of hepatitis B. *Clin. Invest. Med.* 13: 322–328
102 Wong DK, Cheung AM, O'Rourke K, Naylor CD, Detsky AS, Heathcote J (1993) Effect of alpha-interferon treatment in patients with hepatitis B e antigen-positive chronic hepatitis B. A meta-analysis. *Ann Int Med* 119: 312–323
103 Janssen HL, Gerken G, Carreno V, Marcellin P, Naoumov NV, Craxi A, Ring-Larsen H, Kitis G, van Hattum J, se Vries RA et al. (1999) Interferon alpha for chronic hepatitis B infection: increased efficacy of prolonged treatment. The European Concerted Action on Viral Hepatitis (EUROHEP). *Hepatology* 30: 238–243
104 Hoofnagle JH, di Bisceglie AM (1997) The treatment of chronic viral hepatitis. *N Engl J Med* 336: 347–350
105 Kao JH, Wu NH, Chen PI, Lai MY, Chen DS (2000) Hepatitis B genotypes and the response to interferon therapy. *J Hepatol* 33: 998–1002
106 Brook MG, Karayiannis P, Thomas HC (1989) Which patients with chronic hepatitis B virus infection will respond to alpha-interferon therapy? A statistical analysis of predictive factors. *Hepatology* 10: 761–763
107 Main J, McCarron B, Thomas HC (1998) Treatment of chronic viral hepatitis. *Antiviral Chem. Chemother.* 9: 449–460
108 Radecke K, Protzer U, Trippler M, Mayer zum Buschenfelde KH, Gerken G (2000) Selection of

hepatitis B virus variants with amino acid substitutions inside the core antigen during interferon-alpha therapy. *J Med Virol* 62: 479–485
109 Onji M, Lever AM, Saito I, Thomas HC (1989) Defective response to interferons in cells transfected with the hepatitis B virus genome. *Hepatology* 9: 92–96
110 Whitten T, Quets S, Schloemer R (1991) Identification of the hepatitis B factor that inhibits expression of the beta interferon gene. *J Virol* 65: 4699–4704
111 Hilleman MR (2001) Overview of the pathogenesis, prophylaxis and therapeusis of viral hepatitis B, with focus on reduction to practical applications. *Vaccine* 19: 1837–1848
112 Kane M (1998) Status on hepatitis B immunization programmes in 1998. *Vaccine* 16: S104–S108
113 Young MD, Schneider DL, Zuckerman AJ, Du W, Dickson B, Maddrey WC (2001) Adult hepatitis B vaccination using a novel triple antigen recombinant vaccine. *Hepatology* 34: 372–376
114 Margolis HS, Coleman PJ, Brown RE, Mast EE, Sheingold SH, Arevalo JA (1995) Prevention of hepatitis B virus transmission by immunization: an economic analysis of current recommendations. *JAMA* 264: 1201–1208
115 Watson B, West DJ, Chilkatowsky A, Piercy S, Ioli VA (2001) Persistence of immunological memory for 13 years in recipients of a recombinant hepatitis B vaccine. *Vaccine* 19: 3164–3168
116 European Consensus Group on Hepatitis B Immunity (2000) Are booster immunization needed for lifelong hepatitis B immunity? *Lancet* 355: 561–565
117 Confavreux C, Sussia S, Saddier P, Bourdes V, Vukusic S (2001) Vaccinations and the risk of relapse in multiple sclerosis. Vaccines in Multiple Sclerosis Study Group. *N Engl J Med* 344: 319–326
118 Ascherio A, Zhang SM, Hernan MA, Olek MJ, Caplan PM, Brodovicz K, Walker AM (2001) Hepatitis B vaccination and the risk of multiple sclerosis. *N Engl J Med* 344: 327–332
119 Sadovnick AD, Scheifele DW (2000) School-based hepatitis B vaccination programmeme and adolescent multiple sclerosis. *Lancet* 355: 549–550
120 Zipp F, Weil JG, Einhaupl KM (1999) No increase in demyelinating diseases after hepatitis B vaccination. *Nat Med* 5: 964–965
121 Hasley NA, Duclos P, Van Damme P, Margolis H (2000) Hepatitis B vaccine and central nervous system demyelinating diseases. Viral Hepatitis Prevention Board. *Pediatr Infect Dis J* 18: 23–24
122 Duclos P (2003) Safety of immunization and adverse events following vaccination against hepatitis B. *Expert Opin Invest Drugs* 2: 225–231
123 Huang KY, Lin S-R (2000) Nationwide vaccination: a success story in Taiwan. *Vaccine* 18: 535–538
124 Fruttaldo L, Schettino G, Mongio F (1997) Anti-HBV vaccination before therapy with interferon (IFN) in chronic B hepatitis. *Eur Rev Med Pharmacol Sci* 1: 197–201
125 Kaymakoglu S, Demir K, Cakaloglu Y, Tuncer I, Dincer D, Gurel S, Okten A (1999) Combination therapy with hepatitis B vaccine and interferon alfa in chronic hepatitis B. *Am J Gastroenterol* 94: 856–857
126 Couillin I, Pol S, Mancini M, Driss F, Brechot C, Tiollais P, Michel ML (1999) Specific vaccine therapy in chronic hepatitis B: induction of T cell proliferative responses specific for envelope antigens. *J Infect Dis* 180: 15–26

Hepatitis Prevention and Treatment
Edited by J.M. Colacino and B.A. Heinz

Immunology of hepatitis C virus

Eleanor J. Barnes, Nasser Semmo and Paul Klenerman

Nuffield Department of Medicine, University of Oxford, Peter Medawar Building for Pathogen Research, South Parks Road, Oxford OX1 3SY, UK

Hepatitis C virus (HCV) immunology: the challenges

HCV presents many challenges to the immunologist—but one key incentive. A fraction of those infected are able to clear the virus spontaneously; that is, with the help of the host immune system. Understanding the differences between immune responses which are successful and those which fail provides an important reference point in trying to evaluate the role of various aspects of immunity against HCV [1]. This is where the challenges come in. The acute presentation of HCV, where it is clear that the immune responses are having the most critical role, is rare. Those that do present acutely are probably slightly unusual immunologically. The main activity is in the liver, which is difficult to access and allow longitudinal immunological experiments to be performed. The role of specific immune responses in immunopathology, fibrosis and liver failure in the long term—the main clinical concern—is very difficult to assess. Finally, the huge variability in the virus, which is doubtless associated with its capacity to evade host responses, is a major problem for those studying immune responses and those designing vaccines. However, since this is a major pathogen where we need to understand the immunopathogenesis and need a vaccine and better therapies, immunological studies, however flawed, are crucial—and have already given us some important leads

Immune responses to HCV

Following HCV viral infection there are three potential clinical outcomes: viral control or even clearance such that HCV becomes undetectable by sensitive PCR, persistent infection characterized by circulating viraemia with minimal evidence of hepatic inflammation and little fibrosis, or persistent infection associated with hepatic necro-inflammation and progressive fibrosis. These clearly delineated clinical outcomes should in theory enable correlates of protective immunity and viral persistence to be determined. In practice however defining protective immunity in HCV infection is proving particularly challenging.

Host–virus interactions during primary viral infections are likely to play a major role in determining clinical outcome. If T cell, B cell or innate immune responses are ineffective during primary infection then high viral loads will ensue. Once a high viral load is established the attenuation of T cell responses through other mechanisms such as exhaustion and escape may occur (Fig. 1). A cycle of events ultimately resulting in persistent infection with

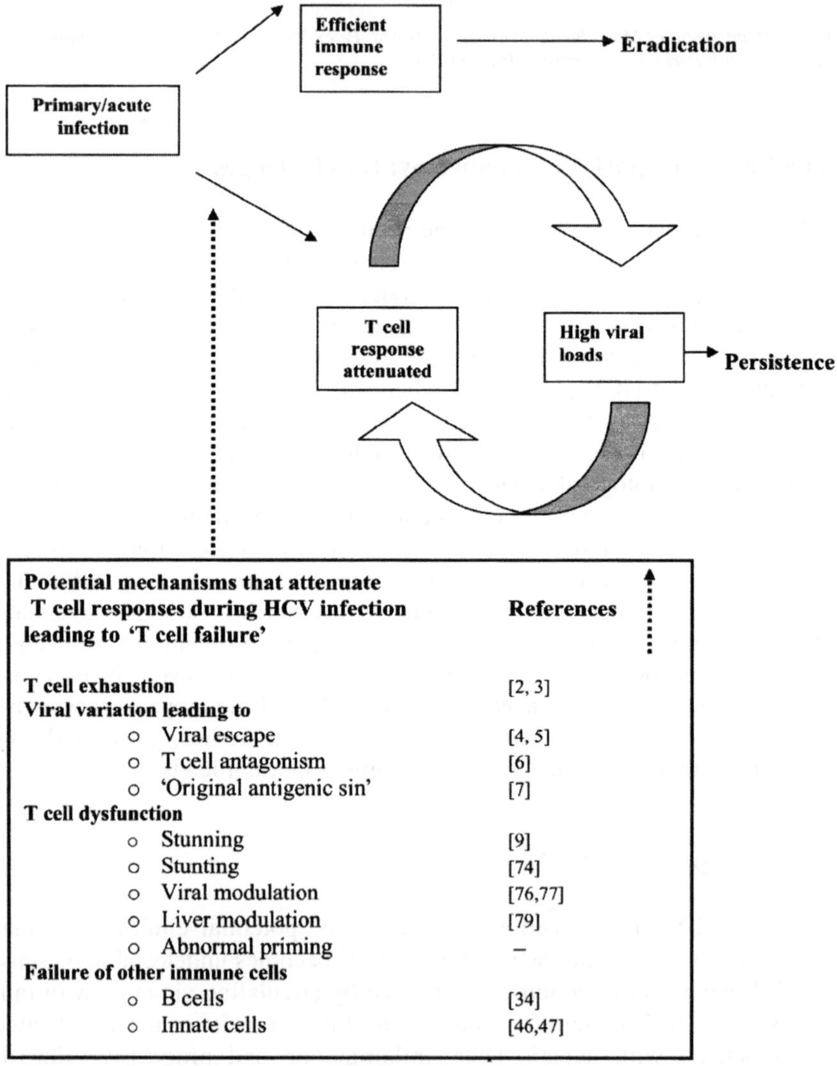

Figure 1. A simple model for T cell failure in HCV infection. During primary acute infection, an efficient immune response will lead to viral control and/or eradication. A variety of mechanisms may subvert this initial response. Once high viral loads are established, T cell exhaustion in conjunction with a variety of other mechanisms may lead to further attenuation of T cell responses and viral persistence.

attenuated T cell responses will result. It is likely then that immune events occurring during primary infection crucially determine outcome.

Background immunology

What are the possible immune parameters that arise during primary infection and which may lead to T cell failure? In lymphochoriomeningitis virus (LCMV) infection, a murine model of persistent viral infection, these are easier to define as individual immune components may be manipulated *in vivo* [1]. Firstly, a brisk multi-specific CD8+ T cell response is associated with sustained viral clearance. However, with high doses of specific LCMV strains (LCMV-DOCILE), CD8+ T cells become exhausted and persistent infection then follows [8]. Secondly, in adoptive-transfer experiments of persistently LCMV-infected mice, CD4+ T helper lymphocytes, antibody and interferon γ (IFNγ) responses are required to clear infection and may do so in the absence of a CD8+ T cell response. CD8+ T cells contribute to an initial reduction in viral titres but alone are insufficient to clear the virus as they become exhausted [10]. Furthermore CD4+ T cell responses are required to maintain a functional CD8+ response with the ability to produce appropriate cytokines and to retain cytotoxic capacity [11]. Finally, the innate immune response must be intact [8, 12]. If all these criteria are met there is long-term viral control with a detectable, sustained population of CD8+ T cells with preserved function. It is likely that similar criteria will apply to the control of HCV infection. However, it should be noted that even small changes in the infecting strain of LCMV, the dose of antigen administered and the mode of antigen administration can alter the outcome of infection [8, 13], underscoring the complex nature of viral pathogenesis in this model and in other settings.

CD8+ T cell responses: primary HCV infection

A large number of potential T cell epitopes for HCV have been reported (reviewed by Ward et al. [14]). However, at the outset it should be noted that a comprehensive analysis of CD8+ T cell responses during primary HCV infection of only a single patient has been published [9]. This reflects the rare clinical presentation of primary HCV and also the fact that comprehensive analysis of CD8+ T cell responses requires very large blood donations. However, CD8+ T cell responses have also been studied using a limited number of peptides in a small number of patients [9, 15, 16] and also more comprehensively in chimpanzee studies [17, 18].

Following HCV inoculation, viral RNA becomes detectable in blood within 2 weeks by PCR. HCV RNA levels rise exponentially thereafter. Liver inflammation as measured by the enzyme alanine aminotransferase (ALT) peaks 7–14 weeks following inoculation. Generally, the peak in ALT coincides with the onset of cellular immune responses, although a fall in viral RNA may occur before this, suggesting that either non-cytolytic mechanisms play a role in viral eradication or that cellular responses are localized to the liver [19]. T cell responses to eight HCV epitopes in humans [9] and nine HCV epitopes in chimpanzees have been recorded [20] during the acute phase of infection. Magnitudes as high as 7% of the total CD8+ T cell population to a single epitope have been documented during acute infection using tetramer staining [9]. This new technology allows for direct *ex vivo* detection of antigen-specific CD8+ T cells with great accuracy. Antigen-specific T cells are identified using a complex of major histocompatibility complex (MHC) class I molecules binding viral peptide and coupled to a fluorochrome. Samples can be stained directly and frequencies analysed using a flow cytometer.

CD4+ T cell responses: primary HCV infection

Several studies have addressed the role of CD4+ T cells in HCV infection. It appears that strong multispecific CD4+ responses arise during primary HCV infection, and several studies support the hypothesis that these responses are associated with viral clearance [19, 21, 22]. In addition, a number of studies have shown a clear association between class II genotype-histocompatibility locus antigen (HLA) DRB1*1101/DQ0301 (these genes are in close linkage disequilibrium) and spontaneous viral control [23]. Patients with this HLA genotype also possess a stronger HCV-specific CD4+ T cell response than do non-DQ0301 controls [24].

T cell responses—persistent HCV infection

Once chronic HCV infection is established, cellular immune responses are rarely detectable using *ex vivo* current technology- and when detectable appear to be directed against a few epitopes only- and are lacking in markers of activation. Furthermore, impaired cytokine secretion has been described for both CD8+ [25] and CD4+ [26] T cells during persistent infection. This is in sharp contrast to other persistent but controlled viral infections such as Epstein–Barr virus (EBV) and cytomegalovirus (CMV) where large numbers of functional viral-specific T cells are found. The T cells can be readily identified after brief re-stimulation *in vitro*, suggesting in many cases that they are still present but at extremely low frequencies. Once persistent infection is established, all of the mechanisms described above that may attenuate the immune response during primary infection may continue to subvert effective immune responses during persistent infection.

It has been suggested that the liver acts as a ‘sink’ for antigen-specific T cells, so that HCV-specific T cells isolated from peripheral blood mononuclear cells (PBMCs) underestimate the number of antigen-specific T cells. In other words, HCV-specific T cells are not lacking in number: we are just looking in the wrong place. This is partially true, in that when HCV-specific cells are found in the peripheral blood, they are usually at a magnitude of 1 log lower than found in the liver [27–29]. However, a comprehensive study of CD8+ T cells from liver biopsies of chronically infected patients failed to identify HCV-specific CD8+ cells in the majority of patients [30].

B cells/antibodies

The lack of an *in vitro* model of productive HCV infection has meant that the characterization of the envelope (E) proteins, and the process of binding of E proteins to cellular receptor(s) followed by viral entry, are poorly understood. Antibody-neutralization studies that test this process have been impeded by the lack of an appropriate model. Attempts have been made to establish neutralization tests that estimate HCV binding *in vitro* [31, 32]. However, the significance of *in vitro* binding to *in vivo* neutralization is not clear.

Studies in chimpanzees or in the clinical setting have shed some light on the role of antibodies in HCV infection *in vivo*. Following HCV infection in humans, antibody seroconversion usually arises in the first 4 weeks. These are directed against both structural and non-structural proteins, but they arise in both individuals who go on to resolve infection as well as those who develop persistent infection, and so should be regarded as a marker of exposure rather than protection. The natural antibody response in experimentally infected chimpanzees is weaker, particularly to the structural proteins [33] and, again, does not afford protection against viral persistence [18].

Although critical in most situations, antibodies appear to have a limited role in containing HCV, as evidenced by the following clinical data: (1) there is no evidence that the time of seroconversion or the pattern of the antibody response correlates with the outcome of primary infection; (2) chimpanzees can be infected repeatedly with both homologous and heterologous strains of HCV despite an antibody response [34]; (3) HCV antibodies in humans are quasi-species-specific as evidenced by a study where serum derived from a patient with acute HCV was used to inoculate chimpanzees which were then infected with the same patient serum obtained 2 and 11 years later after primary infection. Neutralization of the acute-phase virus was obtained using the 2-year, but not the 11-year, serum [35].

Neutralizing antibodies to HCV, as evidenced by either the prevention of infection or a benign disease course, have been demonstrated in chimpanzees [33, 36]. However, in these experiments homologous virus strains were used to infect the chimpanzees. It has yet to be shown that protection can be afforded from natural human HCV infection made up of multiple quasi-species variants. Why then do antibodies 'fail' to control HCV infection during natural infection? It is thought that the hypervariable (HVR1) region, located at the

N-terminus of the E2 protein (27–31 amino acids in length) plays a key role, as seen in chimpanzee studies which have shown that antibodies to this region alone can prevent homologous strain HCV infection [36]. This region forms part of the E2 protein that is located outside of the envelope [37], making it a prime target for antibodies.

Perhaps the best evidence that variation within this region is immune-driven comes again from patients with agammaglobulinaemia, common variable immunodeficiency and human immunodeficiency virus (HIV) infection with low CD4 counts who show a marked conservation of HVR1 sequence [38–40].

Multiple studies have suggested that anti-E2/HVR1 antibodies are not necessarily isolate specific; that is, that antibodies derived from synthetic peptides that conform to a particular HVR1 strain may react with multiple strains of HCV [41–43]. This suggests that these antibodies react against a non-linear epitope that is determined by the three-dimensional structure of the protein. Although the HVR shows remarkable diversity, a number of studies have shown that there are amino acids in the C-terminus [41, 42] that are relatively conserved and that these amino acids are particularly important for the recognition of the HVR by cross-reactive antibodies [44]. Thus, it is likely that these amino acid sites crucially determine HVR three-dimensional structure.

Innate immunity

Natural killer (NK) cells

NK cells are known to play an important role in the early innate immune response against pathogens [45]. The HCV envelope protein (E2) is known to bind to CD81 and the cross-linking of CD81 on NK cells has been shown to inhibit cytotoxicity, proliferation and IFNγ and tumour necrosis factor α (TNFα) secretion [46, 47]. Furthermore, NK cell function is decreased significantly in chronic HCV patients compared to healthy controls [48]. The function of NK cells during primary HCV infection has not been established.

IFNα

This area has drawn much interest, particularly because IFNα is the mainstay of therapy for HCV. Although HCV is resistant to the effects of endogenous IFNα in the majority of infected individuals, clearly this can be overcome by exogenous IFNα if given in sufficient quantity. How then do endogenous and exogenous IFNs exert their effects?

IFNs were originally identified in 1957 [49] as agents produced by virally infected cells which rendered uninfected cells resistant to infection through the expression of multiple IFN-stimulated genes (ISGs) [50]. It is now clear that the IFNs induce an antiviral state through at least three pathways: (1) protein kinase (PKR), (2) the 2', 5'-oligoadenylate synthetase (OAS)/RNase system and (3) Mx proteins.

The PKR pathway was first described in 1976. Acting through the Janus kinase (JAK)/signal transduction and activators of transcription (STAT) pathway, IFNs induce PKR (Fig. 2). Two molecules of PKR then dimerize and bind to activating double-stranded (ds) RNA [51, 52]. Activated PKR induces the phosphorylation and activation of eukaryotic initiation factor (eIF) 2α, which inhibits protein translation and therefore virion production in infected cells (Fig. 2). The OAS pathway is an RNA-decay pathway that activates RNases and so degrades virus [53]. MxA proteins are IFN-induced GTPases that interfere with the replication of negative-strand RNA viruses [54]. However, IFN has substantial residual antiviral effects in mice that are deficient in all of these pathways, suggesting that there are other pathways that are yet to be described [55]. It was this latter observation that has stimulated the search for further IFN pathways.

IFNα is induced by dsRNA—a viral pathogenic pattern recognition molecule—and can be produced by most cell types. Until recently, dsRNA was only known to bind to intracellular targets, such as PKR and 2', 5'-OAS. It is now known that dsRNA also binds to Toll-like receptor-3 (TLR3), at least in murine models, and in this way induces the upregulation of nuclear factor κB (NF-κB), which promotes IFNα transcription [56]. So, although this study identified a new dsRNA receptor (TLR3), the subsequent signalling reactions then occur through NF-κB, a pathway that was identified in the 1980s [57].

The induction of an antiviral state by IFN causes a theoretical problem for antigen-presenting cells (APCs). If professional APCs are resistant to infection then how will they present viral antigen to CD8+ T cells? 'Cross presentation' is one possible way but this may be relatively inefficient. Moreover, dendritic cells (DCs) in the periphery at the site of infection, exposed to high doses of IFN, mature, migrate and produce large quantities of cytokines. How can they do this if IFN induces PKR and eIF-2α and thus inhibits protein translation? The answer may be that in DCs (in contrast to other cells) IFNα seems to increase protein synthesis [58]. This implies that the eIF-2α pathway is inactivated in DCs, although the molecular basis for this has not been explored. Nevertheless, despite the lack of mechanistic detail, the effect is likely to be critical. As the authors of the original report point out, because dsRNA stimulates both maturation and resistance, DCs can act as 'altruistic' APCs, which are capable of priming naïve T cells and polarized Th1 responses.

We have described so far the known antiviral molecular pathways stimulated by IFNα, but IFN does more than this. It has been shown to modulate cellular immune responses and so exerts antiviral effects in this way. Known mechanisms of immune modulation include the upregulation of HLA class I expression on many cell types, enhancement of the terminal differentiation of DCs [59], T cell anti-proliferative effects and the stimulation of IFNγ by CD4+ T cells [60].

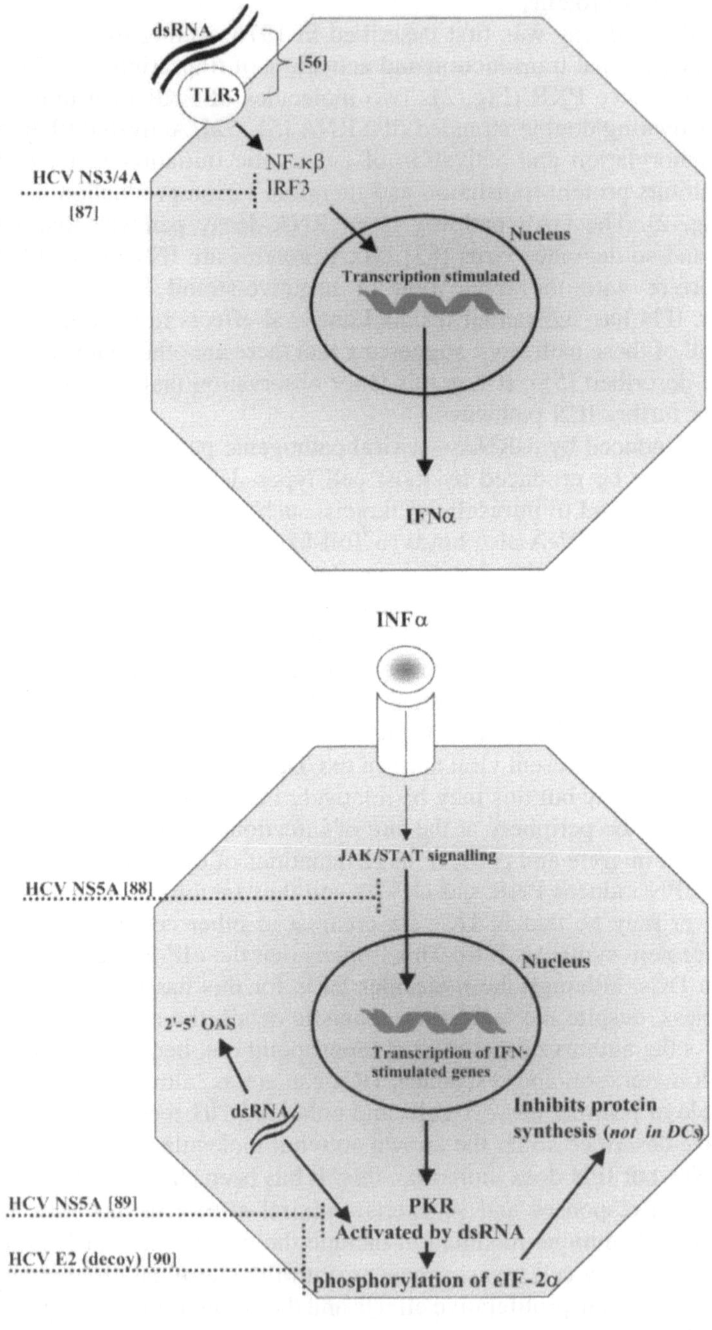

Figure 2. The complex interaction between dsRNA, IFNα and viral resistance. The references refer to the work describing the different pathways (see text for details [51–60, 87–90]). Dendritic cell (DC) function may be uniquely affected by IFN in terms of preservation of protein synthesis for antigen presentation to T cells. For definitions of abbreviations see text.

Viral escape from immune responses

Viral variation

HCV mutations are common, due to the error-prone RNA-dependent RNA polymerase specific to RNA viruses and the very high HCV replication rates [61]. The high viral mutation rates found in both HCV and HIV must in part account for their propensity to persist in immunocompetent hosts.

The infecting inoculum of HCV in any given individual is composed of a complex and diverse quasi-species so that CD8+ T cells that successfully target the dominant species will soon be faced other variants, which they may fail to recognize or do so sub-optimally. In support of this, HCV viral control has been associated with evolutionary stasis of the quasi-species population whereas viral persistence has been linked to HCV genetic mutation arising predominantly within the HVR1 of E2 where dominant B cell [62] and also T cell [5, 14] epitopes are found. The high mutation rate within this region is presumably immune-driven. So, the first problem that the host's immune responses have to face is multiple inoculating viral variants.

Next, within a single quasi-species, mutation of HLA class I restricted epitopes can impair viral clearance by impairing CD8+ T cell recognition of infected hepatocytes presenting variant epitopes. Epitope variants that inhibit CD8+ T cell recognition through the presentation of altered peptide ligands have been observed in both chimpanzee [63] and human [4, 5] studies. CD4+ T cell escape has also been observed in the context of HIV [64] and LCMV [65] infection. Furthermore, the presence of such variants arising during primary infection has been associated with viral persistence in both chimpanzees [66] and humans [5].

Viral variants may not only escape T cell recognition but may also inhibit a T cell response to the wild-type epitope as shown in hepatitis B virus and HIV infection [6, 67], a phenomenon known as 'T cell antagonism'. Alternatively, viral variants may subvert recognition by driving a CD8+ T cell response that targets the wild-type viral strain [7], a phenomenon known as 'original antigenic sin'.

However, factors other than viral variation are likely to contribute to viral persistence, given that regions of the HCV genome are highly conserved, including regions that are targeted by cellular immune responses. It has been observed that, especially during acute disease, one epitope (NS3 1073-81) is commonly recognized in HLA-A2-positive individuals [68]. However, the levels of response to this epitope (and others) are very much lower if persistence is established. Sequence analysis of this epitope over time has shown that in fact it is highly conserved during chronic infection, at a clonal level, and is conserved between individuals. Thus, although variation within the epitope does occur, lack of reactivity in peripheral blood against this epitope in chronic disease, and loss of control of virus, cannot be explained entirely by viral escape. In addition, multiple epitopes would need to mutate simultaneously or sequentially to generate the globally weak CD8+ T cell response found in chronic HCV infection.

T cell exhaustion

In murine models, high viral loads induce T cell exhaustion, leading ultimately to clonal deletion [8, 69, 70]. The mechanism behind T cell exhaustion remains unclear. While T cell exhaustion may arise in HCV infection, it is unlikely to be the sole mechanism responsible for viral persistence as T cell responses can be recovered during combination therapy [71] or after brief restimulation *in vitro* [72].

T cell dysfunction

Stunning

In the search for responses using both tetramer and functional assays during acute infection, major discrepancies in antigen-specific T cell number and function have been found. It is well recognized, using both ELISpot and intracellular cytokine staining (ICS), that not all tetramer-positive cells secrete IFNγ upon peptide stimulation. In a study [9] of a single patient at an early time point during acute HCV infection, a large proportion of tetramer-positive cells were present and highly activated as defined by the phenotypic activation markers CD38 and HLA class II. Nevertheless, at this time point very few HCV-specific cells were able to secrete IFNγ. Our group has termed this phenomenon 'stunning' and it has been similarly observed in a number of subsequent studies in primary HCV infection [17, 19] and in acute simian immunodeficiency virus (SIV) infection [73].

Stunting

The phenomenon of T cell stunting has been described by our group and has been studied in the context of persistent HCV infection. This term refers to a combination of low-frequency responses together with lack of acquisition of 'markers of maturation' in blood. Phenotypic analysis of HCV-specific CD8+ T cells shows a distinct pattern of phenotypic markers that are associated with impaired T cell maturation. This phenotype is different from that found in EBV-, CMV- and HIV-specific T cells [74], and may represent a failure of T cell maturation. The mechanisms behind this are unclear, but may result from failure of generation of mature cells or early death (e.g. in the liver—the 'two loop hypothesis' [75]. In the latter model, T cells are continuously primed or reactivated in lymphoid tissue (positive loop) but also continuously deleted or anergized within liver tissue (negative loop). The net result may be a small population of immature cells *in vivo*, which nevertheless retain *in vitro* proliferative capacity when stimulated with HCV antigens.

Viral modulation
It has been reported that vaccinia constructs incorporating HCV core proteins subvert vaccinia-specific CD8+ T cell responses in a murine model [76]. This suggests that HCV viral proteins may modulate immune responses and contribute to viral persistence. A possible mechanism for this is the binding HCV core to complement (C1q) receptor on T cells which inhibits interleukin-2 and IFNγ production in murine and human antigen-specific T cells [77].

Liver modulation
As long ago as 1969 it was clear that the liver had distinct immune properties and was capable of inducing tolerance in porcine liver transplantation [78]. Anatomically the liver is uniquely constructed to do just this. It has a rich blood supply that passes slowly through hepatic sinusoids and the liver sinusoidal endothelial cells have a fenestrated basement membrane through which naïve T cells might directly contact hepatocytes.

It has been reported that hepatocytes may prime naïve CD8+ T cells to proliferate at magnitudes comparable to those induced by professional APCs *in vivo*. However, thereafter the T cells primed by the hepatocytes die quickly [79]. Liver sinusoidal endothelial cells have also been implicated in T cell tolerance. These have been shown to downregulate Th1 responses through the induction of interleukins 10 and 4 production by CD4+ T cells [80] and to incompletely activate CD8+ T cells. Kupffer cells have also been shown to suppress T cell activation *in vivo* [81]. However, the molecular mechanism of tolerance induction remains unclear.

The preferential death of T helper cells with a type-1, IFNγ-producing phenotype that promotes antiviral effects, compared to T cells with a type-2, tolerance-inducing profile, may further promote viral persistence [82]

Perforin- and granzyme-induced cell death is an important mechanism for the clearance of infected cells by CD8+ T cells. However, hepatocytes in mice have been shown to be poorly susceptible to perforin-dependent pathways [83]. If this is true of human hepatocytes, then T cell responses, even if appropriately generated, may be impotent to achieve full effector status. Finally, 'back-killing' through Fas–Fas-L-interactions in HCV-infected livers may contribute to the killing of effector T cells [84]

Abnormal priming or regulation
It has been suggested that monocyte-derived DCs from infected patients with persistent HCV infection function abnormally in a primary mixed leucocyte reaction and that this is due to direct infection of these cells with HCV [85, 86]. The implication of these studies is that DCs that function abnormally will also fail to prime T cells with appropriate effector function and this may contribute to viral persistence.

Recent years have seen a revival of interest in the role of regulatory T cells —notably the CD4+/CD25+ subset. It is plausible that in HCV infection excessive regulation is involved in the suppression of HCV specific T cell responses. There are currently few published data to support this, but it is fascinating to speculate that T regulatory cells may be triggered in the context of persistent HCV infection and so promote the suppression of T cell responses.

Resistance to IFNα

HCV has developed multiple strategies to subvert these pathways, which are outlined in Figure 2. HCV NS3A/4A block the phosphorylation of IFN-regulatory factor-3 (IRF3) [87]. HCV NS5A subverts both the JAK/STAT signalling pathway of IFNα [88] and prevents phosphorylation and activation of PKR [89]. Finally, HCV E2 has sequence homology to the phosphorylation sites of PKR and eIF-2α and so may act as a decoy for these sites [90].

Challenges for the development of an HCV vaccine

The challenges for vaccine development are multiple, given the degree of variability of the virus and the number of potential escape mechanisms outlined above. Nevertheless the incentive is strong, given that immune responses can clear virus in a proportion of infected patients. The vaccine could be used in two ways.

1. A prophylactic vaccine. This would ideally need to generate neutralizing antibodies in order to provide 'sterilizing' immunity, but the difficulties with this approach are outlined above. Vaccines based on cellular immune response based vaccines cannot provide sterilizing immunity but could push the balance of power in favour of the host. A vaccine generated using a particular genotype would probably be genotype-specific, but since little is known about cross-genotype specificity, it might provide protection in some cases against alternative strains [91]. A standard question would be: which antigens to include? The answer is as many as possible. The epitopes are not clustered in specific viral regions and a broad response is clearly optimal in driving down virus load while avoiding escape mutation. We currently know little about the relative efficacy of responses to various epitopes; some of the responses may provide relatively little protection as they target weakly presented epitopes ('passenger' responses). The prime boost approaches, as used to generate protective immune responses to malaria and HIV, will need to be followed closely if this route is to be taken.
2. A therapeutic vaccine. The schemes of viral escape presented above suggest that attempts to boost cellular immune responses in the face of high viral loads will be ineffective. A more interesting approach might be a combination of treatment and vaccination. In this context, viral load is low and boosting immune responses at the end of treatment might prevent relapse. The additional benefits are that the viral genotype is known and the patients

can be followed closely. We already know that therapy can boost responses but these are often transient, probably as antigen is withdrawn if treatment is prolonged [71, 92]. It is also possible that ribavirin, added routinely with IFNα in current therapies, promotes the switch *in vivo* to an effective antiviral Th1-type immune response [71]. A potential problem with this approach is immune-mediated pathology if a rising viral load at the end of therapy meets a pre-formed immune response. Such approaches would need very careful thought to avoid this potentially serious outcome. However, the principles behind the vaccine design would be identical to those for a prophylactic vaccine.

Thus, generation of a broad T cell response, targeting both conserved and variable virus regions with strong Th1-effector-type memory, would be ideal and likely ultimately successful in providing some 'enhanced' protection. However, much further work is needed to define some critical elements of 'successful' T cell responses and also to reliably generate strong T cell responses in humans before we can embark on this with confidence.

Acknowledgements
We are grateful for support from the Wellcome Trust, the Medical Research Council and the Fritz Thyssen Foundation. Thanks to Rodney Phillips and Geoff Dusheiko for helpful discussions about many issues contained in this chapter.

References

1 Klenerman P, Lucas M, Barnes E, Harcourt G (2002) Immunity to hepatitis C virus: stunned but not defeated. *Microbes Infect* 4: 57–65
2 Moskophidis D, Laine E, Zinkernagel RM (1993) Peripheral clonal deletion of antiviral memory CD8+ T cells. *Eur J Immunol* 23: 3306–11
3 Gallimore A, Glithero A, Godkin A, Tissot AC, Pluckthun A, Elliott T, Hengartner H, Zinkernagel R (1998) Induction and exhaustion of lymphocytic choriomeningitis virus-specific cytotoxic T lymphocytes visualized using soluble tetrameric major histocompatibility complex class I-peptide complexes. *J Exp Med* 187: 1383–93
4 Kaneko T, Moriyama T, Udaka K, Hirioshi K, Kita H, Okamoto H, Yagita H, Okumara K, Imawari M (1997) Impaired induction of CTL by antagonism of a weak agonist borne by a variant HCV epitope. *Eur J Immunol* 27: 1782–87
5 Tsai S, Chen Y, Chen M, Huang C, Sheen I, Yeh L, Huang J, Kuo G, Liaw Y (1998) HCV variants circumventing CTL activity as a mechanism of chronicity. *Gastroenterology* 115: 954–66
6 Bertoletti A, Sette A, Chisari FV, Penna A, Levrero M, De Carli M, Fiaccadori F, Ferrari C (1994) Natural variants of cytotoxic epitopes are T-cell receptor antagonists for antiviral cytotoxic T cells. *Nature* 369: 407–10
7 Klenerman P, Zinkernagel RM (1998) Original antigenic sin impairs cytotoxic T lymphocyte responses to viruses bearing variant epitopes. *Nature* 394: 482–5
8 Moskophidis D, Battegay M, van den Broek M, Laine E, Hoffmann-Rohrer U, Zinkernagel RM (1995) Role of virus and host variables in virus persistence or immunopathological disease caused by a non-cytolytic virus. *J Gen Virol* 76: 381–91
9 Lechner F, Wong DK, Dunbar PR, Chapman R, Chung RT, Dohrenwend P, Robbins G, Phillips R, Klenerman P, Walker BD (2000) Analysis of Successful Immune Responses in Persons Infected with Hepatitis C Virus. *J Exp Med* 191: 1499–1512
10 Planz O, Ehl S, Furrer E, Horvath E, Brundler M, Hengartner H, Zinkernagel R (1997) A critical role for neutralising antibody-producing B cells, CD4+ cells and interferons in persistent infec-

tions of mice with LCMV: implications for adoptive immunotherapy of virus carriers. *Proc Natl Acad Sci USA* 94: 6874–6879
11 Zajac A, Blattman J, Murali-Krishna K, Sourdive D, Suresh M, Altman J, Ahmed R (1998) Viral immune evasion due to persistence of activated T cells without effector function. *J Exp Med* 188: 2205–13
12 Ou R, Zhou S, Huang L, Moskophidis D (2001) Critical role for alpha/beta and gamma interferons in persistence of lymphocytic choriomeningitis virus by clonal exhaustion of cytotoxic T cells. *J Virol* 75: 8407–23
13 Kalams SA, Walker BD (1998) The critical need for CD4 help in maintaining effective cytotoxic T lymphocyte responses. *J Exp Med* 188: 2199–204
14 Ward S, Lauer G, Isba R, Walker B, Klenerman P (2002) Cellular immune responses against hepatitis C virus: the evidence base 2002. *Clin Exp Immunol* 128: 195–203
15 Sobao Y, Tomiyama H, Nakamura S, Sekihara H, Tanaka K, Takiguchi M (2001) Visual demonstration of hepatitis C virus-specific memory CD8(+) T-cell expansion in patients with acute hepatitis C. *Hepatology* 33: 287–94
16 Lechner F, Gruener NH, Urbani S, Uggeri J, Santantonio T, Kammer AR, Cerny A, Phillips R, Ferrari C, Pape GR et al. (2000) CD8+ T lymphocyte responses are induced during acute hepatitis C virus infection but are not sustained. *Eur J Immunol* 30: 2479–87
17 Shoukry NH, Grakoui A, Houghton M, Chien DY, Ghrayeb J, Reimann KA, Walker CM (2003) Memory CD8+ T cells are required for protection from persistent hepatitis C virus infection. *J Exp Med* 197: 1645–55
18 Cooper S, Erickson AL, Adams EJ, Kansopon J, Weiner AJ, Chien DY, Houghton M, Parham P, Walker CM (1999) Analysis of a successful immune response against hepatitis C virus. *Immunity* 10: 439–49
19 Thimme R, Oldach D, Chang KM, Steiger C, Ray SC, Chisari FV (2001) Determinants of viral clearance and persistence during acute hepatitis C virus infection. *J Exp Med* 194: 1395–406
20 Cooper D, Gold J, Maclean P, Donovan B, Finlayson R, Barnes T, Michelmore H, Brooke P, Penny R (1985) Acute AIDS retrovirus infection. Definition of a clinical illness associated with seroconversion. *Lancet* i: 537–540
21 Gerlach J, Diepolder H, Jung M-C, Gruener N, Schraut W, Zachoval R, Hoffman R, Schirren C, Santantonio T, Pape G (1999) Recurrence of HCV after loss of virus specific CD4+ T cell response in acute Hepatitis C. *Gastroenterology* 117: 933–41
22 Missale G, Bertoni R, Lamonaca V, Valli A, Massari M, Mori C, Rumi M, Houghton M, Fiaccadori F, Ferrari C (1996) Different clinical behaviours of acute HCV infection are associated with different vigor of the anti-viral T cell response. *J Clin Invest* 98: 706–14
23 Thursz M, Yallop R, Goldin R, Trepo C, Thomas HC (1999) Influence of MHC class II genotype on outcome of infection with hepatitis C virus. The HENCORE group. Hepatitis C European Network for Cooperative Research. *Lancet* 354: 2119–24
24 Harcourt G, Hellier S, Bunce M, Satsangi J, Collier J, Chapman R, Phillips R, Klenerman P (2001) Effect of HLA class II genotype on T helper lymphocyte responses and viral control in hepatitis C virus infection. *J Viral Hepat* 8: 174–9
25 Gruener N, Lechner F, Jung M-C, Diepolder H, Gerlach T, Lauer G, Walker B, Sullivan J, Phillips R, Pape G et al. (2001) Sustained dysfunction of antiviral CD8+ T lymphocytes after infection with HCV. *J Virol* 75: 5550–8
26 Godkin A, Jeanguet N, Thursz M, Openshaw P, Thomas H (2001) Characterization of novel HLA-DR11-restricted HCV epitopes reveals both qualitative and quantitative differences in HCV-specific CD4+ T cell responses in chronically infected and non-viremic patients. *Eur J Immunol* 31: 1438–46
27 He XS, Rehermann B, Boisvert J, Mumm J, Maecker HT, Roederer M, Wright TL, Maino VC, Davis MM, Greenberg HB (2001) Direct functional analysis of epitope-specific CD8+ T cells in peripheral blood. *Viral Immunol* 14: 59–69
28 He X-S, Rehermann B, Lopez-Labrador F, Boisvert J, Cheung R, Mumm J, Wedermeyer H, Berenguer M, Wright T, Davis M (1999) Quantitative analysis of HCV-specific CD8+ T cells in peripheral blood and liver using peptide-MHC tetramers. *Proc Natl Acad Sci USA* 96: 5692–7
29 Grabowska AM, Lechner F, Klenerman P, Tighe PJ, Ryder S, Ball JK, Thomson BJ, Irving WL, Robins RA (2001) Direct *ex vivo* comparison of the breadth and specificity of the T cells in the liver and peripheral blood of patients with chronic HCV infection. *Eur J Immunol* 31: 2388–94
30 Wong DK, Dudley DD, Afdhal NH, Dienstag J, Rice CM, Wang L, Houghton M, Walker BD,

Koziel MJ (1998) Liver-derived CTL in hepatitis C virus infection: breadth and specificity of responses in a cohort of persons with chronic infection. *J Immunol* 160: 1479–88
31 Zibert A, Schreier E, Roggendorf M (1995) Antibodies in human sera specific to hypervariable region 1 of hepatitis C virus can block viral attachment. *Virology* 208: 653–61
32 Rosa D, Campagnoli S, Moretto C, Guenzi E, Cousens L, Chin M, Dong C, Weiner AJ, Lau JY, Choo QL et al. (1996) A quantitative test to estimate neutralizing antibodies to the hepatitis C virus: cytofluorimetric assessment of envelope glycoprotein 2 binding to target-cells. *Proc Natl Acad Sci USA* 93: 1759–63
33 Choo QL, Kuo G, Ralston R, Weiner A, Chien D, Van Nest G, Han J, Berger K, Thudium K, Kuo C et al. (1994) Vaccination of chimpanzees against infection by the hepatitis C virus. *Proc Natl Acad Sci USA* 91: 1294–8
34 Farci P, Alter HJ, Govindarajan S, Wong DC, Engle R, Lesniewski RR, Mushahwar IK, Desai SM, Miller RH, Ogata N et al. (1992) Lack of protective immunity against reinfection with hepatitis C virus. *Science* 258: 135–40
35 Farci P, Alter HJ, Wong DC, Miller RH, Govindarajan S, Engle R, Shapiro M, Purcell RH (1994) Prevention of hepatitis C virus infection in chimpanzees after antibody-mediated *in vitro* neutralization. *Proc Natl Acad Sci USA* 91: 7792–6
36 Farci P, Shimoda A, Wong D, Cabezon T, De Gioannis D, Strazzera A, Shimizu Y, Shapiro M, Alter HJ, Purcell RH (1996) Prevention of hepatitis C virus infection in chimpanzees by hyperimmune serum against the hypervariable region 1 of the envelope 2 protein. *Proc Natl Acad Sci USA* 93: 15394–15399
37 Weiner AJ, Geysen HM, Christopherson C, Hall JE, Mason TJ, Saracco G, Bonino F, Crawford K, Marion CD, Crawford KA (1992) Evidence for immune selection of hepatitis C virus (HCV) putative envelope glycoprotein variants: potential role in chronic HCV infections. *Proc Natl Acad Sci USA* 89: 3468–72
38 Kumar U, Monjardino J, Thomas HC (1994) Hypervariable region of hepatitis C virus envelope glycoprotein (E2/NS1) in an agammaglobulinemic patient. *Gastroenterology* 106: 1072–5
39 Booth JC, Kumar U, Webster D, Monjardino J, Thomas HC (1998) Comparison of the rate of sequence variation in the hypervariable region of E2/NS1 region of hepatitis C virus in normal and hypogammaglobulinemic patients. *Hepatology* 27: 223–7
40 Toyoda H, Fukuda Y, Koyama Y, Takamatsu J, Saito H, Hayakawa T (1997) Effect of immunosuppression on composition of quasispecies population of hepatitis C virus in patients with chronic hepatitis C coinfected with human immunodeficiency virus. *J Hepatol* 26: 975–82
41 Zhou YH, Moriyama M, Esumi M (1999) Multiple sequence-reactive antibodies induced by a single peptide immunization with hypervariable region 1 of hepatitis C virus. *Virology* 256: 360–70
42 Shang D, Zhai W, Allain JP (1999) Broadly cross-reactive, high-affinity antibody to hypervariable region 1 of the hepatitis C virus in rabbits. *Virology* 258: 396–405
43 Esumi M, Ahmed M, Zhou YH, Takahashi H, Shikata T (1998) Murine antibodies against E2 and hypervariable region 1 cross-reactively capture hepatitis C virus. *Virology* 251: 158–64
44 Li C, Candotti D, Allain JP (2001) Production and characterization of monoclonal antibodies specific for a conserved epitope within hepatitis C virus hypervariable region 1. *J Virol* 75: 12412–12420
45 Biron CA, Nguyen KB, Pien GC, Cousens LP, Salazar-Mather TP (1999) Natural killer cells in antiviral defense: function and regulation by innate cytokines. *Annu Rev Immunol* 17: 189–220
46 Crotta S, Stilla A, Wack A, D'Andrea A, Nuti S, D'Oro U, Mosca M, Filliponi F, Brunetto RM, Bonino F et al. (2002) Inhibition of natural killer cells through engagement of CD81 by the major hepatitis C virus envelope protein. *J Exp Med* 195: 35–41
47 Tseng CT, Klimpel GR (2002) Binding of the hepatitis C virus envelope protein E2 to CD81 inhibits natural killer cell functions. *J Exp Med* 195: 43–9
48 Corado J, Toro F, Rivera H, Bianco NE, Deibis L, De Sanctis JB (1997) Impairment of natural killer (NK) cytotoxic activity in hepatitis C virus (HCV) infection. *Clin Exp Immunol* 109: 451–7
49 Isaacs A, Lindenmann J (1957) Virus interference. I. The interferon. *Proc R Soc Lond B Biol Sci* 147: 258–67
50 Der SD, Zhou A, Williams BR, Silverman RH (1998) Identification of genes differentially regulated by interferon alpha, beta, or gamma using oligonucleotide arrays. *Proc Natl Acad Sci USA* 95: 15623–15628
51 Clemens MJ, Elia A (1997) The double-stranded RNA-dependent protein kinase PKR: structure and function. *J Interferon Cytokine Res* 17: 503–24

52 Roberts WK, Hovanessian A, Brown RE, Clemens MJ, Kerr IM (1976) Interferon-mediated protein kinase and low-molecular-weight inhibitor of protein synthesis. *Nature* 264: 477–80
53 Kerr IM, Brown RE (1978) pppA2'p5'A2'p5'A: an inhibitor of protein synthesis synthesized with an enzyme fraction from interferon-treated cells. *Proc Natl Acad Sci USA* 75: 256–60
54 Arnheiter H, Frese M, Kambadur R, Meier E, Haller O (1996) Mx transgenic mice–animal models of health. *Curr Top Microbiol Immunol* 206: 119–47
55 Zhou A, Paranjape JM, Der SD, Williams BR, Silverman RH (1999) Interferon action in triply deficient mice reveals the existence of alternative antiviral pathways. *Virology* 258: 435–40
56 Alexopoulou L, Holt AC, Medzhitov R, Flavell RA (2001) Recognition of double-stranded RNA and activation of NF-kappaB by Toll- like receptor 3. *Nature* 413: 732–8
57 Visvanathan KV, Goodbourn S (1989) Double-stranded RNA activates binding of NF-kappa B to an inducible element in the human beta-interferon promoter. *EMBO J* 8: 1129–38
58 Cella M, Salio M, Sakakibara Y, Langen H, Julkunen I, Lanzavecchia A (1999) Maturation, activation, protection of dendritic cells induced by double-stranded RNA. *J Exp Med* 189: 821–9
59 Luft T, Pang KC, Thomas E, Hertzog P, Hart DN, Trapani J, Cebon J (1998) Type I IFNs enhance the terminal differentiation of dendritic cells. *J Immunol* 161: 1947–53
60 Brinkmann V, Geiger T, Alkan S, Heusser CH (1993) Interferon alpha increases the frequency of interferon gamma-producing human CD4+ T cells. *J Exp Med* 178: 1655–63
61 Neumann A, Lam N, Dahari H, Gretch D, Wiley T, Layden T, Perelson A (1998) HCV Viral dynamics *in vivo* and the efficacy of Interferon alpha therapy. *Science* 282: 103–7
62 Farci P, Shimoda A, Coiana A, Diaz G, Peddis G, Melpolder JC, Strazzera A, Chien DY, Munoz SJ, Balestrieri A et al. (2000) The outcome of acute hepatitis C predicted by the evolution of the viral quasispecies. *Science* 288: 339–44
63 Weiner A, Erickson AL, Kansopon J, Crawford K, Muchmore E, Hughes AL, Houghton M, Walker CM (1995) Persistent hepatitis C virus infection in a chimpanzee is associated with emergence of a cytotoxic T lymphocyte escape variant. *Proc Natl Acad Sci USA* 92: 2755–9
64 Harcourt GC, Garrard S, Davenport MP, Edwards A, Phillips RE (1998) HIV-1 variation diminishes CD4 T lymphocyte recognition. *J Exp Med* 188: 1785–93
65 Ciurea A, Hunziker L, Martinic MM, Oxenius A, Hengartner H, Zinkernagel RM (2001) CD4+ T-cell-epitope escape mutant virus selected *in vivo*. *Nat Med* 7: 795–800
66 Erickson AL, Kimura Y, Igarashi S, Eichelberger J, Houghton M, Sidney J, McKinney D, Sette A, Hughes AL, Walker CM (2001) The outcome of hepatitis C virus infection is predicted by escape mutations in epitopes targeted by cytotoxic T lymphocytes. *Immunity* 15: 883–95
67 Klenerman P, Rowland-Jones S, McAdam S, Edwards J, Daenke S, Lalloo D, Koppe B, Edwards A, Phillips RE, McMichael AJ (1994) Naturally ocurring HIV-1 gag variants antagonise cytotoxic T-cell activity. *Nature* 369: 403–406
68 Kantzanou M, Lucas M, Barnes E, Komatsu H, Dusheiko G, Ward S, Harcourt G, Klenerman P (2003) Viral escape and T cell exhaustion in hepatitis C virus infection analysed using Class I peptide tetramers. *Immunol Lett* 85: 165–71
69 Gallimore A, Glitheroe A, Godkin A, Tissot A, Hengartner H, Elliott T, Zinkernagel R (1998) Induction and exhaustion of LCMV-specific CTL visualised using soluble tetrameric MHC-peptide complexes. *J Exp Med* 187: 1383–93
70 Moskophidis D, Lechner F, Pircher H, Zinkernagel RM (1993) Virus persistence in acutely infected immunocompetent mice by exhaustion of antiviral cytotoxic effector T cells. *Nature* 362: 758–61
71 Barnes E, Harcourt G, Brown D, Lucas M, Phillips R, Dusheiko G, Klenerman P (2002) The Dynamics of T-Lymphocyte Responses During Combination Therapy for Chronic Hepatitis C Virus Infection. *Hepatology* 36: 743–754
72 Lauer GM, Ouchi K, Chung RT, Nguyen TN, Day CL, Purkis DR, Reiser M, Kim AY, Lucas M, Klenerman P et al. (2002) Comprehensive analysis of CD8(+)-T-cell responses against hepatitis C virus reveals multiple unpredicted specificities. *J Virol* 76: 6104–13
73 Vogel TU, Allen TM, Altman JD, Watkins DI (2001) Functional impairment of simian immunodeficiency virus-specific CD8+ T cells during the chronic phase of infection. *J Virol* 75: 2458–61
74 Appay V, Dunbar PR, Callan M, Klenerman P, Gillespie GM, Papagno L, Ogg GS, King A, Lechner F, Spina CA et al. (2002) Memory CD8+ T cells vary in differentiation phenotype in different persistent virus infections. *Nat Med* 8: 379–85
75 Willberg C, Barnes E, Klenerman P (2003) HCV immunology-Death and the maiden T cell. *Cell Death Differ* 10 (suppl 1): S39–S47

76 Large MK, Kittlesen DJ, Hahn YS (1999) Suppression of host immune response by the core protein of hepatitis C virus: possible implications for hepatitis C virus persistence. *J Immunol* 162: 931–8
77 Yao ZQ, Ray S, Eisen-Vandervelde A, Waggoner S, Hahn YS (2001) Hepatitis C virus: immunosuppression by complement regulatory pathway. *Viral Immunol* 14: 277–95
78 Calne RY, Sells RA, Pena JR, Davis DR, Millard PR, Herbertson BM, Binns RM, Davies DA (1969) Induction of immunological tolerance by porcine liver allografts. *Nature* 223: 472–6
79 Bertolino P, Trescol-Biemont MC, Rabourdin-Combe C (1998) Hepatocytes induce functional activation of naive CD8+ T lymphocytes but fail to promote survival. *Eur J Immunol* 28: 221–36
80 Limmer A, Ohl J, Kurts C, Ljunggren HG, Reiss Y, Groettrup M, Momburg F, Arnold B, Knolle PA (2000) Efficient presentation of exogenous antigen by liver endothelial cells to CD8+ T cells results in antigen-specific T-cell tolerance. *Nat Med* 6: 1348–54
81 Crispe IN (2003) Hepatic T cells and liver tolerance. *Nat Rev Immunol* 3: 51–62
82 Zhang X, Brunner T, Carter L, Dutton RW, Rogers P, Bradley L, Sato T, Reed JC, Green D, Swain SL (1997) Unequal death in T helper cell (Th)1 and Th2 effectors: Th1, but not Th2, effectors undergo rapid Fas/FasL-mediated apoptosis. *J Exp Med* 185: 1837–49
83 Kafrouni MI, Brown GR, Thiele DL (2001) Virally infected hepatocytes are resistant to perforin-dependent CTL effector mechanisms. *J Immunol* 167: 1566–74
84 Dennert G (2002) Elimination of virus-specific cytotoxic T cells in the liver. *Crit Rev Immunol* 22: 1–11
85 Bain C, Fatmi A, Zoulim F, Zarski JP, Trepo C, Inchauspe G (2001) Impaired allostimulatory function of dendritic cells in chronic hepatitis C infection. *Gastroenterology* 120: 512–24
86 Kanto T, Hayashi N, Takehara T, Tatsumi T, Kuzushita N, Ito A, Sasaki Y, Kasahara A, Hori M (1999) Impaired allostimulatory capacity of peripheral blood dendritic cells recovered from hepatitis C virus-infected individuals. *J Immunol* 162: 5584–91
87 Foy E, Li K, Wang C, Sumpter R Jr, Ikeda M, Lemon SM, Gale M Jr (2003) Regulation of interferon regulatory factor-3 by the hepatitis C virus serine protease. *Science* 300: 1145–8
88 He Y, Katze MG (2002) To interfere and to anti-interfere: the interplay between hepatitis C virus and interferon. *Viral Immunol* 15: 95–119
89 Gale MJ Jr, Korth MJ, Katze MG (1998) Repression of the PKR protein kinase by the hepatitis C virus NS5A protein: a potential mechanism of interferon resistance. *Clin Diagn Virol* 10: 157–62
90 Taylor DR, Shi ST, Romano PR, Barber GN, Lai MM (1999) Inhibition of the interferon-inducible protein kinase PKR by HCV E2 protein. *Science* 285: 107–10
91 Harcourt GC, Lucas M, Godkin AJ, Kantzanou M, Phillips RE, Klenerman P (2003) Evidence for lack of cross-genotype protection of CD4+ T cell responses during chronic hepatitis C virus infection. *Clin Exp Immunol* 131: 122–9
92 Komarova NL, Barnes E, Klenerman P, Wodarz D (2003) Boosting immunity by antiviral drug therapy: A simple relationship among timing, efficacy, and success. *Proc Natl Acad Sci USA100: 1855–1860*

Hepatitis Prevention and Treatment
Edited by J.M. Colacino and B.A. Heinz

Current treatment of patients with chronic hepatitis B virus infection

Nathaniel A. Brown and Richard E. Boehme

Idenix Pharmaceuticals, 60 Hampshire Street, Cambridge, MA 02139, USA

Epidemiology and natural history of hepatitis B virus infection

Hepatitis B virus (HBV) infection causes both acute and chronic hepatitis and is a major global health problem. Worldwide, it is estimated that 50 million new HBV infections occur annually, and approximately 350–400 million people are chronically HBV-infected [1–2]. In parts of Asia and Africa, as many as 7–20% of people are chronically infected with HBV, i.e. are chronic HBV 'carriers' [3]. In North America and northern Europe, HBV chronic carrier rates are much lower, usually below 1%. Intermediate HBV carrier rates of 1 to 7% are found in southern and eastern Europe, parts of central and South America, the Middle East, and parts of Japan [4]. It has been estimated that, worldwide, more than 1 million people die annually due to HBV-related end-stage disease manifestations such as cirrhosis and hepatocellular carcinoma (HCC) [4].

Primary HBV infection during adolescence and adult years is most common in non-endemic regions, such as North America and Europe, and is usually due to sexual transmission or to parenteral exposures to HBV-contaminated devices or blood products [5]. More than 90% of adults who become acutely infected with HBV recover from the infection without apparent clinical sequelae and develop immunity to the virus. However, 2–10% of HBV-infected adults become chronic HBV carriers [1, 3–5]. The clinical manifestations of this chronic infection are variable, ranging from an asymptomatic viral carrier state with negligible liver damage, to progressive liver inflammation with end-stage complications such as cirrhosis or HCC [1, 2].

In contrast, in endemic regions, such as Asia and parts of Africa, HBV infection is most commonly acquired perinatally, from HBV-infected mothers, or in early childhood, and most (70–95%) perinatally infected infants remain chronically infected. Because of the high rate of persistent infection in infants, perinatal and early childhood infection is responsible for most HBV-associated liver disease in adults in populations where HBV infection is endemic [1, 5].

The early stage of chronic HBV infection is characterized by the presence of several serologic markers; i.e. there is persistent detectability of hepatitis

B surface antigen (HBsAg), with initially high serum levels of markers of ongoing HBV replication, e.g. hepatitis B e antigen (HBeAg), and HBV DNA, the latter corresponding to the presence of HBV virions (Dane particles) in the bloodstream [1, 6]. HBsAg and hepatitis B core antigen (HBcAg), comprising envelope and capsid epitopes of HBV, are detectable in HBV-infected hepatocytes in the liver. Because HBsAg circulates in the bloodstream in very high amounts in HBV-infected individuals, HBsAg is the serologic marker of HBV infection that has been easiest to detect in routine laboratory testing. In many or most chronic HBV carriers, high-level HBV replication probably subsides before the onset of appreciable chronic liver disease. Such individuals have persistent detectability of HBsAg, but lose detectable HBeAg and exhibit lower serum HBV DNA levels (typically in the range of 2–5 $\log_{10}$ copies/ml). As long as HBV replication remains at low levels, these HBV carriers are thought to be at low risk for liver disease and are termed 'inactive' carriers.

Chronic hepatitis B

The term 'chronic hepatitis B' refers to patients who are chronically HBV-infected and have evidence of chronic HBV-associated liver disease, i.e. chronic necroinflammatory liver injury [1, 6]. In HBV carriers, chronic liver inflammation is typically reflected serologically by persistently, albeit sometimes irregularly, elevated serum aminotransferase [alanine aminotransferase (ALT), aspartate amino transferase (AST)] levels. The pathogenesis of liver damage related to HBV infection does not appear to be due to a direct cytopathic effect of the virus but rather to an interplay between viral replication and the host's cell-mediated immune response to HBV-infected hepatocytes [7].

For unknown reasons, patients with chronic hepatitis B often have a waxing and waning disease course, characterized by occasional 'flares' in disease activity evidenced by exacerbations of elevated serum ALT and AST levels, histologic worsening in the liver, and occasionally signs of hepatic decompensation [1, 6–8]. Sometimes these disease flares are associated with clearance of HBV replication, signaled by loss of detectable serum HBeAg and eventual gain of detectable serum anti-HBe (in most cases), a process called 'spontaneous HBeAg seroconversion'. HBeAg loss and seroconversion can also occur in the absence of ALT/AST flare activity. Importantly, HBeAg loss and seroconversion, whether occurring spontaneously or in response to antiviral treatment, are usually associated with substantial, multi-log reductions in ongoing HBV replication and reduced liver inflammation [9, 10]. Hepatitis B patients who experience HBeAg loss or seroconversion often maintain such states of reduced viral replication indefinitely, and in the absence of reactivation of HBV replication such patients are thought to have reduced risk for progressive liver disease if HBeAg seroconversion has occurred prior to the advent of cirrhosis. Without durable seroconversion, however, disease flares related to incomplete immune clearance of HBV infection can lead to histologic worsening and the possibility of progressive cirrhosis, liver failure, and death.

In some patients, HBeAg loss and seroconversion are followed by persistent liver disease with detectable HBV viremia—i.e. with appreciable persisting HBV DNA levels—despite the loss of detectable HBeAg [11–13]. The endogenous emergence of so-called pre-core mutant HBV strains can usually be demonstrated in such circumstances [14]. With emergence of these HBV variants, HBeAg secretion from infected hepatocytes is abrogated or reduced by pre-core or core gene mutations in HBV genomic templates, but HBV virion particles continue to be produced nearly unabated since the structural components of the HBV virion are unaffected by the mutations. Pre-core mutant HBV strains are especially prevalent in adult hepatitis B patients in southern Europe (Italy, Spain, and Greece) and certain parts of Asia [13, 15, 16]. In such HBV-endemic populations, pre-core mutant HBV infection is occasionally transmitted from person to person and does not always arise through viral quasispecies evolution in HBV-infected individuals. However, it is generally thought that in most patients with HBeAg-negative chronic hepatitis B, pre-core mutant HBV species have evolved as a later virologic stage of the chronic HBV infection, through viral quasispecies evolution, decades after initial infection of the patient with wild-type HBV [1, 13].

Criteria for a diagnosis of chronic hepatitis B have recently been described in a consensus Practice Guideline from the American Association for the Study of Liver Diseases (AASLD) [1, 17]. Briefly, patients are said to have chronic hepatitis B if they have chronic HBs antigenemia, elevated ALTs, chronic hepatitis on biopsy, and HBV viremia detectable in 'unamplified' (i.e. direct hybridization) assays [1, 2, 6]. The latter is generally taken to correspond to viremia levels of >5 $\log_{10}$ copies/ml [1]. In this AASLD Practice Guideline, both virologic phases of chronic hepatitis B, i.e. HBeAg-positive and HBeAg-negative, are said to be defined by HBV viremia levels detectable by unamplified hybridization assays.

A number of studies have indicated that chronic HBV carriers with persistently high levels of HBV replication, e.g. with persistently detectable serum HBeAg and/or HBV DNA levels detectable in conventional hybridization assays, are at greatest risk for the development of progressive necroinflammatory liver disease and hepatocellular carcinoma [18–22]. In both HBeAg-positive and HBeAg-negative hepatitis B, liver disease progression tends to correlate with the level of ongoing HBV replication, i.e. with serum HBV DNA levels. However, there may be differences in the range of viremia levels for untreated HBeAg-positive and HBeAg-negative patients. Based on the characteristics of patient populations enrolled in multicenter trials of antiviral agents, it appears that patients with histologically active, HBeAg-positive hepatitis B typically have serum HBV DNA levels of approximately 6–10 $\log_{10}$ genome copies/ml [1, 2, 23, 24]. In large multicenter trials, mean/median viremia levels in HBeAg-positive patients have typically been about 8–9 $\log_{10}$ genome copies/ml [23, 24]. In contrast, it appears that the range of HBV DNA levels in HBeAg-negative patients with active liver disease may be broader, with some HBeAg-negative patients having active disease with viremia levels as low as 4 $\log_{10}$ genome copies/ml [13, 14, 16]. Nonetheless, HBeAg-negative patients enrolled in multicenter clinical trials have typically had mean/median HBV DNA levels of approximately 7–8 $\log_{10}$ genome copies/ml [25, 26].

Based on the range of findings reported in the literature with various assays for HBV DNA, in the viremia range of 3–6 $\log_{10}$ genome copies/ml there appears to be a mixture of HBV carriers with active disease (especially, some HBeAg-negative patients) and HBV carriers with inactive disease [1, 2, 13]. Trials of antiviral agents for hepatitis B have generally been targeted to capturing the higher-viremic patients at greatest risk for progressive liver disease, e.g. patients with HBV DNA levels of 5–6 $\log_{10}$ genome copies/ml or more [1, 2, 22–26].

Among HBV carriers with the markers of viral replication described above, it has been estimated that 15–20% may develop cirrhosis within 5 years [21, 27]. For patients with chronic HBV infection and the histologic pattern of chronic active hepatitis with cirrhosis, 5-year mortality rates of 45–86% have been reported, depending on whether clinical signs of hepatic decompensation were evident at the start of the observation period [21, 22, 27].

In summary, although there has been progress in global HBV vaccination programs, chronic HBV infection and its end-stage complications of cirrhosis and HCC still comprise a significant global disease burden with regard to morbidity and mortality. The course of HBV-associated liver disease is variable, with some patients spontaneously seroconverting to lower levels of HBV replication and to negligible risk for subsequent progression of HBV-associated liver disease, while others progress to cirrhosis or liver cancer. Because of this variable disease course, optimal patient management involves a thorough knowledge of the scientific and clinical issues relevant to hepatitis B as well as familiarity with, and individualization of care to, each individual patient. Since current antiviral therapy for hepatitis B is rarely eradicative, antiviral treatment for hepatitis B should be targeted to patients with well-documented chronic HBV infection, with evidence of significant underlying necroinflammatory liver disease, and with relatively high-level persistent HBV replication, since these are the individuals who are most likely to suffer progressive liver disease.

Current antiviral therapy for chronic hepatitis B

The goal of antiviral therapy for hepatitis B is to reduce patients' risks for progressive liver disease through prolonged suppression or eradication of HBV infection and consequent arrest or amelioration of HBV-related necroinflammatory liver damage [1, 2].

Only three antiviral agents currently have widespread regulatory approvals for the treatment of patients with chronic hepatitis B: interferon α, which is administered parenterally, generally by subcutaneous injection; lamivudine, an orally bioavailable nucleoside analog; and adefovir dipivoxil, an orally bioavailable nucleotide. Adefovir dipivoxil has recently been approved in the USA and several European and Asian countries, but has not yet achieved widespread use comparable to lamivudine and interferon. The characteristics of each of these three antiviral treatment modalities are summarized below.

Interferon α

During the 1980s and early 1990s, numerous clinical trials were undertaken with various interferon α products, utilizing treatment periods of typically 4–6 months and study periods of 10–24 months, and typically involving relatively high-viremic HBeAg-positive adult patients with chronic hepatitis B and compensated liver disease. The results of these trials indicated that HBeAg loss and seroconversion (gain of anti-HBe) were significantly more frequent with interferon treatment compared to no treatment, and that durable clinical benefits (ALT normalization and longer-term histologic improvement) were common in the minority of patients with such serologic responses (HBeAg loss/seroconversion) [28]. The major pre-treatment factors favorably associated with HBeAg responses were elevated serum ALT levels (e.g. $\geq 2 \times$ the upper limit of normal (ULN)) and low serum HBV DNA levels (<100–200 pg/ml by an unamplified solution hybridization assay) [1, 28, 29].

In association with HBeAg loss/seroconversion, interferon-treated patients' HBV viremia is usually still detectable by sensitive DNA amplification methods (e.g. PCR) [1, 7]. Quantitatively, however, with HBeAg clearance serum HBV DNA levels typically drop 100–1000-fold, to levels of 2–5 $\log_{10}$ copies/ml, similar to the viremia levels found in 'inactive' HBV carriers or in HBeAg-positive patients who have spontaneously seroconverted. In endemic areas for precore mutant HBV strains, however, some patients' HBeAg loss post-treatment has been found to be due to the coincidental emergence of precore mutant HBV strains [30–32]. Such patients can often be detected by their persisting high serum levels of HBV DNA (>5–6 $\log_{10}$ copies/ml), in contrast to the shift to lower viremia levels observed with true HBeAg seroconversion.

After serum HBeAg loss, some patients eventually lose detectable serum HBsAg, usually months or years after the loss of detectable HBeAg [1, 2, 33, 34]. At the time of HBsAg loss, serum HBV DNA usually becomes non-detectable (even by PCR methods), but HBV DNA remains detectable in patients' livers, probably mostly in the form of covalently closed circular DNA (cccDNA) but perhaps also with some HBV DNA integrants [7, 35]. Thus, true 'virologic cures' are rare or non-existent, even among long-term 'complete' responders to interferon therapy. Probably the same holds true for long-term responders to nucleoside/nucleotide therapy, although long-term follow-up data for lamivudine and adefovir are not yet adequate for inferences in this regard.

With regard to the possible long-term benefits of interferon therapy, although still somewhat controversial it appears likely that therapeutic induction of HBeAg loss/seroconversion in pre-cirrhotic patients is associated with a reduced long-term risk for decompensated cirrhosis [33, 34, 36] Also, one retrospective follow-up study suggested a beneficial effect of interferon treatment on the incidence of HCC, at least in serologic responders to interferon; however, this observation needs confirmation in additional studies [36].

It is uncertain as to whether perinatally infected patients (e.g. most Asians) respond as well as Caucasians to interferon therapy. Early reports suggested reduced responsiveness to interferon in Asians with chronic hepatitis B, thought to possibly relate to the predominance of perinatal transmission and residual partial immune tolerance to HBV infection in such patients [37]. Later reports, however, indicated that treatment responses to lamivudine may be similar for Asian and Caucasian patients, after adjustment for pre-treatment ALT levels [38, 39].

Interferon therapy has also been attempted in additional HBV patient groups, other than the HBeAg-positive patient population. In early trials, interferon treatment of patients with HBeAg-negative chronic hepatitis B (i.e. patients with precore mutant HBV infection) appeared potentially beneficial, with ALT normalization and histologic responses in some patients; however, most responders relapsed after 4–12 months of interferon treatment [40–42]. Therefore, the role of interferon in treating patients with HBeAg-negative chronic hepatitis B remains uncertain, although some experts recommend a trial of long-term maintenance therapy in such patients if they can tolerate interferon's side effects [43].

Interferon treatment of hepatitis B patients with decompensated cirrhosis was found to be risky, with some treated patients experiencing leukopenia-related bacterial sepsis and some experiencing disease flares with hepatic decompensation [44, 45]. Therefore, interferon is considered to be relatively contraindicated in patients with decompensated cirrhosis [1, 2].

In summary, in patients with HBeAg-positive chronic hepatitis B and elevated ALT levels (e.g. $>2 \times$ ULN), treatment with interferon can produce HBeAg loss in about 20–35% of patients, assessed 6–12 months after completion of treatment [28]. Long-term HBV suppression in non-seroconverted patients has not been well studied with interferon, perhaps due to issues of tolerability and inconvenience for long-term interferon treatment. Interferon therapy is associated with significant side effects, which can sometimes be severe [1, 2, 28]. Additionally, patients with a number of medical conditions, or with advanced disease (i.e. decompensated cirrhosis), are poor candidates for interferon therapy due to the potential for severe complications in such patients [1, 2, 44–45].

In a US multicenter trial of interferon α, improvements in liver histology correlated with clearance of detectable HBV DNA in an unamplified assay, regardless of whether HBeAg seroconversion was achieved [29]. Thus, it became apparent that the therapeutic goal of abrogating progression of HBV-associated liver disease can be accomplished in two ways, i.e.: by inducing HBeAg loss or HBeAg seroconversion, with associated persistent reduction of HBV replication or, in the absence of HBeAg loss/seroconversion, through persistent suppression of HBV replication (reflected by suppression of HBV DNA levels), which is associated with ALT normalization and amelioration of underlying liver inflammation as long as the suppression is maintained. This observation from interferon trials, regarding an association between treatment-related reductions in HBV DNA levels and reduced liver inflammation, provided important rationale for subsequent antiviral investigations in hepatitis B with 'direct' antiviral agents (nucleoside and nucleotide agents), as discussed in the following sections.

Lamivudine

Large phase III trials involving HBeAg-positive hepatitis B patients in Asia, North America, and Europe indicated that treatment with lamivudine induces HBeAg loss in 16–33% of patients at 1 year, and lamivudine can provide continuing HBV suppression in patients who have not seroconverted during the first year of therapy, with a potential for additional HBeAg seroconversions during subsequent years of treatment [23, 46–50]. However, after the first 6–12 months of treatment, maintenance therapy in non-seroconverters can be problematic due to the emergence of lamivudine-resistant HBV strains, which is cumulative with prolonged therapy and results in variable loss of therapeutic response, as discussed in more detail below [51, 52].

As with interferon, therapy responses that are durable post-treatment can be observed in lamivudine-treated HBeAg-positive patients who achieve HBeAg loss/seroconversion. The post-treatment durability of such virologic responses appears to be better in patients who have been treated with lamivudine for a minimum of 1 year, and in whom HBeAg clearance has been evident for 6 months or more before treatment discontinuation [53–55]. Pretreatment factors associated with HBeAg responses to lamivudine therapy include elevated ALT levels and greater liver inflammation (higher histologic activity index score), while pretreatment serum HBV DNA levels appear to have less influence on treatment responsiveness to lamivudine compared to interferon [56].

A phase III trial of lamivudine in patients with HBeAg-negative hepatitis B indicated consistent HBV suppression and normalization of serum ALT levels during the first year of treatment [26]. However, disease relapse was nearly universal post-treatment, when lamivudine was arbitrarily discontinued by protocol after 1 year. Unfortunately, in contrast to the situation for HBeAg-positive patients, a treatment discontinuation endpoint is not yet apparent for HBeAg-negative patients [1, 2]. Many clinicians have attempted long-term maintenance therapy with lamivudine in HBeAg-negative patients. However, as is discussed further below, the cumulative development of lamivudine-resistant HBV can limit long-term lamivudine treatment in this patient population [17, 26, 57].

Lamivudine therapy has been generally well tolerated, and exhibits a safety profile similar to placebo during one-year study treatment periods [34–37, 49]. Post-treatment, occasional hepatitis flares have been reported after lamivudine therapy, including ALT elevations with or without hepatic decompensation [23, 46, 58]. Controlled phase III trial data suggest that the incidence of grade 3–4 ALT elevations is higher after discontinuation of lamivudine compared to placebo, but severe post-treatment flares (e.g. ALT elevations with bilirubin elevations or with clinical signs of hepatic decompensation) were rare in the controlled clinical trials and occurred with indistinguishable frequency for lamivudine and placebo recipients [23, 46–48]. In light of such observations, it is recommended that patients discontinuing lamivudine be monitored for several months after treatment discontinuation [1, 2].

Although lamivudine was an important treatment advance for patients with chronic hepatitis B and can confer significant treatment benefit as long as viral suppression is maintained, the limiting feature of lamivudine therapy is a propensity for the cumulative emergence of lamivudine-resistant HBV variants during prolonged treatment. In the first year of lamivudine therapy, HBV strains with mutations in the YMDD motif of the viral polymerase gene (which confer reduced susceptibility to lamivudine) were detectable by PCR methods in the sera of about 15–30% of phase III patients [23, 26, 46–48]. Prolonged lamivudine therapy, e.g. beyond 1 year, is associated with cumulatively increasing detectability of such lamivudine-resistant HBV variants; after 3 years of lamivudine treatment, a majority of phase III patients were found to have detectable YMDD-mutant HBV [50]. It appears that the evolution of lamivudine-resistant HBV is inversely proportional to the degree of HBV DNA suppression during early treatment [51]. And while some patients with YMDD-mutant HBV maintain good therapeutic responses, patients with YMDD-mutant HBV more commonly develop variable return of appreciable serum HBV DNA levels and variably elevated serum ALT levels, reflecting diminished therapeutic responses [51]. Furthermore, after the development of YMDD-mutant HBV, patients have lower rates of HBeAg seroconversion, no greater than placebo recipients. Finally, clinical disease progression, including death, has occasionally been observed in patients who lose response to lamivudine due to the emergence of YMDD-mutant HBV [51, 69]. Severe disease progression appears to be more likely in patients with advanced hepatitis B or in the liver-transplant setting [59]. Disease progression after the development of lamivudine resistance appears to be primarily related to the quantitative return of HBV replication, i.e. to loss of control of HBV replication, rather than to any structural changes in YMDD-mutant HBV strains [51, 59].

As noted above, one important difference for nucleoside therapy, compared to interferon therapy, is that patients who do not achieve HBeAg clearance can nonetheless achieve therapeutic responses to maintained nucleoside treatment. If HBV suppression is successfully maintained (e.g. serum HBV DNA levels < 4–5 $\log_{10}$ copies/ml), patients' ALT levels and underlying liver histology tend to remain improved [49, 50]. Conversely, if HBV resistance develops and

HBV suppression is lost, the return of appreciable serum HBV DNA levels is associated with return of variably elevated ALT levels and histologic disease activity.

Adefovir dipivoxil

While interferon and lamivudine can each produce clinically valuable responses in hepatitis B patients, non-response to interferon is common and resistance to lamivudine develops with cumulatively increasing frequency after the first 6 months of treatment. Therefore there is a need for additional treatment options and continuing optimization of antiviral therapy for patients with chronic hepatitis B.

Adefovir dipivoxil (ADV; Hepsera™) recently completed phase III registration trials and achieved regulatory approval by the US Food and Drug Administration (FDA) for use in hepatitis B patients. At present ADV use is not as widespread as interferon or lamivudine therapy, but ADV is currently in the process of achieving more widespread regulatory approvals, worldwide.

ADV is an oral prodrug of PMEA, a phosphonate nucleotide analogue of adenosine monophosphate which exhibits antiviral activity in laboratory assays against hepadnaviruses, retroviruses, and herpesviruses [60]. ADV is active *in vitro* against both wild-type and lamivudine-resistant (YMDD-mutant) HBV strains [60, 61].

Two multicenter phase III trials of ADV were recently completed, involving adults with chronic hepatitis B and compensated liver disease [24, 25]. These principal registration trials for ADV are summarized below, together with product label recommendations [67].

One phase III study of ADV, known as the GS-98-437 protocol, was a controlled assessment of ADV treatment in HBeAg-positive patients, at either of two ADV dosing levels (10 or 30 mg/day), compared to placebo [24]. This study enrolled 515 patients, with the following randomized treatment assignments: ADV at 30 mg/day ($n = 173$ patients), ADV 10 mg/day ($n = 172$), or placebo ($n = 170$).

The 48-week GS-98-437 study results indicated significant benefit for ADV treatment compared to placebo, but also indicated a higher frequency of safety issues for ADV dosing at 30 mg/day, compared to the 10 mg/day dose [24]. Nephrotoxicity, evidenced by an increase in serum creatinine levels of 0.5 mg/dl or more, was evident in 8% of patients in the ADV 30 mg/day group. Therefore, product registration focused on efficacy and safety data for the ADV 10 mg/day dose. In the GS-98-437 trial, on the primary endpoint of protocol-defined histologic response, improvement was observed significantly more frequently in patients receiving ADV 10 mg/day compared to patients receiving placebo (53 *versus* 25%; $P < 0.001$). There was also evidence of treatment benefit on secondary efficacy endpoints. HBeAg seroconversion was observed in 12% of patients treated with ADV 10 mg/day for 48 weeks compared with 6% of patients on placebo ($P = 0.049$). Change in HBV viral load, as measured by the Roche Amplicor® HBV Monitor™ Test (PCR), was assessed. At week 48, patients in the ADV 10 mg/day arm had a median reduction in serum HBV DNA level of 3.52 $\log_{10}$ copies/ml, compared with a reduction of 0.55 $\log_{10}$ copies/ml in patients receiving placebo ($P < 0.001$). Additionally, following 48 weeks of treatment with ADV 10 mg/day, 48% of patients treated with ADV 10 mg/day achieved normalization of serum ALT levels, compared with 16% of patients receiving placebo ($P < 0.001$). Beyond week 48, longer-term ADV treatment data and post-treatment data (efficacy, safety, resistance) are unfortunately not available.

The overall safety profile for the 10 mg/day ADV dose was satisfactory in the GS-98-437 trial [24]. Through the first 48 weeks of treatment, the discontinuation rate was similar between the ADV 10 mg/day and placebo arms, with 7% of patients receiving ADV 10 mg/day and 8% receiving placebo discontinuing the study, and the rates of various clinical adverse events and laboratory abnormalities appeared similar between these two treatment groups. Notably, in contrast to the observations for ADV 30 mg/day in the same study, no patients in either the ADV 10 mg/day or placebo groups had confirmed increases in serum creatinine of ≥0.5 mg/dl from baseline or a serum phosphorus level <1.5 mg/dl, laboratory markers of renal function.

The other phase III registration trial for ADV, designated GS-98-438, involved patients with HBeAg-negative chronic hepatitis B [25]. This study enrolled 185 patients, randomized 2:1 to treatment with ADV 10 mg/day ($n = 123$ patients) or placebo ($n = 61$). The 30 mg/day ADV dose was not investigated in this latter study. As in the GS-98-437 study, the primary efficacy endpoint was protocol-defined histologic response, and again treatment benefit was evident for the ADV recipients: histologic improvement was observed significantly more frequently in patients receiving ADV 10 mg/day compared with patients receiving placebo (64 *versus* 33%; $P < 0.001$). Benefits of ADV treatment were found for several secondary efficacy evaluations. At week 48 the median decrease in serum HBV DNA level was 3.91 $\log_{10}$ copies/ml for ADV-treated patients compared to 1.35 $\log_{10}$ copies/ml for placebo-treated patients ($P < 0.001$). HBV DNA levels became non-detectable by PCR (lower limit of detection; 400 copies/ml) in 51% of ADV recipients compared with 0% of placebo recipients. Finally, in concert with the HBV suppression and histologic responses, ALT normalization was observed in 72% of ADV recipients compared with 29% of the placebo group ($P < 0.001$) [25].

In the GS-98-438 study involving HBeAg-negative patients, the overall safety observations for the ADV 10 mg/day treatment group were similar to those for placebo recipients—as in the GS-98-437 trial—although headaches and abdominal pain were reported to be marginally more common in the ADV 10 mg/day group [25]. Two types of safety observation from the ADV clinical trial program necessitated boxed warnings on the US product label for Hepsera™: a potential for nephrotoxicity with long-term use, especially in patients who might have other circumstances associated with renal insufficiency or renal toxicity; and a relatively high rate (25%) of severe (Grade 4) post-treatment ALT flares, in patients discontinued from ADV treatment prior to HBeAg seroconversion [62].

Based on the phase III results, after 1 year of treatment, the antiviral and clinical efficacy of ADV appear to be roughly comparable to lamivudine. The efficacy assessments in the phase III trials for lamivudine and ADV were conducted using slightly different methods, but at 1 year the rates of histologic response and ALT normalization appeared quite similar for these two anti-HBV agents. Although the rate of full HBeAg seroconversion (12%) with ADV was lower than any of the corresponding results for lamivudine (which were ≥16–18% in three phase III trials), the rate of HBeAg loss appeared roughly comparable (ca. 23% for both agents at 1 year).

In summary, with regard to the roles of lamivudine and ADV in treating hepatitis B patients, lamivudine appears to have an overall edge with regard to potential safety concerns, while ADV appears to have an edge with regard to less frequent emergence of drug-resistant HBV. In contrast to lamivudine, analyses of patient sera during the first year of ADV treatment in the phase III trials did not suggest an identifiable pattern of emergence of ADV-resistant HBV strains. However, ADV-resistant HBV has been reported to emerge in some patients during subsequent prolonged treatment [63–65]. Three-year clinical data from the GS-98-438 study were recently presented, with a reported cumulative HBV resistance rate of 5.9% at 3 years due to emergence of viral mutants with an Asn–236–Thr mutation in the polymerase gene as well as mutants with an Ala–181–Val polymerase mutation [66]. Thus initial data suggest that the ADV resistance profile offers substantial improvements over lamivudine for the primarily HBeAg-negative patients and liver transplant recipients who were the subject of the ADV long-term (>1 year) resistance analysis. However, quantitative comparisons of resistance rates for these agents are hampered by the lack of long-term resistance data for ADV pertaining to the high-viremic, HBeAg-positive patients from which the long-term lamivudine data were gathered [63–66]. This consideration is important insofar as high pretreatment HBV DNA level is the strongest risk factor for the development of drug-resistant HBV strains [51, 52]. More data (beyond 1 year) are needed to fully establish the long-term resistance advantage pertaining to ADV treatment of HBeAg-positive patients.

Prospects for further improvements in antiviral therapy for chronic hepatitis B

Interferon α, lamivudine, and ADV allow the clinical management of many patients, and patients' prospects for reduced morbidity and mortality have therefore improved considerably with these treatment advances over the past 15 years. However, antiviral therapy for hepatitis B remains problematic with regard to optimizing durable response rates with safe, well-tolerated agents. Fortunately the expanding marketplace for hepatitis B drugs is now driving pharmaceutic companies' increasing interest in the discovery and development of new treatments for this global scourge.

Two therapeutic goals remain desirable: more rapid and extensive induction of durable HBeAg and HBsAg seroconversions, and better long-term suppression of HBV replication and liver disease in non-seroconverting patients; i.e. more effective and safe 'maintenance' therapy, especially for patients with HBeAg-negative hepatitis B and for cirrhotic patients (in whom treatment discontinuation, with risk of aggressive post-treatment relapse, may be deemed too risky). It is likely that achievement of these goals will be aided by: (a) the development of more potent antiviral agents allowing greater inhibition of HBV replication; (b) the development of new treatment strategies with anti-HBV agents (e.g. combination or sequential treatment strategies, aimed at maintaining maximal HBV suppression); and (c) a better understanding of HBV-specific immune phenomena associated with HBeAg seroconversion and other forms of durable response. Better delineation of HBV-specific immune may facilitate the rational development of specific immunomodulatory agents that can augment the durable response rates achievable with optimized antiviral therapy. The development of HBV-specific immunomodulatory agents, probably to be used in combination regimens with nucleoside/nucleotide agents, could well be important in achieving optimal durable response rates with finite treatment regimens, and could render the non-specific activities and considerable toxicities of interferon obsolete.

Among new anti-HBV agents in current clinical testing, both entecavir and telbivudine (LdT) have considerable promise as more potent anti-HBV agents, based on supportive phase II head-to-head data *versus* concurrent lamivudine-treated cohorts [67, 68]. Also, exploratory analyses of a 104-patient database from a phase IIb telbivudine trial supported the importance of maximizing patients' antiviral responses in the first 6 months of treatment, to achieve better clinical efficacy at 1 year and beyond [69]. Results from the phase IIb telbivudine trial, and two earlier reports, suggest that reducing patients' serum HBV DNA to levels below 3–4 $\log_{10}$ copies/ml in the first six months of treatment is likely to be associated with better HBeAg seroconversion and ALT normalization rates, with reduced viral breakthrough [52, 69, 70].

Conclusions

With three antiviral agents now available (interferon, lamivudine, and ADV), sufficient progress has been made to allow the clinical management of many patients with chronic hepatitis B. Some patients can achieve responses that are durable post-treatment, while others are currently treated with maintenance therapy regimens with varying degrees of success. The AASLD, a professional society that includes many American and foreign hepatitis experts, recently promulgated recommendations to guide the evaluation and treatment of hepatitis B patients [1, 17]. These and other expert recommendations from various sources serve to highlight that, with the current anti-HBV armamentarium, therapy is not yet sufficiently eradicative of HBV infection to recommend treatment for all HBV carriers. Instead, antiviral treatment can currently be recommended for patients with chronic HBV infection (>6 months), persisting high levels of HBV replication (e.g. HBeAg-positive, and/or serum HBV DNA >4–5 $\log_{10}$ copies/ml), and persistently elevated serum aminotransferase levels (and/or evidence of a consistent pattern of chronic inflammation in a liver biopsy, spreading beyond the peri-portal areas). Considerations of initial treatment choice, among the currently available anti-HBV agents, are controversial but are well-addressed in the recent AASLD guidelines [1, 17]. Due to the remarkable and unpredictable variation in clinical course among individuals with chronic hepatitis B, regular patient monitoring is essential during antiviral treatment, regardless of treatment choice; and periodic long-term follow-up remains important, even in patients who appear to achieve durable post-treatment responses.

More potent, orally bioavailable anti-HBV agents appear to be on the near horizon, but phase III data are still to be gained to conclusively demonstrate the superiority of these new nucleoside agents. If the immune pathophysiology of durable responses can be better understood, it may be possible to fully optimize early achievement of durable responses to anti-HBV therapy, with combinations of nucleoside/nucleotide agents and specific immunomodulatory therapies.

References

1 Lok ASF, McMahon BJ (2001) Chronic hepatitis B. *Hepatology* 34: 1225–1241

2 Liaw YF, Leung N, Guan R, Lau GK, Merican I (2003) Asia Pacific consensus statement on the management of chronic hepatitis B: an update. *J Gastroenterol Hepatol* 18: 239–245

3 Andre F (2000) Hepatitis B epidemiology in Asia, the Middle East and Africa. *Vaccine* 18: S20–S22

4 Mast EE, Alter MJ, Margolis HS (1999) Strategies to prevent and control hepatitis B and C virus infections: a global perspective. *Vaccine* 17: 1730–1733

5 Alter MJ (2003) Epidemiology of hepatitis B in Europe and worldwide. *J Hepatol* 39 (suppl 1): S64–S69

6 Lee W (1997) Hepatitis B virus infection. *N Engl J Med* 337: 1733–1745

7 Ganem D, Prince AM (2004) Hepatitis B virus infection–natural history and clinical consequences. *N Engl J Med* 350: 1118–1129

8 Lok AS, Lai CL (1990) Acute exacerbations in Chinese patients with chronic hepatitis B (HBV) virus infection: incidence, predisposing factors and etiology. *J Hepatol* 10: 29–34

9 Hoofnagle JH, Dusheiko GM, Seeff LB, Jones EA, Waggoner JG, Bales ZB (1981) Seroconversion from hepatitis B e antigen to antibody in chronic type B hepatitis. *Ann Intern Med* 94: 744–748

10 Liaw YF, Chu CM, Su IJ, Huang MJ, Lin DY, Chang-Chien CS (1983) Clinical and histological events preceding hepatitis B e antigen seroconversion in chronic type B hepatitis. *Gastroenterology* 84: 216–219

11 Fattovich G, Brollo L, Alberti A, Pontisso P, Giustina G, Realdi G (1988) Long term follow-up of anti-HBe-positive chronic active hepatitis B. *Hepatology* 8: 1651–1654

12 Zarski JP, Marcellin P, Cohard M, Lutz JM, Bouche C, Rais A (1994) Comparison of anti-HBe-positive and HBe-antigen-positive chronic hepatitis B in France. French Multicentre Group. *J Hepatol* 20: 636–640
13 Hadziyannis S (1995) Hepatitis B e antigen negative chronic hepatitis B: from clinical recognition to pathogenesis and treatment. *Viral Hepatitis Rev* 1: 7–36
14 Grandjacques C, Pradat P, Stuyver L, Chevallier M, Chevallier P, Pichoud C, Maisonnas M, Trepo C, Zoulim F (2000) Rapid detection of genotypes and mutations in the pre-core promoter and the pre-core region of hepatitis B virus genome: correlation with viral persistence and disease severity. *J Hepatol* 33: 430–439
15 Tu H, Ziong SD, Trepo C, Wen YM (1997) Frequency of hepatitis B virus e-minus mutants varies among patients from different areas of China. *J Med Virol* 51: 85–90
16 Chan HLY, Leung NWY, Hussain M, Wong ML, Lok ASF (2000) Hepatitis B e antigen-negative chronic hepatitis B in Hong Kong. *Hepatology* 31: 763–768
17 Lok ASF, McMahon BJ (2004) Chronic hepatitis B: an update of recommendations. *Hepatology* 39: 857–861
18 Yang HI, Lu SN, Liaw YF, You SL, Sun CA, Wang LY, Hsiao CK, Chen PJ, Chen DS, Chen CJ (2002) Hepatitis B e antigen and the risk of hepatocellular carcinoma. *N Engl J Med* 347: 168–174
19 Evans AA, O'Connell AP, Pugh JC, Mason WS, Shen FM, Chen GC, Lin WY, Dia A, M'Boup S, Drame B et al. (1998) Geographic variation in viral load among hepatitis B carriers with differing risks of hepatocellular carcinoma. *Cancer Epidemiol Biomarker Prev* 7: 559–565
20 Ohkubo K, Kato Y, Ichikawa T, Kajiya Y, Takeda Y, Higashi S, Hamasaki K, Nakao K, Nakata K, Eguchi K (2002) Viral load is a significant prognostic factor for hepatitis B virus-associated hepatocellular carcinoma. *Cancer* 94: 2663–2668
21 De Jongh FE, Janssen HLA, De Man RA, Hop WC, Schalm SW, van Blankenstein M (1992) Survival and prognostic indicators in hepatitis B surface antigen-positive cirrhosis of the liver. *Gastroenterology* 103: 1630–1635
22 Fattovich G, Brollo L, Giustina G, Noventa F, Pontisso P, Alberti A, Realdi G, Ruol A (1991) Natural history and prognostic factors for chronic hepatitis type B. *Gut* 32: 294–298
23 Lai CL, Chien RN, Leung NW, Chang TT, Guan R, Tai DI, Ng KY, Wu PC, Dent JC, Barber J et al. (1998) A one-year trial of lamivudine for chronic hepatitis B. *N Engl J Med* 339: 61–68
24 Marcellin P, Chang TT, Lim SG, Tong MJ, Sievert W, Shiffman ML, Jeffers L, Goodman Z, Wulfsohn MS, Xiong S et al. (2003) Adefovir dipivoxil for the treatment of hepatitis B e antigen-positive chronic hepatitis B. *N Engl J Med* 348: 808–816
25 Hadziyannis S, Tassopoulos N, Heathcote EJ, Chang TT, Kitis GG, Rizetto M, Marcellin P, Lim SG, Goodman Z, Wulfsohn MS et al. (2003) Adefovir dipivoxil for the treatment of HBeAg-negative chronic hepatitis B. *N Eng J Med* 348: 800–807
26 Tassopoulos NC, Volpes R, Pastore G, Heathcote J, Buti M, Goldin RD, Hawley S, Barber J, Condreay L, Gray DF (1999) Efficacy of lamivudine in patients with hepatitis B e antigen-negative/hepatitis B virus DNA-positive (precore mutant) chronic hepatitis B. Lamivudine Precore Mutant Study Group. *Hepatology* 29: 889–896
27 Weissberg JI, Andres LL, Smith CI, Weick S, Nichols JE, Garcia G, Robinson WS, Merigan TC, Gregory PB (1984) Survival in chronic hepatitis B. An analysis of 379 patients. *Ann Intern Med* 101: 613–616
28 Wong DK, Cheung AM, O'Rourke K, Naylor CD, Detsky A, Heathcote J (1993) Effect of alpha-interferon treatment in patients with hepatitis B e antigen-positive chronic hepatitis B. *Ann Intern Med* 119: 312–323
29 Perrillo RP, Schiff ER, Davis GL, Bodenheimer HC, Lindsay K, Payne J, Dienstag JL, O'Brien C, Tamburro C, Jacobson IM et al. (1990) A randomized controlled trial of interferon alfa-2b alone and after prednisone withdrawal for the treatment of chronic hepatitis B. *N Engl J Med* 323: 295–301
30 Brunetto MR, Giarin M, Saracco G, Oliveri F, Calvo P, Capra G, Randone A, Abate ML, Manzini P, Capalbo M et al. (1993) Hepatitis B virus unable to secrete e antigen and response to interferon in chronic hepatitis B. *Gastroenterology* 105: 845–850
31 Zhang X, Zoulim F, Habersetzer F, Xiong S, Trepo C (1996) Analysis of hepatitis B virus genotypes and pre-core region variability during interferon treatment of HBe antigen negative chronic hepatitis B. *J Med Virol* 48: 8–16
32 Brunetto MR, Rodriguez UA, Bonino F (1999) Hepatitis B virus mutants. *Intervirology* 42: 69–80
33 Niederau C, Heintges T, Lange S, Goldmann G, Niederau C, Mohr L, Häussinger D (1996) Long-

term follow-up of HBeAg-positive patients treated with interferon alpha for chronic hepatitis B. *N Engl J Med* 334: 1422–1427
34 Korenman J, Baker B, Waggoner J, Everhart JE, Di Bisceglie AM, Hoofnagle JH (1991) Long-term remission of chronic hepatitis B after alpha-interferon therapy. *Ann Intern Med* 114: 629–634
35 Pollicino T, Squadrito G, Cerenzia G, Cacciola I, Raffa, Crax A, Farinati F, Missale G, Smedile A, Tiribelli C et al. (2004) Hepatitis B virus maintains its pro-oncogenic properties in the case of occult HBV infection. *Gastroenterology* 126: 102–110
36 Lin SM, Sheen IS, Chien RN, Chu CM, Liaw YF (1999) Long-term beneficial effect of interferon therapy in patients with chronic hepatitis B virus infection. *Hepatology* 29: 971–975
37 Lok ASF, Lai CL, Wu PC, Lau JYN, Leung EKY, Wong LS (1989) Treatment of chronic hepatitis B with interferon: experience in Asian patients. *Sem Liver Dis* 9: 249–253
38 Lok ASF, Wu PC, Lai CL, Lau JYN, Leung EKY, Wong LS, Ma OC, Lauder IJ, Ng CP, Chung HT (1992) A controlled trial of interferon with or without prednisone priming for chronic hepatitis B. *Gastroenterology* 102: 2091–2097
39 Guan R (2000) Interferon monotherapy in chronic hepatitis B. *J Gastroenterol Hepatol* 15 (suppl):E34–E40
40 Brunetto MR, Oliveri F, Rocca G, Criscuolo D, Chiaberge E, Capalbo M, David E, Verme G, Bonino F (1989) Natural course and response to interferon of chronic hepatitis B accompanied by antibody to hepatitis B e antigen. *Hepatology* 10: 198–202
41 Hadziyannis S, Bramou T, Makris A, Moussoulis G, Zignego L, Papaioannou C (1990) Interferon alpha-2b treatment of HBeAg negative/serum HBV DNA positive chronic hepatitis type B. *J Hepatol* 1: S133–S136
42 Fattovich G, Farci P, Rugge M, Brollo L, Mandes A, Pontisso P, Giustina G, Lai ME, Belussi F, Busatto G et al. (1991) A randomized controlled trial of lymphoblastoid IFN-alpha in patients with chronic hepatitis B lacking HBeAg. *Hepatology* 15: 584–589
43 Lampertico P, Del Ninno E, Manzin A, Donato MF, Rumi MG, Lunghi G, Morabito A, Clementi M, Colombo M (1997) A randomized controlled trial of a 24-month course of interferon alfa 2b in patients with chronic hepatitis B who had hepatitis B virus DNA without hepatitis B e antigen in serum. *Hepatology* 26: 1621–1625
44 Perrillo R, Tamburro C, Regenstein F, Balart L, Bodenheimer H, Silva M, Schiff E, Bodicky C, Miller B, Denham C et al. (1995) Low-dose, titratable interferon alfa in decompensated liver disease caused by chronic infection with hepatitis B virus. *Gastroenterology* 109: 908–916
45 Hoofnagle JH, Di Bisceglie AM, Waggoner JG, Park Y (1993) Interferon alfa for patients with clinically apparent cirrhosis due to chronic hepatitis B. *Gastroenterology* 104: 1116–1121
46 Dienstag JL, Schiff ER, Wright TL, Perrillo RP, Hann HW, Goodman Z, Crowther L, Condreay LD, Woessner M, Rubin M, Brown NA (1999) Lamivudine as initial treatment for chronic hepatitis B in the United States. *N Engl J Med* 341: 1256–1263
47 Schalm SW, Heathcote J, Cianciara J, Farrell G, Sherman M, Willems B, Dhillon A, Moorat A, Barber J, Gray DF (2000) Lamivudine and alpha interferon combination treatment of patients with chronic hepatitis B infection: a randomised trial. *Gut* 46: 562–568
48 Schiff ER, Dienstag JL, Karayalcin S, Grimm IS, Perrillo RP, Husa P, De Man RA, Goodman Z, Condreay LD, Crowther LM et al. (2003) Lamivudine and 24 weeks of lamivudine/interferon combination therapy for hepatitis B e antigen-positive chronic hepatitis B in interferon nonresponders. *J Hepatol* 38: 818–826
49 Liaw TJ, Leung NWY, Chang TT, Guan R, Tai DI, Ng KY, Chien RN, Dent J, Roman L, Edmundson S (2000) Effects of extended lamivudine therapy in Asian patients with chronic hepatitis B. *Gastroenterology* 119: 172–180
50 Leung NWY, Lai CL, Chang TT, Guan R, Lee CM, Ng KY, Lim SG, Wu PC, Dent JC, Edmundson S et al. (2001) Extended lamivudine treatment in patients with chronic hepatitis B enhances hepatitis B e antigen seroconversion rates: results after 3 years of therapy. *Hepatology* 33: 1527–1532
51 Lai CL, Dienstag JL, Schiff ER, Leung NWY, Atkins M, Hunt C, Brown N, Woessner M, Boehme R, Condreay L (2003) Prevalence and clinical correlates of YMDD variants during lamivudine therapy for patients with chronic hepatitis B. *Clin Infect Dis* 36: 687–696
52 Yuen MF, Sablon E, Hui CK, Yuan HJ, Decraemer H, Lai CL (2001) Factors associated with hepatitis B virus DNA breakthrough in patients receiving prolonged lamivudine therapy. *Hepatology* 34: 785–791
53 Dienstag JL, Cianciara J, Karayalcin S, Kowdley KV, Willems B, Plisek S, Woessner M, Gardner

S, Schiff E (2003) Durability of serologic response after lamivudine treatment of chronic hepatitis B. *Hepatology* 37: 748–755

54 Chien RN, Yeh CT, Tsai SL, Chu CM, Liaw YF (2003) Determinants for sustained HBeAg response to lamivudine therapy. *Hepatology* 38: 1267–1273

55 Ryu SH, Chung YH, Choi MH, Kim JA, Shin JW, Jang MK, Park NH, Lee HC, Lee YS, Suh DJ et al. (2003) Long-term additional lamivudine therapy enhances durability of lamivudine-induced HBeAg loss: a prospective study. *J Hepatol* 39: 614–619

56 Perrillo RP, Lai CL, Liaw YF, Dienstag JL, Schiff ER, Schalm SW, Heathcote EJ, Brown NA, Atkins M, Woessner M et al. (2002) Predictors of HBeAg loss after lamivudine treatment for chronic hepatitis B. *Hepatology* 36: 186–194

57 Hadziyannis SJ, Papatheodoridis GV, Vassilopoulos D (2003) Treatment of HBeAg-negative chronic hepatitis B. *Sem Liver Dis* 20: 81–88

58 Lok ASF, Lai CL, Leung N, Yao GB, Cui ZY (2003) Long-term safety of lamivudine treatment in patients with chronic hepatitis B. *Gastroenterology* 125: 1714–1722

59 Ben-Ari Z, Mor E, Tur-Kaspa R (2003) Experience with lamivudine therapy for hepatitis B virus infection before and after liver transplantation, and review of the literature. *J Intern Med* 253: 544–552

60 Dando T, Plosker G (2003) Adefovir dipivoxil: a review of its use in chronic hepatitis B. *Drugs* 63: 2215–2234

61 Xiong X, Flores C, Yang H, Toole JJ, Gibbs CS (1998) Mutations in hepatitis B DNA polymerase associated with resistance to lamivudine do not confer resistance to adefovir *in vitro*. *Hepatology* 28: 1669–1673

62 Hepsera product label. Gilead Inc. www.gilead.com/pdf/hepsera_pi.pdf

63 Angus P, Vaughan R, Xiong S, Yang H, Delaney W, Gibbs C, Brosgart C, Colledge D, Edwards R, Ayres A (2003) Resistance to adefovir dipivoxil therapy associated with the selection of a novel mutation in the HBV polymerase. *Gastroenterology* 125: 292–297

64 Yang H, Westland CE, Delaney WE, Heathcote EJ, Ho V, Fry J, Brosgart C, Gibbs CS, Miller MD, Xiong S (2002) Resistance surveillance in chronic hepatitis B patients treated with adefovir dipivoxil for up to 60 weeks. *Hepatology* 36: 464–473

65 Bartholomeusz A, Locarnini S, Ayres A, Thompson G, Edwards R, Colledge D, Angus P et al. (2003) Molecular modeling and functional studies of adefovir resistant mutations in the hepatitis B virus polymerase selected during therapy. [Abstract]. *Hepatology* 38: 273A

66 Hadziyannis S, Tassopoulos N, Chang TT, Heathcote EJ, Kitis G, Rizzetto M, Marcellin P, Lim SG, Chen SS, Arteburn S et al. (2004) Three year study of adefovir dipivoxil (ADV) shows sustained efficacy in presumed precore mutant chronic hepatitis B (CHB) patients in a long-term safety and efficacy study (LTSES). [abstract] Gastroenterology 126(4 suppl 7): A50

67 Lai CL, Rosmawati M, Lao J, Van Vlierberghe H, Anderson FH, Thomas N, Dehertogh D (2002) Entecavir is superior to lamivudine in reducing hepatitis B virus DNA in patients with chronic hepatitis B infection. *Gastroenterology* 123: 1831–1838

68 Lai CL, Leung NWY, Teo EK, Tong M, Wong F, Hann HW, Itan S, Poynard T, Myers M, Chao G et al. (2003) Results of A One-Year International Phase IIb Comparative Trial of Telbivudine, Lamivudine, and the Combination, in Patients with Chronic Hepatitis B. [abstract] *Hepatology* 38 (4 suppl 1): 262A

69 Poynard T, Lai CL, Leung NWY, Teo EK, Tong M, Wong F, Hann HW, Han S, Myers M, Chao G et al. (2004) Improving antiviral therapy in chronic hepatitis B: maximal viral suppression at Week 24 correlates with better clinical efficacy at one year. *J Hepatol* 40(suppl.1): 130A

70 Gauthier J, Bourne EJ, Lutz M, Crowther LM, Dienstag JL, Brown NA, Condreay LD (1999) Quantitation of hepatitis B viremia and emergence of YMDD variants in patients with chronic hepatitis B treated with lamivudine. *J Infect Dis* 180: 1757–1762

Hepatitis Prevention and Treatment
Edited by J.M. Colacino and B.A. Heinz

Current standard of care in hepatitis C virus infection

Anouk Dev, Keyur Patel and John G. McHutchison

Duke Clinical Research Institute and Division of Gastroenterology, Duke University Medical Centre, PO Box 17969, Durham, NC 27715, USA

Introduction

Since the identification of the hepatitis C virus (HCV) in the 1980s [1] and determination of the HCV genome in 1991 [2], there has been considerable progress in the management of chronic hepatitis C (CHC) in terms of achieving viral eradication and improving histology. Interferons (IFNs) were among the first agents used in the mid 1980s for the treatment of non-A, non-B hepatitis, a term then used to describe the yet-to-be identified HCV. Initially, the use of short courses of IFN monotherapy resulted in sustained losses of detectable virus in fewer than 10% of patients with only modest improvements associated with a longer duration of IFN therapy [3–5]. The major advance in treatment, in recent times, has been the addition of the oral nucleoside analog ribavirin to IFN, achieving an overall sustained virological response (SVR) in 30–40% of patients [6, 7] and, more recently, in over 50% of patients treated with the longer-acting PEGylated IFN and ribavirin combination [8, 9]. However, these drug regimens are expensive, require prolonged treatment (6 or 12 months), are associated with significant side effects, and clearly are not suited for all patients [6–9]. The accurate targeting of therapy to those most likely to develop progressive liver disease, thus avoiding exposure of low-risk patients to potential adverse effects of antiviral treatment, remains a significant challenge. In this chapter, the efficacy, adverse events and limitations of combination IFN and ribavirin therapy as well as newer IFN and immunomodulator agents will be discussed.

Current standard of care: IFN and ribavirin therapy

The aim of therapy for HCV infection is to achieve sustained, long-term eradication of the virus, thereby decreasing or preventing progressive liver damage, cirrhosis, liver failure or hepatocellular carcinoma. Type I IFNs, a group of naturally occurring cytokines that exhibit antiviral, antiproliferative and immunomodulatory effects, are the backbone of current antiviral treatment regimens. They function through a series of complex signaling pathways resulting in the expression of IFN-inducible genes, which mediate a variety of immunoregulatory actions including an increase in class I major histocompatibility complex (MHC) antigens, promotion of lymphoid cell differentiation and function, activation of macrophages and natural killer cells, and upregulation of cytotoxic Th1 T helper cells. Type I IFNs may also suppress cell proliferation and inhibit virus replication in infected cells [10].

The classification of IFNs is based on the distinct cell-surface receptor to which they bind. Type I IFNs include IFNα (IFNα-2a and -2b are approved for therapeutic use in CHC patients), β, κ and ω. Type II IFNs include IFNγ [11]. Recent advances include the PEGylation of IFN, which involves the covalent attachment of polyethylene glycol (PEG) to the native compound, resulting in reduced drug clearance and allowing once-weekly dosing. PEG-IFNα-2a has a 40 kDa branched-chain moiety compared to the 12 kDa linear-chain moiety in PEG-IFNα-2b. Each drug has different comparative pharmacokinetics [12, 13] (Tab. 1). The effectiveness of these two PEGylated compounds has not been compared directly, but data from published trials suggest that these compounds have similar response rates and similar adverse events [8, 9].

Ribavirin, although ineffective as monotherapy in inducing sustained HCV clearance or reductions in HCV RNA concentrations, markedly potentiates the clinical effect of IFN, resulting in enhanced sustained response rates. The

Table 1. Characteristics of PEG-IFNs compared with standard IFN

Characteristics	PEG-IFNα-2a	PEG-IFNα-2b	Standard recombinant IFN
Type of dosing	Fixed	Weight-based	Fixed
Dose interval	Once/week	Once/week	Three times/week
Molecular weight (kDa)			
IFN molecule	20	19	19–20
PEG	40	12	NA
Total	60	31	19–20
Volume of distribution (l)	8	20	70–80
Elimination half-life (h)	80	40	5–8
C_{max} (hours after dose)	72–96	15–44	3–12
Trough concentration (pg/ml at 48 weeks)	16 000	320	0
Peak-to-trough ratio	1.3	6	Infinity
Clearance when CrCl < 50 (% of normal clearance)	55–75%	50%	50% [19]
Dosing in patients with cirrhosis	No change	No change	No change

Adapted from product information: Pegasys™ (PEG-IFNα-2a), Hoffmann-La Roche Inc, Nutley, NJ, USA (2002) and PEG-Intron™ (PEG-IFNα-2b), Schering Corporation, Kenilworth, NJ, USA (2001). NA, not applicable.

underlying mechanisms for improvement in response rates with combination therapy are not completely understood. However, several possible mechanisms invoked to explain the beneficial clinical actions of ribavirin have been proposed. These include immune-mediated alteration of the host Th1/Th2 cytokine balance favoring Th2 antiviral effects, inhibition of the host inosine monophosphate dehyrogenase (IMPDH; a rate-limiting enzyme in guanine nucleotide synthesis), activity against viral RNA-dependent RNA polymerase, and as an RNA viral mutagen associated with a loss of genome viability and infectivity [14]. Ribavirin may also act as an IFN-enhancing agent, as *in vitro* studies indicate that it acts in part on virus-infected cells, making them more responsive to IFN by activating IFN signaling pathways [15]. In combination, these two drugs provide the current most effective therapy for the management of HCV infection.

Defining a response to antiviral therapy

An important outcome of the National Institutes of Health Consensus Development Conference in 1997 was the standardization of definitions of response to antiviral therapy in HCV infection [16]. An end-of-treatment response (ETR) is thus defined as an absence of detectable HCV RNA at the time of cessation of a course of therapy, and an SVR, as the absence of detectable virus 6 months after the end of treatment. These definitions have become widely accepted as the best indicators of a beneficial treatment effect in both the short and long term [17]. More recently, a failure to achieve an early virological response (EVR), defined as greater than or equal to a 2-log reduction in HCV RNA levels from baseline to week 12 of therapy, has been shown to be highly predictive of eventual non-response, particularly in patients infected with genotype 1 [18]. A relapse is defined as an HCV RNA which becomes detectable again following either an ETR or SVR (the risk of late relapse after achieving SVR is 0.25–0.75% per year) [19]. Some treated patients may never demonstrate a normal alanine aminotransferase (ALT) or undetectable HCV RNA during therapy, categorized as a primary non-response. Others may initially normalize their ALTs and have an undetectable HCV RNA during treatment followed by a rise in ALT values and reappearance of HCV RNA that is referred to as a breakthrough. In considering these different categories, it is most practical to consider relapse, breakthrough and primary non-response as manifestations of unsuccessful therapy (a so-called 'non- response').

Efficacy of treatment

In 1998, the results of two large international studies involving over 1700 previously untreated CHC patients treated with standard IFN and ribavirin or IFN monotherapy indicated that SVR was significantly higher in the combination-therapy groups (33–41% *versus* 6–16%) [6, 7]. These studies also demonstrated conclusively the importance of viral load and viral genotype in achieving a therapeutic response. Patients with genotype 1 infection required a 12-month duration of therapy to achieve a 29% SVR [20, 21]. Similarly designed studies evaluating combination therapy and IFN monotherapy in patients who had relapsed following IFN monotherapy indicated that re-treatment with combination therapy achieved a higher SVR (49% *versus* 7%). Additionally, a SVR occurred in 95% of patients with genotypes 2 and 3 compared with 24% of genotype 1-infected patients [21].

The improved pharmacokinetic and pharmacodynamic profiles of the PEG-IFNs over the standard formulation were anticipated to improve SVR. The results of PEG-IFNα-2a and -2b monotherapy studies, which are now of a historical importance, indicated that given alone for 48 weeks they were approximately twice as effective as their standard IFN counterpart [21–23]. This finding underscores the importance of ribavirin when combined with IFN in improving the number of patients who eradicate virus during therapy and preventing relapse [6, 7, 20].

Two pivotal international studies, involving previously untreated CHC patients treated with combination PEG-IFN and ribavirin and standard IFN and ribavirin, have recently been reported. Manns and colleagues [8] evaluated PEG-IFNα-2b (0.5ug/kg or 1.5 μg/kg dosed weekly) plus ribavirin (800 –1200 mg/day) for 12 months compared with standard IFNα2-b and ribavirin. An overall 54% SVR was achieved in the combination PEG-IFN arm (*versus* 47% SVR in the standard IFN and ribavirin arm; $P = 0.01$) with 42% of genotype 1-infected patients and 82% of genotype 2- and 3-infected patients achieving a SVR. This advantage was most beneficial for the often difficult-to-treat genotype 1-infected patient group, where there was a 9% enhanced sustained response rate over the 33% SVR observed in the standard IFN plus ribavirin arm. Additionally, PEG-IFNα-2b at 1.5 μg/kg/week, the current approved dose, was the most effective in this trial. Retrospective analyses also indicated that weight-based dosing of ribavirin incrementally enhanced sustained response rates to 61% in patients receiving more than 10.6 mg/kg of body weight (equivalent to greater than 800 mg/day for a 75-kg person). More recently and in agreement with this analysis, Hadziyannis and colleagues [24] have prospectively demonstrated that higher doses of ribavirin enhance response rates in genotype 1-infected patients. By contrast, genotype 2- and 3-infected patients should be treated with ribavirin at 800 mg/day for 24 weeks, as there is no additional benefit by prolonging therapy or increasing the doses of ribavirin in these patients.

In the second major study, Fried and colleagues [9] evaluated a fixed dose of PEG-IFNα-2a (180 μg/kg dosed weekly) plus ribavirin (1000–1200 mg/daily) for 12 months compared with standard IFNα 2b and ribavirin. Patients in the PEG-IFN arm achieved an overall 56% SVR (*versus* 45% SVR in the standard IFN and ribavirin arm; $P = 0.01$) with 46% of genotype 1-infected patients and 76% of genotype 2- and 3-infected patients attaining a SVR. Genotype 1-infected patients achieved a similar incremental benefit with the PEG-IFN combination of 9% over the standard IFN combination (46 *versus* 37%). Furthermore, the authors were able to evaluate early response rates. The patients in this study who failed to achieve an EVR were eventual non-responders 97% of the time.

Post hoc analyses of both these studies indicate that certain patient baseline characteristics are important in achieving a SVR. Patients with HCV RNA levels of <2 million copies/ml had higher SVR rates than those with higher viral loads (56 *versus* 49% in IFNα–2a/ribavirin groups and 75 *versus* 41% for IFNα–2b/ribavirin groups). Response rates were also lower in patients with advanced fibrosis compared to patients with minimal or no fibrosis (41–44% *versus* 54–55%). However, the negative impact of fibrosis was small compared with previously observed results with standard combination therapy [6, 7]. Advancing age was also shown to negatively impact response. Sustained response rates were highest in patients younger than 35 years compared to patients over 45 years (>60% *versus* <50%). Similarly, pretreatment weight influenced response rates with 55–70% of patients under 75 kg achieving SVR compared with 35–50% of patients over 75 kg. The greatest negative impact on response occurred in patients weighing over 98 kg. Analysis of the influence of race on therapeutic response in the PEG-IFNα-2a and ribavirin study confirmed previous findings of a significant difference in response between white and black patients (54 *versus* 22%) [9].

Although the findings from these two studies indicate similar overall sustained response rates, a direct comparison between the drug regimens is problematic for a number of reasons and results must be interpreted with this in mind. Differences exist in the study designs, doses of PEG-IFN and ribavirin, patient characteristics, management of adverse events, and criteria for dose reduction and early discontinuation. Of particular importance is the difference in baseline demographic characteristics of patients in these two trials, many of which have been shown to influence treatment response. It is probable that although the overall response rates were similar, comparisons of response rates between individual subgroups may demonstrate a significant difference. This underscores the need for a head-to-head comparative trial of the two currently available regimens.

Retreatment of non-responders and relapsers

Preliminary results indicate that non-responders to previous IFN monotherapy achieved a higher SVR than IFN/ribavirin-tested non responders (26 *versus* 11%) [25] when retreated with PEG-IFN and ribavirin. Furthermore, the rate of SVR was significantly higher in genotype 2- and 3-infected patients compared with genotype 1-infected patients (52 *versus* 14%) [25]. Relapsing patients also demonstrate a better response when retreated with combination PEG-IFN therapy. In one ongoing study, 60% of relapsers achieved a SVR when retreated with PEG-IFN and ribavirin [26]. Determining the appropriate treatment for patients who have failed to respond to PEG-IFN and ribavirin continues to be a challenge. For those patients who have minimal hepatic fibrosis (Metavir Fibrosis scores 0 and 1) careful observation and reserving retreatment until more effective therapies are available is a prudent option. For non-responders with more advanced fibrosis (Metavir Fibrosis scores 2, 3 and 4) who are at risk of developing complications, continuing low-dose PEG-IFN as maintenance therapy has been proposed on the basis of an observed histological response in approximately 40% of non-responders with some level of viral-load reduction during treatment [27]. Ongoing, large multicenter trials evaluating the efficacy of these maintenance antifibrotic approaches are underway.

Predictors of a response to antiviral therapy

As stated previously, the factors associated with a favorable response to combination IFN and ribavirin therapy are genotype 2 or 3 infection, lower baseline HCV RNA levels, absence of bridging fibrosis or cirrhosis, younger age, normal weight, absence of co-infection with hepatitis B and/or human immunodeficiency virus (HIV), absence of steatosis and non-black race [6–9, 28–31]. Additionally, factors that favor a response to retreatment include previous relapse rather than primary non-response, a partial virological response during previous treatment and previous treatment with IFN monotherapy [25, 26] (Tab. 2).

The ability to tolerate and comply with therapy is also important, as patients in whom the dose of IFN or ribavirin has to be modified or discontinued are

Table 2. Factors predicting a response to IFN and ribavirin therapy

Factor	Treatment-naïve patients	Patients retreated with antiviral therapy
Age	<40 years	<40 years
Stage of disease	Fibrosis stage 0–1	Fibrosis stage 0–1
Gender	Female	Female
HCV genotype	2 and 3	2 and 3
HCV viral load	<3.5 million copies/ml	<3.5 million copies/ml
Race	Non-black	Non-black
Absence of steatosis		IFN monotherapy
Previous treatment		Partial virological response
Previous on-treatment response		(<100 000 copies/ml)
Previous post-treatment response		Relapse

Adapted from [6, 7, 25, 26].

less likely to achieve a durable response. Genotype 1-infected patients who are able to tolerate 80% of the prescribed drug doses for at least 80% of the duration of treatment have significantly higher SVR values compared with those receiving less therapy [32]. Maintaining the optimal dose of ribavirin appears to be particularly important in preventing relapse [20].

Long-term benefits of antiviral therapy

Those patients who achieve sustained viral eradication have also been shown to have improvements in their health-related quality of life and work productivity [33–35]. SVR is also accompanied by a decrease in hepatic inflammation. If hepatic fibrosis is a result of the host immune response targeted against the virus, then eradication of the virus should lead to a diminution of this inflammatory response, a subsequent reduction in 'new' fibrosis, and the potential reversal of existing fibrosis. In a recent study and secondary analysis, treatment with PEG-IFNα-2b and ribavirin was associated with reducing the rate of fibrosis progression in 73% of patients and the reversal of cirrhosis in 49% of patients with established cirrhosis [36]. This effect was most prominent in patients who achieved SVR. A number of other studies have associated a response to IFN therapy with a decreased prevalence of HCC (3–4% of IFN-treated patients *versus* 12–38% of untreated patients with a follow up of 2–7 years) [37, 38]. Sustained responders also achieve long-term clinical benefits by a reduction in liver-related death [39].

Large-scale, prospective studies of IFN-based therapies have yet to conclusively demonstrate the prevention of cirrhosis, a delayed time to liver transplantation and improvement in long-term morbidity and mortality.

Adverse effects of antiviral therapy

The adverse effects of combination PEG-IFN and ribavirin therapy are well described and are listed in Table 3. The hematological and neuropsychiatric side effects deserve special attention as these events are often responsible for dose reduction, non-compliance and early and late withdrawal of treatment.

Depression, anxiety, acute psychosis, suicidal ideation, confusion, memory loss and in rare cases seizures have been observed in patients receiving IFN-based therapies. The proposed mechanisms of IFN-induced depression include IFN effects on the hypothalamic-pituitary-adrenal axis and a reduction of synaptic serotonin concentrations [40]. The successful use of selective serotonin re-uptake inhibitors (SSRIs) for the treatment of IFN-induced depression lends credence to the involvement of serotonin pathways in this process. Pretreatment with an SSRI significantly diminished the development of major depression and prevented treatment discontinuation in patients receiving high-dose IFN for melanoma [41]. Similar studies have not been performed in

Table 3. Frequency of adverse effects associated with combination IFN and ribavirin therapy

Adverse effect	PEG-IFNα-2a and ribavirin (%)*	PEG-IFNα-2b and ribavirin (%)	IFNα-2b and ribavirin (%)‡
Fatigue	54	64	55–60
Headache	47	62	52–58
Pyrexia	43	46	33–56
Myalgia	42	56	50
Rigors	24	48	35–41
Insomnia	37	40	39–41
Nausea	29	43	33
Alopecia	28	36	32–34
Irritability	24	35	28
Arthralgia	27	34	25–27
Anorexia/weight loss	21	32/29	20–22
Dermatitis	21	31	18
Depression	22	58	30–34
Injection-site reaction		58	36
Injection-site inflammation		25	18

Uncommon serious adverse events of IFN and ribavirin

Acute psychosis	Photosensitivity	Autoimmune disease
Severe depression	Severe skin rash	Hypothyroidism
Suicidal ideation	Local abscess formation	Hyperthyroidism
Confusion	Acute renal failure	Type I diabetes mellitus
Memory loss	Interstitial pneumonitis	Celiac disease
Retinopathy	Stroke	Hemolytic anemia
Neuropathy	Cardiac arrhythmias	Thrombocytopenic purpura
Seizures	Ischemic heart disease	Addison's disease
		Myasthenia gravis
		Lupus-like syndrome
		Autoimmune hepatitis

Adapted from product information: Pegasys™ (PEG-IFNα-2a), Hoffmann-La Roche Inc, Nutley, NJ, USA (2002) and PEG-Intron™ (PEG-IFNα-2b), Schering Corporation, Kenilworth, NJ, USA (2001). *Adapted from [9]. Adapted from [8]. ‡Adapted from [8, 9].

patients with CHC. Depressive symptoms can also be effectively treated with SSRIs during IFN and ribavirin therapy once depression has developed [42, 43]. More recently, the addition of a noradrenergic agent to an SSRI has been shown to improve both the rate and depth of response and is more effective than SSRIs alone in the acute remission of depression [44, 45]. The value of this two-drug regimen in HCV-infected patients needs to be studied further. At the more severe end of the spectrum, suicidal behaviour remains a significant concern, albeit a rare one. Patients should be adequately screened prior to the commencement of and regularly while on therapy to determine who is at most risk from the development of neuropsychiatric effects. While the majority of patients may be able to continue antiviral therapy with the aid of antidepressants or anxiolytics, antiviral therapy may need to be terminated in patients with severe symptoms. Patients who have a history of severe psychiatric disturbance should also be counselled against receiving antiviral therapy.

The administration of both standard and PEG-IFNs also causes inhibition of progenitor proliferation in the bone marrow of hematopoietic lineages resulting in anemia, neutropenia and thrombocytopenia, which are not completely counteracted by the production of endogenous hemopoietic growth factors [46]. Additionally, ribavirin causes a dose-dependent, reversible hemolytic anemia. The sequestration of ribavirin metabolites by erythrocytes and the associated depletion of adenosine triphosphate in red blood cells result in membrane oxidative damage and extravascular hemolysis [47]. Other proposed mechanisms of ribavirin-induced anemia include marrow erythroid hypoplasia, histiocytic phagocytosis and vacuolization of erythroid precursors [48].

Hemoglobin levels fall within the first 2–4 weeks of PEG-IFN and ribavirin therapy, with a mean maximal decrease of 3 g/dl, resulting in the need for dose reductions in 9% of patients [6–8]. Ribavirin dose reduction is suggested at hemoglobin levels between 8.5 g/dl and 10 g/dl and cessation of therapy at levels less than 8.5 g/dl. Recently, interest has focused on the use of recombinant human erythropoietin (r-HuEPO) as an alternative to ribavirin dose alteration. Dieterich and colleagues [49] have reported higher mean hemoglobin levels in IFN- and ribavirin-treated patients who received r-HuEPO compared with patients who underwent ribavirin dose reduction (13.8 g/dl *versus* 11.4 g/dl; $P < 0.0001$) with fewer ribavirin dose reductions in the r-HuEPO arm. It is yet to be determined if the maintenance of ribavirin doses improves both early and sustained response rates. Alternatives to ribavirin that have similar immunomodulatory effects but potentially lower incidences and severity of hemolytic anemia such as levovirin (an L-isomer of ribavirin) and viramidine (a carboxamide prodrug converted to ribavirin in the liver) are also under investigation in combination with PEG-IFNs [50].

PegIFN-associated neutropenia (neutrophil levels decrease on average to approximately 45% of baseline) is seen as early as 2 weeks after the commencement of therapy. However, at this stage, there appears to be no association between absolute neutrophil count and serious bacterial infections in these patients receiving combination therapy [9, 51]. IFN dose reduction is generally recommended at neutrophil counts less than 500–750 cells/mm^3. Granulocyte colony-stimulating factor (GCSF) has been used successfully in the treatment of neutropenia associated with cancer chemotherapy. However, there are limited data available regarding its use in IFN-induced neutropenia in CHC and its efficacy, with respect to preventing bacterial infections, improving therapeutic adherence or enhancing SVR, is largely unknown.

IFN-induced thrombocytopenia is a result of bone marrow suppression and is commonly seen in cirrhotic patients secondary to hypersplenism. Patients treated with PEG-IFN and ribavirin experience a 30% decrease in median platelet count; however, this is rarely associated with bleeding [8]. Most patients can be managed by IFN dose reduction at platelet counts below 50 000/mm^3 with cessation of therapy at counts below 25 000/mm^3. Thrombopoietic growth factors such as interleukin 11 have been approved for use in cancer chemotherapy-associated thrombocytopenia. However, this medication is associated with numerous serious cardiovascular and pulmonary side effects and limited data are available regarding its use in CHC [52].

Limitations of antiviral therapy

The use of IFN-based therapies is complicated by significant side effects which can be problematic in certain patients groups including those with advanced liver disease, those who have not responded to currently available therapies and those in whom therapy is contraindicated.

Patients with cirrhosis, those most in need of therapy, are at greatest risk from developing HCV-related complications. When treated with a combination of PEG-IFN and ribavirin, sustained response rates of approximately 40% were reported in these patients. However, therapy can be complicated by cytopenias and the benefits of treatment need to be weighed against risks [22]. For all these reasons, treatment of patients with decompensated cirrhosis is complex. The efficacy of therapy in advanced liver disease is considerably less than in compensated cirrhosis [53, 54]. Lower doses of IFN and ribavirin have been used in an attempt to offset these hematological adverse events, but the rates of achieving a SVR will more than likely be reduced.

Treatment recommendations derive largely from trials that have selected specific patient populations in which patients with co-morbidities were excluded and in which there was often a low representation of ethnic minorities. Consequently, there are limited data in populations including those with advanced and decompensated liver disease, children, the elderly, ethnic minorities, the incarcerated and in patients with ongoing or recent alcohol and substance abuse, renal disease, immunosuppression, severe psychiatric or neurologic illness, autoimmune disorders and solid organ transplant. Efforts should focus on including these groups in future studies so that there can be a better understanding of drug safety and efficacy in these and other populations.

Other interferons

In an attempt to improve therapeutic efficacy, drug development has more recently focused on the modification of existing IFN compounds.

Modified IFN compounds

Consensus IFNs (CIFNs) such as IFN-alfacon 1 are cytokines engineered to contain the specific amino acid sequences found in other IFNα subtypes. Randomized control trials of CIFNs have yielded variable results. At high doses, CIFN is associated with a higher SVR compared with standard IFN [55], but at equivalent doses both IFNs were equally effective [56]. Other IFNs have been produced by 'gene shuffling' of IFNα DNA encoding sequences resulting in markedly increased antiviral activity *in vitro*, though they have yet to be tested in human populations. Recently, interleukin-28 and interleukin-29, a newly discovered family of novel class II cytokines distantly related to IFNα and interleukin 10 have been evaluated in a bovine diarrhoeal virus (BVDV) model (surrogate viral system for HCV) [57]. Interleukins 28 and 29 demonstrate inhibition of BVDV-dependent cell death and may potentially have therapeutic value against HCV in human patients. Additionally, fusion products such as IFN and serum albumin (Albuferon) have been developed, resulting in molecules with superior pharmacokinetics and pharmacodynamics when compared to standard IFN *in vitro* [58, 59]. Other IFNs such as IFNω and IFNγ have been shown to inhibit protein synthesis and RNA replication in HCV replicon systems and are also being evaluated in humans with CHC [60]. IFNγ also has antifibrotic properties and is currently in phase II trials in chronic hepatitis C patients with advanced fibrosis. As a means of improving the pharmacokinetic profile of IFNs, novel delivery systems including disposable infusion pumps, controlled-release formulations for parenteral use and liposome encapsulation are presently under development. While many other IFNs are being developed, the safety and efficacy of these agents in combination with ribavirin and how they compare to the current standard of care requires evaluation.

Immunomodulatory agents

Oral IFN inducers

Oral IFN inducers are compounds that may affect the immune response by inducing or modulating the release of host cytokines. Oral agents including the imidazoquinolones imiquimod and resimiquod, and nucleoside analogs ANA245 and ANA971, induce the innate IFN response and are currently being evaluated in patients with CHC [61]. ANA245 induces numerous cytokines in addition to IFNα and has been shown to have activity against a broad range of viruses. However, reduced bioavailability has led to the development of an intravenous form that is currently in early clinical trials. ANA971 is a novel molecule that efficiently delivers ANA245 after oral administration in animal models.

IMPDH inhibitors

In addition to levovirin and viramidine, other IMPDH inhibitors include VX-497, an orally bioavailable agent that has an additive inhibitory effect on viral replication when combined with IFN in the HCV replicon system and in humans [62], and mycophenylate mofetil (MMF), commonly used as an immunosuppressant following organ transplantation. Although ineffective when combined with standard IFNα in achieving a SVR, mycophenylate mofetil (MMF) and VX-497 are currently under evaluation in combination with PEGylated IFN [63].

Other immunomodulators

Thymosin α 1, a synthetic nonglycosylated peptide derived from thymus-gland extracts, stimulates the production of cytokines including IFNγ and interleukins 2 and 3 and promotes T cell maturation and natural killer cell activity. When combined with IFNα it is well tolerated and is currently under investigation in combination with PEG-IFN in phase II/III studies of cirrhotic and noncirrhotic patients who have not responded to previous therapy [64]. Histamine dihydrochloride acts via H_2-type histamine receptors to inhibit the production of oxygen radicals by phagocytes, and therefore protects natural killer cells and T cells from functional inhibition and apoptosis. This mechanism may be important in enhancing the immune response when histamine is combined with IFN. Patients treated with histamine dihydrochloride and IFNα-2b achieved an overall 31–38% SVR rate with 18–42% SVR rate in genotype 1-infected patients [65]. The drug was well tolerated and is currently being evaluated in combination with PEGylated IFN and ribavirin in phase II studies. Amantadine is an antiviral agent that has been primarily used for the treatment of influenza A virus. Amantadine monotherapy has little effect on HCV viral levels, but trials of IFN, ribavirin and amantadine have yielded mixed results with significant relapse rates [66, 67].

The efficacy and safety of these 'modified' IFNs and immonomodulators in HCV patients when given alone or combined with ribavirin must now be established through carefully controlled clinical trials.

Conclusion

Significant advances have been made in developing antiviral therapies for HCV infection. However, a number of challenges remain for the development of more effective, cost-saving, simplified and safer drug therapies. Patients at high risk of developing progressive disease and associated complications especially require improved therapeutic options. Additionally, groups who have previously been excluded should have the opportunity to participate in future therapeutic trials. With the optimization of IFN-based regimens for different HCV populations currently ongoing and a number of promising therapies in clinical trials, it is anticipated that more-effective combination regimens will be available in future years.

References

1 Choo QL, Kuo G, Weiner AJ, Overby LR, Bradley DW, Houghton M (1989) Isolation of a cDNA clone derived from a blood-borne non-A, non-B viral hepatitis genome. *Science* 244: 359–362
2 Choo QL, Richman KH, Han JH, Berger K, Lee C, Dong C, Gallegos C, Coit D, Medina-Selby R, Barr PJ et al. (1991) Genetic organization and diversity of the hepatitis C virus. *Proc Natl Acad Sci USA* 88: 2451–2455
3 Hoofnagle JH, Mullen KD, Jones DB, Rustgi V, Di Bisceglie A, Peters M, Waggoner JG, Park Y, Jones EA (1986) Treatment of chronic non-A, non-B hepatitis with recombinant human alpha interferon. A preliminary report. *N Engl J Med* 315(25): 1575–1578
4 Di Bisceglie AM, Martin P, Kassianides C, Lisker-Melman M, Murray L, Waggoner J, Goodman Z, Banks SM, Hoofnagle JH (1989) Recombinant interferon alfa therapy for chronic hepatitis C. A randomized, double-blind, placebo-controlled trial. *N Engl J Med* 321(22): 1506–1510
5 Poynard T, Leroy V, Cohard M, Thevenot T, Mathurin P, Opolon P, Zarski JP (1996) Meta-analysis of interferon randomized trials in the treatment of viral hepatitis C: effects of dose and duration. *Hepatology* 24(4): 778–789
6 McHutchison JG, Gordon SC, Schiff ER, Shiffman ML, Lee WM, Rustgi VK, Goodman ZD, Ling MH, Cort S, Albrecht JK (1998) Interferon alfa-2b alone or in combination with ribavirin as initial treatment for chronic hepatitis C. *N Engl J Med* 339(21): 1485–1492
7 Poynard T, Marcellin P, Lee SS, Niederau C, Minuk GS, Ideo G, Bain V, Heathcote J, Zeuzem S, Trepo C, Albrecht J (1998) Randomised trial of interferon alpha-2b plus ribavirin for 48 weeks or for 24 weeks *versus* interferon alpha-2b plus placebo for 48 weeks for treatment of chronic infection with hepatitis C virus. *Lancet* 352: 1426–1432
8 Manns MP, McHutchison JG, Gordon SC, Rustgi VK, Shiffman M, Reindollar R, Goodman ZD, Koury K, Ling M, Albrecht JK (2001) Peginterferon alfa-2b plus ribavirin compared with interferon alfa-2b plus ribavirin for initial treatment of chronic hepatitis C: a randomised trial. *Lancet* 358: 958–965
9 Fried MW, Shiffman ML, Reddy KR, Smith C, Marinos G, Goncales FL Jr, Haussinger D, Diago M, Carosi G, Dhumeaux D et al. (2002) Peginterferon alfa-2a plus ribavirin for chronic hepatitis C virus infection. *N Engl J Med* 347(13): 975–982
10 Zein NN (1998) Interferons in the management of viral hepatitis. *Cytokines Cell Mol Ther* 4(4): 229–241
11 Grossberg SE, Taylor JL, Kushnaryov VM (1989) Interferon receptors and their role in interferon action. *Experientia* 45(6): 508–513
12 Glue P, Fang JW (2000) Pegylated interferon alfa-2b: Pharmacokinetics, Pharmacodynamics, safety and preliminary efficacy data. *Clin Pharmacol Ther* 68: 556–567
13 Hadziyannis SJ, Papatheodoridis GV (2003) Peginterferon-alpha2a (40 kDa) for chronic hepatitis C. *Expert Opin Pharmacother* 4(4): 541–551
14 Lau JY, Tam RC, Liang TJ, Hong Z (2002) Mechanism of action of ribavirin in the combination treatment of chronic HCV infection. *Hepatology* 35: 1002–1009
15 Zhang Y, Jamaluddin M, Wang S, Tian B, Garofalo RP, Casola A, Brasier AR (2003) Ribavirin treatment up-regulates antiviral gene expression via the interferon-stimulated response element in respiratory syncytial virus-infected epithelial cells. *J Virol* 77(10): 5933–5947
16 United States National Institutes of Health Consensus Development Conference Panel Statement (1997) Management of hepatitis C. *Hepatology* 26 (suppl 1): 2S–10S
17 Marcellin P, Boyer N, Gervais A, Martinot M, Pouteau M, Castelnau C, Kilani A, Areias J, Auperin A, Benhamou JP et al. (1997) Long-term histologic improvement and loss of detectable intrahepatic HCV RNA in patients with chronic hepatitis C and sustained response to interferon-alpha therapy. *Ann Intern Med* 127(10): 875–881
18 Davis GL, Wong JB, McHutchison JG, Manns MP, Harvey J, Albrecht J (2003) Early virological

response to treatment with pegylated interferon alfa-2b plus ribavirin in patients with chronic hepatitis C. *Hepatology* 38(3): 645–652
19 McHutchison JG, Davis GL, Esteban-Mur R (2001) Durability of sustained viral response in patients with chronic hepatitis C after treatment with interferon alfa2-b alone or in combination with ribavirin. *Hepatology* 34: 244A
20 Davis GL, Esteban-Mur R, Rustgi V, Hoefs J, Gordon SC, Trepo C, Shiffman ML, Zeuzem S, Craxi A, Ling MH, Albrecht J (1998) Recombinant interferon alfa-2b alone or in combination with ribavirin for retreatment of interferon relapse in chronic hepatitis C. *N Engl J Med* 339: 1493–1496
21 Lindsay KL, Trepo C, Heintges T, Shiffman ML, Gordon SC, Hoefs JC, Schiff ER, Goodman ZD, Laughlin M, Yao R, Albrecht JK for the Hepatitis Interventional Therapy group (2001) A randomized, double-blind trial comparing pegylated interferon alfa-2b to interferon alfa-2b as initial treatment for chronic hepatitis C. *Hepatology* 34: 395–403
22 Heathcote EJ, Shiffman ML, Cooksley WG, Dusheiko GM, Lee SS, Balart L, Reindollar R, Reddy RK, Wright TL, Lin A et al. (2000) Peginterferon alfa-2a in patients with chronic hepatitis C and cirrhosis. *N Engl J Med* 343: 1673–1680
23 Zeuzem S, Feinman SV, Rasenack J, Heathcote EJ, Lai MY, Gane E, O'Grady J, Reichen J, Diago M, Lin A et al. (2000) Peginterferon alfa-2a in patients with chronic hepatitis C. *N Engl J Med* 343: 1666–1672
24 Hadziyannis SJ, Sette H Jr, Morgan TR, Balan V, Diago M, Marcellini P, Ramadori G, Bodenheimer H Jr, Bernstein D, Rizzetto M et al. (2004) Peginterferon-α-2a and ribavirin combination therapy in chronic hepatitis C. A randomized study of treatment duration and ribavirin dose. *Ann Intern Med* 140: 346–355
25 Jacobson I (2002) Pegylated interferon alfa-2b plus ribavirin in patients with chronic C: Atrial in prior nonresponders to interferon monotherapy or combination therapy and in combination therapy nonresponders. *Gastroenterology* 122: A626
26 Shiffman ML (2002) Retreatment of HCV non-responders with peginterferon and ribavirin: results from the lead-in phase of the hepatitis C antiviral long term treatment against cirrhosis (HALT-C) trial. *Hepatology* 36: 295A
27 Shiffman ML, Hofmann CM, Thompson EB, Ferreira-Gonzalez A, Contos MJ, Koshy A, Luketic VA, Sanyal AJ, Mills AS, Garrett CT (1997) Relationship between biochemical, virological, and histological response during interferon treatment of chronic hepatitis C. *Hepatology* 26(3): 780–785
28 Perez-Olmeda M, Nunez M, Romero M, Gonzalez J, Castro A, Arribas JR, Pedreira J, Barreiro P, Garcia-Samaniego J, Martin-Carbonero L et al. (2003) Pegylated IFN-alpha2b plus ribavirin as therapy for chronic hepatitis C in HIV-infected patients. *AIDS* 17(7): 1023–1028
29 Villa E, Grottola A, Buttafoco P, Colantoni A, Bagni A, Ferretti I, Cremonini C, Bertani H, Manenti F (2001) High doses of alpha-interferon are required in chronic hepatitis due to coinfection with hepatitis B virus and hepatitis C virus: long term results of a prospective randomized trial. *Am J Gastroenterol* 96(10): 2973–2977
30 Reddy KR, Hoofnagle JH, Tong MJ, Lee WM, Pockros P, Heathcote EJ, Albert D, Joh T (1999) Racial differences in responses to therapy with interferon in chronic hepatitis C. Consensus Interferon Study Group. *Hepatology* 30(3): 787–793
31 McHutchison J, Poynard T, Pianko S, Gordon S, Reid A, Dienstag J, Morgan T, Yao R, Albrecht J (2000) The impact of interferon plus ribavirin on response to therapy in black patients with chronic hepatitis C. *Gastroenterology* 119: 1317–1323
32 McHutchison JG, Manns M, Patel K, Poynard T, Lindsay KL, Trepo C, Dienstag J, Lee WM, Mak C, Garaud JJ et al. (2002) Adherence to combination therapy enhances sustained response in genotype-1 infected patients with chronic hepatitis C. *Gastroenterology* 123(4): 1061–1069
33 Bonkovsky HL, Woolley JM (1999) Reduction of health-related quality of life in chronic hepatitis C and improvement with interferon therapy. The Consensus Interferon Study Group. *Hepatology* 29(1): 264–270
34 Ware JE Jr, Bayliss MS, Mannocchia M, Davis GL (1999) Health-related quality of life in chronic hepatitis C: impact of disease and treatment response. The Interventional Therapy Group. *Hepatology* 30(2): 550–555
35 McHutchison JG, Ware JE Jr, Bayliss MS, Pianko S, Albrecht JK, Cort S, Yang I, Neary MP (2001) Hepatitis Interventional Therapy Group. The effects of interferon alpha-2b in combination with ribavirin on health related quality of life and work productivity. *J Hepatol* 34(1): 140–147

36 Poynard T, McHutchison J, Manns M, Trepo C, Lindsay K, Goodman Z, Ling MH, Albrecht J (2002) Impact of pegylated interferon alfa-2b and ribavirin on liver fibrosis in patients with chronic hepatitis C. *Gastroenterology* 122(5): 1303–1313
37 Nishiguchi S, Kuroki T, Nakatani S, Morimoto H, Takeda T, Nakajima S, Shiomi S, Seki S, Kobayashi K, Otani S (1995) Randomised trial of effects of interferon-alpha on incidence of hepatocellular carcinoma in chronic active hepatitis C with cirrhosis. *Lancet* 346(8982): 1051–1055
38 Yoshida H, Shiratori Y, Moriyama M, Arakawa Y, Ide T, Sata M, Inoue O, Yano M, Tanaka M, Fujiyama S (1999) Interferon therapy reduces the risk for hepatocellular carcinoma: national surveillance program of cirrhotic and noncirrhotic patients with chronic hepatitis C in Japan. IHIT Study Group. Inhibition of Hepatocarcinogenesis by Interferon Therapy. *Ann Intern Med* 131(3): 174–181
39 Yoshida H, Arakawa Y, Sata M, Nishiguchi S, Yano M, Fujiyama S, Yamada G, Yokosuka O, Shiratori Y, Omata M (2002) Interferon therapy prolonged life expectancy among chronic hepatitis C patients. *Gastroenterology* 123(2): 483–491
40 Trask PC, Esper P, Riba M, Redman B (2000) Psychiatric side effects of interferon therapy: prevalence, proposed mechanisms, and future directions. *J Clin Oncol* 18(11): 2316–2326
41 Musselman DL, Lawson DH, Gumnick JF, Manatunga AK, Penna S, Goodkin RS, Greiner K, Nemeroff CB, Miller AH (2001) Paroxetine for the prevention of depression induced by high-dose interferon alfa. *N Engl J Med* 344(13): 961–966
42 Hauser P, Khosla J, Aurora H, Laurin J, Kling MA, Hill J, Gulati M, Thornton AJ, Schultz RL, Valentine AD et al. (2002) A prospective study of the incidence and open-label treatment of interferon-induced major depressive disorder in patients with hepatitis C. *Mol Psychiatry* 7(9): 942–947
43 Kraus MR, Schafer A, Faller H, Csef H, Scheurlen M (2002) Paroxetine for the treatment of interferon-alpha-induced depression in chronic hepatitis C. *Aliment Pharmacol Ther* 16(6): 1091–1099
44 DeBattista C, Solvason HB, Poirier J, Kendrick E, Schatzberg AF (2003) A prospective trial of bupropion SR augmentation of partial and non-responders to serotonergic antidepressants. *J. Clin Psychopharmacol* 23(1): 27–30
45 Thase ME, Entsuah AR, Rudolph RL (2001) Remission rates during treatment with venlafaxine or selective serotonin reuptake inhibitors. *Br J Psychiatry* 178: 234–241
46 Peck-Radosavljevic M, Wichlas M, Homoncik-Kraml M, Kreil A, Hofer H, Jessner W, Gangl A, Ferenci P (2002) Rapid suppression of hematopoiesis by standard or pegylated interferon-alpha. *Gastroenterology* 123(1): 141–151
47 De Franceschi L, Fattovich G, Turrini F, Ayi K, Brugnara C, Manzato F, Noventa F, Stanzial AM, Solero P, Corrocher R (2000) Hemolytic anemia induced by ribavirin therapy in patients with chronic hepatitis C virus infection: role of membrane oxidative damage. *Hepatology* 31 (4): 997–1004
48 Cosgriff TM, Hodgson LA, Canonico PG, White JD, Kastello MD, Donovan JC, Ross PE (1984) Morphological alterations in blood and bone marrow of ribavirin-treated monkeys. *Acta Haematol* 72(3): 195–200
49 Dieterich DT, Wasserman R, Brau N, Hassanein TI, Bini EJ, Bowers PJ, Sulkowski MS (2003) Once-weekly epoetin alfa improves anemia and facilitates maintenance of ribavirin dosing in hepatitis C virus-infected patients receiving ribavirin plus interferon alfa. *Am J Gastroenterol* 98(11): 2491–2499
50 Watson J (2002) Prospects for hepatitis c virus therapeutics: levovirin and viramidine improved derivatives of ribavirin. *Curr Opin Investing Drugs* 3: 680–683
51 Soza A, Everhart JE, Ghany MG, Doo E, Heller T, Promrat K, Park Y, Liang TJ, Hoofnagle JH (2002) Neutropenia during combination therapy of interferon alfa and ribavirin for chronic hepatitis C. *Hepatology* 36(5): 1273–1279
52 Artz AS, Ershler WB, Rustgi V (2001) Interleukin-11 for thrombocytopenia associated with hepatitis C. *J Clin Gastroenterol* 33(5): 425–426
53 Crippin JS, Sheiner P, Terrault NA, McCashland T, Charlton M (2002) A pilot study of the tolerability and efficacy of antiviral therapy in patients awaiting liver transplantation for hepatitis C. *Liver Transpl* 8: 350–355
54 Samuel D, Bizollon T, Feray C, Roche B, Ahmed SN, Lemonnier C, Cohard M, Reynes M, Chevallier M, Ducerf C et al. (2003) Interferon-alpha 2b plus ribavirin in patients with chronic hepatitis C after liver transplantation: a randomized study. *Gastroenterology* 124(3): 642–650

55 Tong MJ, Blatt LM, Resser KJ, Klein MC, Cruickshank SE, Figueroa T, Sayadzadeh K (1998) Treatment of chronic hepatitis C virus infection with recombinant consensus interferon. *J Interferon Cytokine Res* 18(2): 81–86
56 Haydon GH, Mutimer DJ (2002) Genetically modified interferon: is there a consensus yet? *Eur. J Gastroenterol Hepatol* 14(5): 471–473
57 Nelson M (2003) Proceedings from Hep DART, Frontiers in Drug Development for Viral Hepatitis, 14–18 December 2003. Kauai, Hawaii
58 Chang CC, Chen TT, Cox BW, Dawes GN, Stemmer WP, Punnonen J, Patten PA (1999) Evolution of a cytokine using DNA family shuffling. *Nat Biotechnol* 17(8): 793–797
59 Osborn BL, Olsen HS, Nardelli B, Murray JH, Zhou JX, Garcia A, Moody G, Zaritskaya LS, Sung C (2002) Pharmacokinetic and pharmacodynamic studies of a human serum albumin-interferon-alpha fusion protein in cynomolgus monkeys. *J Pharmacol Exp Ther* 303(2): 540–548
60 Frese M, Schwarzle V, Barth K, Krieger N, Lohmann V, Mihm S, Haller O, Bartenschlager R (2002) Interferon-gamma inhibits replication of subgenomic and genomic hepatitis C virus RNAs. *Hepatology* 35(3): 694–703
61 Tompkins WA (1999) Immunomodulation and therapeutic effects of the oral use of interferon-alpha: mechanism of action. *J Interferon Cytokine Res* 19(8): 817–828
62 Mc Hutchison JG, Cheung R, Shiffman ML (2001) A 4-week trial of VX 497 (an IMPDH inhibitor) combined with interferon in previously untreated patients with chronic hepatitis C. *Hepatology* 34 (Suppl): 329A
63 Afdhal N, Flamm S, Imperial JC (2001) Analyses of 40 KDA peginterferon alfa-2a in combination with ribavirin, mycophenolate mofetil, amantadine or amantadine plus ribavirin in patients that relapsed or did not respond to Rebetron therapy: a report of two randomized, multicenter, efficacy and safety studies. *Hepatology* 34 (Suppl.): 243A
64 Pockros PJ (2002) Developments in treatment of chronic hepatitis C. *Expert Opin Invest Drugs* 11: 1–14
65 Lurie Y, Nevens F, Aprosina ZG, Fedorova TA, Kalinin AV, Klimova EA, Ilan Y, Maevskaya MV, Warnes TW, Yuschuk ND et al. (2002) A multicentre, randomized study to evaluate the safety and efficacy of histamine dihydrochloride and interferon-alpha-2b for the treatment of chronic hepatitis C. *J Viral Hepat* 9(5): 346–353
66 Berg T, Kronenberger B, Hinrichsen H, Gerlach T, Buggisch P, Herrmann E, Spengler U, Goeser T, Nasser S, Wursthorn K et al. (2003) Triple therapy with amantadine in treatment-naive patients with chronic hepatitis C: a placebo-controlled trial. *Hepatology* 37(6): 1359–1367
67 Thuluvath PJ, Pande H, Maygers J (2003) Combination therapy with interferon-alpha(2b), ribavirin, and amantadine in chronic hepatitis C nonresponders to interferon and ribavirin. *Dig Dis Sci* 48(3): 594–597

Hepatitis Prevention and Treatment
Edited by J.M. Colacino and B.A. Heinz

Viral response to therapy: viral dynamics

Craig S. Gibbs

Gilead Sciences, 333 Lakeside Drive, Foster City, CA 94404, USA

Introduction: validation of viral markers as measures of response and emergence of the field of viral dynamics

The response to therapy of chronic viral hepatitis B and C can be measured by a variety of parameters; virological (plasma levels of viral nucleic acids and antigens), biochemical (plasma markers of liver injury, function or fibrosis such as alanine aminotransferase (ALT), aspartate aminotransferase (AST), albumin, bilirubin, prothrombin time, platelets, α2-macroglobulin, haptoglobin, apolipoprotein A2, γ-glutamyltransferase), immunological (plasma levels of antibodies against viral antigens or the magnitude and breadth of the viral-specific T cell response), histological (Knodell and Ishak scores of liver inflammation, necrosis and fibrosis) and clinical (Childs–Pugh score, hepatocellular carcinoma, survival). Although all these parameters are of value, they all have relative strengths and weakness with regard to sensitivity, dynamic range, specificity and accessibility.

In recent years large controlled clinical trials of increasingly effective antiviral therapies for chronic hepatitis B [1–3] and hepatitis C [4, 5] have allowed correlations to be established between the parameters used to measure efficacy. The preponderance of data has solidified the thesis that active viral replication is the underlying cause of disease, despite the recognition that a large degree of the pathogenesis associated with chronic hepatitis B and C is mediated through the immune response. Specifically, repression of viral replication by antiviral therapy has been shown to precede improvements in the other markers of disease and to be necessary but not necessarily sufficient to achieve sustained disease remission following cessation of therapy. The major remaining caveat to the statement above is that with a few exceptions most of the studies have been conducted over periods of a few months or years; relatively short periods in the lifetime of a patient with a chronic viral infection. Data are only just emerging to indicate that these observed remissions result in improved long-term outcomes such as survival and incidence of hepatocellular carcinoma.

The increasing acceptance of viral markers as surrogate markers for disease has also been driven by recognition of the exquisite specificity and sensitivity of PCR-based assays for hepatitis B virus (HBV) DNA and hepatitis C virus (HCV) RNA in plasma or serum [6, 7]. Currently available commercial assays for HCV RNA and HBV DNA have lower limits of quantification of 25–50 i.u./ml and 100 copies/ml, respectively. The dynamic range of these assays is as high as 10^5 and can be extended even further by dilution of samples to extend the upper limit of quantification. Furthermore, since serum or plasma samples can be obtained relatively non-invasively by phlebotomy, it is possible to obtain frequent samples allowing close monitoring of patients over time. Consequently there has been motivation to maximize the information that can be obtained using the new assay technology for viral nucleic acids and the readily available plasma samples and this has lead to the burgeoning field of viral dynamics: the study of the changes in plasma viral load over time.

Viral dynamics

The objective of viral dynamics analysis is to maximize the informational value of the most sensitive, dynamic and convenient parameter: plasma viral load. The approach was pioneered in the analysis of the viral response to antiretroviral therapy of human immunodeficiency virus (HIV) infection [8, 9] and has been readily applied to both HBV [10, 11] and HCV [12, 13]. The basic assumption of viral dynamics is that during a chronic viral infection the total viral load in the body, as reflected in the plasma levels of virus, reaches a steady state that is driven by a balance between the rates of viral production and viral clearance. These are in turn dependent on the rates of infection of uninfected cells and clearance or curing of infected cells. During chronic infection with HBV or HCV this steady state can persist for relatively long periods of time. The recent advent of potent antiviral agents provides a means to perturb this steady state and analyze the resulting changes in viral load over time. Two broad categories of analysis have emerged. In the first, mathematical models based upon a basic understanding of the biology of the virus and its interaction with the host have been deployed to calculate numerical values for the fundamental parameters that govern the steady state and the changes observed after initiation of therapy. These parameters include the rate constants of viral production and clearance, the rate constants of infection and clearance or curing of infected cells and the efficacy of the antiviral agent. This approach requires deriving a mathematical solution to a series of differential equations and fitting the solution to the observed data by non-linear regression. The second category of analysis is based on the recognition that the viral-load changes observed are not necessarily uniform over extended periods of time and can display quite complex patterns. Furthermore quite different patterns can be observed in individual patients treated with the same agent. This recognition has lead to analyses where pattern-recognition algorithms have been employed to categorize patients with similar patterns of viral clearance. In both categories of analysis, the next step has been to attempt to identify correlations between the observed parameters or patterns with preclinical observations, baseline patient characteristics and measures of clinical response. Such analyses have provided additional insight into the life cycle of the viruses, the natural history of infection, the mechanism of antiviral agents, the interaction between the virus and host and may provide insight into the factors affecting the emergence of antiviral resistance. Furthermore such analyses have provided a more sophisticated means for comparing different drugs, doses, dosing regimens and combinations and may permit better definition of the predictors of sustained response to therapy.

HBV dynamics

The basic biphasic model of HBV dynamics

The basic model for HBV dynamics, first enumerated by Nowak et al. [14], has three variables; the number of virus particles (v), the number of uninfected cells (x) and the number of infected cells (z). Uninfected cells are assumed to be generated at a constant rate (γ) and they die at a constant per cell-rate (δx), which is thought to be quite slow ($t_{1/2}$, approximately 6 months). Uninfected cells are infected at a rate (bvx) that depends on both the concentration of uninfected cells and free virus and are cleared regardless of mechanism [curing of covalently closed circular DNA (cccDNA) from cells or apoptotic or cytolytic death of infected cells] at a per-cell rate (ay). Free viruses are produced from infected cells at a rate (ky) and are cleared at a rate (uv) through the reticuloendothelial system, via immune complex formation with antibodies or via phagocytosis. Schematically the model is represented in Figure 1. Mathematically the model can be represented as the following series of differential equations.

$$dx/dt = \lambda - \delta x - bvx$$

Change in uninfected cells = generation rate – death rate – infection rate

$$dy/dt = bvx - ay$$

Change in infected cells = infection rate – clearance rate

$$dv/dt = ky - uv$$

Change in virus = production rate – clearance rate

In patients with chronic hepatitis B, a pretherapy steady state is assumed where viral production equals viral clearance ($ky = uv$). When antiviral therapy with nucleoside analogues, whose presumed mechanism is inhibition of viral DNA production by the HBV polymerase, is introduced a biphasic decline in HBV viral load is observed over the first few weeks of therapy with

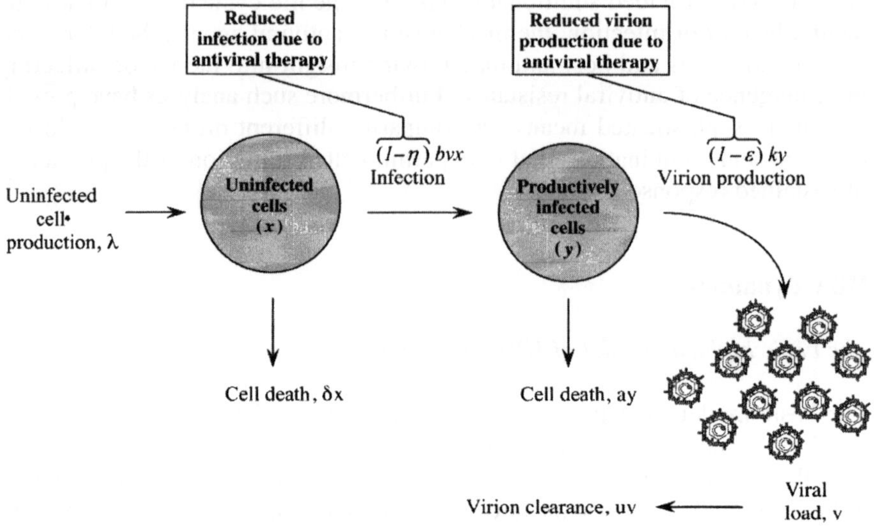

Figure 1. Schematic representation of the basic model of viral dynamics. Adapted from [11].

an initial steep decline of greater than 2 log observed during the first week. This transitions to a second slower phase of decline in subsequent weeks that varies considerably in slope from patient to patient Figure 2. This biphasic decline was first analyzed by Tsiang et al. [15] in HBV patients treated in a phase 2 study of adefovir dipivoxil and was accommodated in the model by providing a term for efficacy (ε) for partial inhibition of viral production during therapy. By making the simplifying assumption that during antiviral therapy infection of new cells would be zero ($bvx = 0$), the system of differential equations was solved and the resulting function was fitted to the viral-decay curves by non-linear regression in order to derive estimates of the various parameters. In this analysis the biphasic pattern of viral decline was explained by identifying the first phase as the clearance of viral particles present in plasma prior to the initiation of therapy with a rapid half-life of 1.1 days and by identifying the second phase as the clearance of viral particles produced during therapy due to incomplete inhibition of viral production. Therefore the decrease in viral load observed during therapy reflects the clearance of infected cells with a mean half-life of 18 days but with considerable variation observed between patients ranging from 11 to 30 days.

Figure 2. Biphasic clearance of HBV DNA from the serum of 10 different patients treated with adefovir dipivoxil, 30 mg daily for 12 weeks. Serum HBV DNA was determined by Roche Amplicor assay at weeks 0, 1, 2, 4, 8, 10 and 12. The solid lines represent the best fit of the equation derived from solving the three differential equations that describe the basic model of viral dynamics to the individual patient data (●). Adapted from [15].

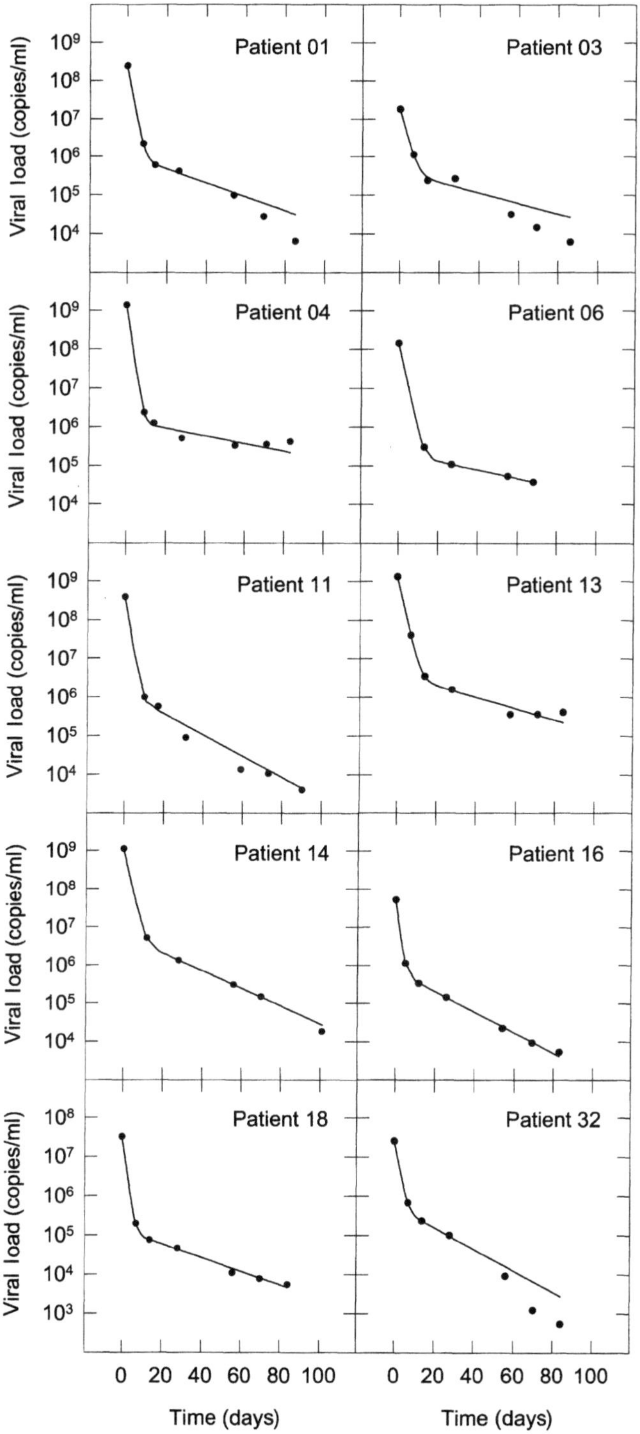
Patient 01
Patient 03
Patient 04
Patient 06
Patient 11
Patient 13
Patient 14
Patient 16
Patient 18
Patient 32
Viral load (copies/ml)
10⁹
10⁸
10⁷
10⁶
10⁵
10⁴
10³
0 20 40 60 80 100
Time (days)

Delays in HBV clearance kinetics

Intuitively, it is unlikely that the viral decline begins immediately upon initiation of antiviral therapy with nucleoside inhibitors of HBV polymerase since time is required for absorption and distribution of the drug to the target organs and for metabolism to the active triphosphate form. Furthermore, inhibition of viral DNA synthesis should not affect assembly and egress of viral particles containing viral genomes synthesized prior to initiation of therapy. These factors are likely to contribute to a lag in viral decline that can cause the slope of the first phase, and therefore the half-life of the free virus, to be underestimated unless very early and frequent sampling is employed. Analysis of a group of hepatitis B patients participating in a study of the pharmacokinetics of the 10-mg dose of adefovir dipivoxil where serum samples were drawn very frequently after the first dose (every 30 min during the first 3 h and at hours 4, 6, 8, 12, 24, 36 and 48) revealed a mean delay of 9.9 h before the classical biphasic viral decline was observed [16]. The detection of this delay quantified for the first time the period required for drug distribution, metabolism to the active diphosphate form within cells and exit from the cell of viral particles containing pre-synthesized viral DNA to leave the cell. Furthermore it permitted a more accurate estimate of the first phase slope giving a new mean estimate of 16.3 h for the half-life of virus particles in plasma. In contrast, administration of an agent with a different mechanism of action than the nucleoside analogues had a quite a different effect on HBV-clearance kinetics [17]. Infusion of hepatitis B patients with Hepex-B, a mixture of two monoclonal antibodies specific for the hepatitis B surface (HBs) antigen resulted in an initial decrease in viral load that was apparent within 30 min of the initiation of the infusion. The decrease was more rapid than that observed during antiviral therapy with nucleoside/nucleotide analogues giving an apparent half-life of 0.9 h, reaching a maximum decrease of 2.5 logs after 8 h. Based on this analysis the investigators concluded that the primary mechanism for Hepex-B was likely to be an approximately 24-fold acceleration in the clearance of free virus from plasma by the antibodies rather than the blocking of virus entry or infection of uninfected cells.

The effect of antiviral dose and antiviral combinations on HBV clearance kinetics

In the initial study of the biphasic clearance of HBV DNA during adefovir dipivoxil therapy, the efficacy of inhibition of viral production of the 30-mg daily dose of adefovir dipivoxil was determined to be 0.993 [15]. In other words, the residual replication observed during therapy was 0.7% of that occurring before treatment was initiated. The effect of varying the efficacy of inhibition of viral production (ε) was simulated and predicted to modulate the magnitude and duration of the first phase of viral decline but not to affect the slope of the first or second phase. These results were confirmed in a more-detailed analysis where data were analyzed from patients receiving 5-, 10-, 30- and 60-mg daily doses of adefovir dipivoxil [16]. This analysis confirmed that increasing dose and efficacy increased the magnitude of the first phase but had no effect on the second phase. The same phenomenon was observed in the first published study of antiviral combination therapy of hepatitis B where patients receiving 150 mg of lamivudine daily were compared with patients receiving combination therapy of 150 mg of lamivudine daily plus 500 mg of famciclovir three times daily [18]. The efficacy of inhibition of viral production (ε) was determined to be 0.94 for the monotherapy group and 0.988 for the combination group, a 5-fold difference in residual replication during therapy between the two groups. The viral declines were biphasic for both groups and the observed effect of the increased efficacy of the combination regimen was an increased duration and magnitude of the first phase of clearance in a manner identical to that observed for increasing doses of adefovir dipivoxil.

Correlation of HBV dynamics parameters with baseline characteristics

Patient characteristics have a significant impact on the response to antiviral therapy of chronic hepatitis B. For example, baseline ALT has been shown to dramatically influence the rate of seroconversion of patients with hepatitis B e antigen positive (HBeAg+) hepatitis B in response to antiviral therapy with interferon α (IFNα), lamivudine and adefovir dipivoxil. The more-sophisticated analysis of the decline in serum HBV DNA offered by viral-dynamics analysis has provided additional insights into the etiology of this effect. All studies of HBV viral-clearance kinetics to date have noted the wide variation in the slope of the second phase between patients. Analysis of the response of HBeAg+ patients receiving various doses of adefovir dipivoxil [16] or lamivudine [19] indicated that the slope of the second phase was positively correlated with the pre-treatment serum ALT level and further strengthening the suggestion that the second phase reflects the rate of loss of productively infected cells and is therefore likely dependent on the immunological response of the patient.

Analysis of patients randomized to placebo in a clinical study of adefovir dipivoxil demonstrated that, as has been established for patients HBeAg– (precore mutant) hepatitis B, the viral load in untreated HBeAg+ patients undergoes spontaneous fluctuations over time [16]. In HBeAg+ patients receiving placebo monitored over 12 months, 47% of patients displayed spontaneous declines in serum HBV DNA of greater than 1 log and 50% of these patients demonstrated subsequent rebound within the 12-month period of observation. Therefore a large proportion of untreated patients displayed viral oscillations of >1 log amplitude with a period of 2–9 months. These oscillations have been attributed to periods of increased immunological pressure followed by viral escape through mutation. In a retrospective examination of the serum HBV kinetics in the month immediately preceding the initiation of adefovir dipivoxil therapy Neumann et al. [16] noted that pre-treatment decline was positively correlated with a faster second phase slope (median $t_{1/2} = 8.6$ days compared to 34.7 days for patients with no pre-treatment decline; $P < 0.01$). This invoked the provocative hypothesis that better responses might result if initiation of treatment was timed to coincide with a period of spontaneous viral decline. However this hypothesis has yet to be tested in prospective analyses.

Refinements to the basic biphasic model of HBV clearance

In the basic biphasic models of HBV clearance during antiviral therapy discussed so far the slower second phase of clearance has been somewhat cryptically referred to as the loss of productively infected cells regardless of the actual mechanism(s) by which this process might occur. However our current understanding of the clearance of infected cells during resolution of acute HBV infection of chimpanzees invokes at least two mechanisms by which this process might occur [20]. Both processes, infected cell killing presumably mediated by HBV-specific cytotoxic T lymphocytes and non-cytolytic curing of HBV cccDNA from infected cells mediated by T helper 1 (Th1) cytokines, appear to be relevant *in vivo*. The basic model does not discriminate between these two mechanisms and assumes a single integrated rate of infected cell loss. Recently, Lewin et al. [21] developed a more sophisticated model for HBV viral dynamics that accommodates these two processes. In this refined model, the rate constant of infected cell loss is the sum of the rate constants of infected cell death and infected cell curing. Consequently, the number of uninfected cells can increase proportionately to the rate of infected cell curing. Lewin et al. [21] also introduced a further modification to the basic model by allowing partial inhibition of infection of target-cells by providing an efficacy term (η) to modify the infection rate constant (b) (Fig. 1). While these more-complex models are undoubtedly more scientifically correct, the additional insights provided by these more-complex models are limited in practice since the actual measurements that serve as input into the model remain constrained to only two variables; serum viral load and time. As techniques evolve to provide independent direct measurements of additional variables (e.g. infected cell number, death rate and curing rate), the true benefit of these more-complex models can be realized.

Non-biphasic HBV clearance kinetics

In a study of 15 hepatitis B patients undergoing treatment with lamivudine monotherapy or the combination of lamivudine plus famciclovir for 12 weeks, Lewin et al. [21] noticed that although the majority of patients displayed a classical biphasic decline in serum HBV DNA, five displayed more complex patterns with an accelerated phase of clearance during the second phase leading to a triphasic staircase pattern of decline. This observation was confirmed in a much larger study of patients receiving therapy for more extended periods by Neumann et al. [22]. Some 342 patients with HBeAg+ chronic hepatitis B were treated with 10 mg of adefovir dipivoxil or placebo for 48 weeks. A pattern-recognition algorithm (Classifase) was used to categorize HBV DNA clearance for each patient into different patterns. Patients receiving placebo displayed three distinct patterns: (1) flat (53%); (2) delayed spontaneous decline of >1 log (unrelated in time to the initiation of therapy) then flat (20%), and (3) oscillatory pattern; delayed spontaneous decline of >1 log, then flat, then rebound (18%). Patients receiving adefovir displayed four major patterns: (1) biphasic, immediate decline (>1 log) then slow decline (42%); (2) biphasic, immediate decline (>1 log) then flat (24%); (3) triphasic, immediate decline (>1 log) then flat, then transient rebound (11%), and (4) triphasic, immediate decline (>1 log), then flat, then acclerated decline (8%). The two biphasic patterns represent the patients with second-phase slopes at the two extremes of the range. Of the two triphasic patterns, the patients displaying transient rebound were shown to be those with treatment interruptions or periods of poor compliance. The third accelerated phase observed in 8% of patients has been suggested to be the result of reactivation of the HBV-specific immune response as a result of viral-load suppression. An alternate hypothesis is that this pattern represents the spontaneous declines, observed in the placebo group as part of the natural history of infection, superimposed over the viral decline induced by adefovir dipivoxil treatment.

Limitations of HBV viral dynamics analyses

Although the current mathematical models for HBV clearance appear complex they remain substantially over-simplified descriptions of a complex biological system. The current models do not account for changes in the rate constants over time to accommodate phenomena such as the observed accelerations in the second-phase slope due to immune reactivation or changes in efficacy due to emergence of drug resistance. Furthermore, Werle et al. [23] have recently shown that 1 year of therapy with adefovir dipivoxil can reduce intra-hepatic cccDNA in patients by almost 1 log. Since the bulk of the decrease appeared to be due to a decrease in the number of copies of cccDNA per cell rather than a decrease in the total number of infected cells it follows that the rate constant for viral production should also change over time. However, as discussed in a previous section the value of more complex models is limited unless experimental estimates of additional parameters, such as infected cell number and cccDNA copy, can be included into the model. Finally, when it comes to correlating viral dynamics parameters or patterns with baseline characteristics and clinical outcomes it is desirable to utilize data from large clinical studies that involve extended periods of dosing so that sufficient statistical power is generated. However, this requirement for larger studies generates a paradox since due to cost constraints large studies generally do not employ the early and frequent sampling required to generate accurate estimates of the viral dynamics parameters and unequivocal categorization of patients using pattern-recognition algorithms.

HCV dynamics

The basic biphasic model of HCV dynamics

HCV viral-clearance kinetics observed during antiviral therapy can be described by the same or similar mathematical models described above for HBV. However some notable differences exist. Firstly unlike HBeAg+ and especially HBeAg– chronic hepatitis B, the viral load in patients with chronic hepatitis C does not change significantly over time in the absence of therapy [24]. Although a biphasic decline (Fig. 3) is also observed during therapy with IFNα the slope of the first phase of clearance is much steeper; a 1–2 log decline is observed over the first 24–48 h giving an estimated $t_{1/2}$ for the free virus of 2.7 h, compared to 16–24 h for HBV. The slope of the second slower phase is approximately 0.3 log per week and like HBV is variable from patient to patient, giving a $t_{1/2}$ for infected cell clearance of 1.7–70 days [25].

HCV clearance kinetics during IFNα therapy

Unlike the anti-HBV nucleoside analogues, where the mechanism of action has been demonstrated comprehensively to be through inhibition of viral DNA synthesis by the HBV polymerase, the precise mechanism for the clinical efficacy of IFNα in hepatitis C patients has not been pinpointed from studies in model systems. In preclinical models, IFNα has demonstrated both direct antiviral effects and immunomodulatory effects. The observed biphasic decline in hepatitis C patients treated with IFNα, following a short lag period of 8–10 h was explained by models wherein the primary mechanism of IFNα was to inhibit viral production with partial efficacy [25]. Alternate hypotheses where the primary mechanism of IFNα was postulated to be inhibition of target-cell infection or enhanced clearance of infected cells by immune mechanisms were less consistent with the short lag period observed before viral decline. Following the pattern described for the anti-HBV agents increasing IFNα dose did not increase the slope of the first or second phases of HCV clearance but increased the magnitude and duration of the first phase [25]. Frequent sampling during standard three-times-daily dosing of IFN α and once weekly (qw) dosing of the long-half-life PEGylated IFNα (IFNα treated with polyethylene glycol; PEG-IFNα) revealed a pattern of viral

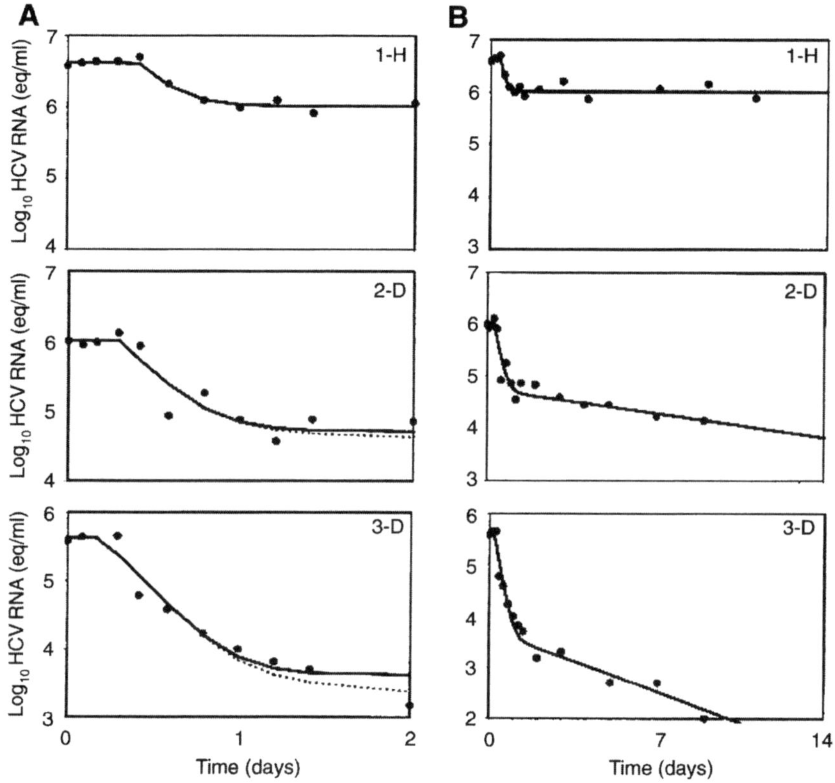

Figure 3. Biphasic clearance of HCV RNA from serum of three different patients 1-H, 2-D, and 3-D, receiving three different daily doses of IFNα (5 , 10 and 15 mi.v.) respectively, for 2 weeks. Panel (A) shows the initial delay and biphasic decline during the first 2 days only. Panel (B) shows the overall biphasic decline over 14 days. The solid lines and the dotted lines show the best fit of the model to the viral load data with and without the simplifying assumption that the number of infected cells remains constant. Adapted from [25].

decline and partial rebound prior to the next dose. These oscillations can be minimized by co-administration of ribavirin [13].

HCV clearance kinetics during ribavirin therapy

When combined with IFNα or PEG- IFNα, ribavirin clearly increases sustained response rates in hepatitis C patients with most of the effect stemming from a reduced rate of relapse when therapy is stopped [5]. However, the mechanistic basis for the clinical efficacy of ribavirin in hepatitis C patients remains unclear. From preclinical studies a multitude of mechanisms have been proposed including direct inhibition of the HCV polymerase by ribavirin triphosphate, inhibition of host IMP dehydrogenase, immune modulation by enhancing the Th1 cytokine response and induction of an error catastrophe secondary to incorporation of ribavirin monophosphate into viral RNA. The dynamics of HCV clearance in response to ribavirin have been analyzed in order to provide some insight into the potential mechanism. Ribavirin monotherapy was initially observed to have no effect on HCV viral load but did reduce plasma ALT [26–28]. However frequent early sampling revealed a significant but transient decrease in HCV viral load of approximately 0.5 log in the first 2 days of therapy [29, 30]. As mentioned above, co-administration of ribavirin can block the partial rebounds observed between doses of IFNα. In a recent study, ribavirin combined with PEG-IFNα appeared to increase the rate of clearance of infected cells by approximately 2-fold compared to PEG-IFNα alone [31]. However in another study in the same issue of *Hepatology*, co-administration of ribavirin with IFNα did not significantly change any viral dynamics parameters (delay, efficacy of inhibition of viral production, rate of infected cell loss, rate of free virus clearance) in either white or African-American patients [32]. Therefore, in contrast to IFNα, analysis of viral dynamics has not provided significant insight into the likely mechanism for ribavirin *in vivo*.

HCV clearance kinetics during direct antiviral therapy with BILN 2061 and NM283

More recently, HCV-clearance data have been published from some of the pilot clinical studies evaluating novel direct antiviral agents in patients with chronic hepatitis C. The first such study was the evaluation of the direct HCV NS3/NS4A serine protease inhibitor, BILN 2061, at 25, 200 or 500 mg twice daily for 2 days in patients with genotype 1 HCV [33, 34]. Rapid decreases in serum HCV RNA reminiscent of but greater than those observed with IFNα occurred, with all patients in the 200- and 500-mg-dose groups displaying a >2 log reduction after 2 days and 3/8 and 7/8 displaying >3 log reductions, respectively. Closer analysis of the viral kinetics [35], revealed that even with only 2 days of dosing, a biphasic pattern of clearance was observed with the rapid first phase lasting approximately 1 day, although HCV RNA was already below the limit of detection for many patients after 24 h. The efficacy of inhibition of viral production was >0.999 for both the 200- and 500-mg-dose groups in patients with no minimal fibrosis but was significantly lower (0.9978) for patients with Childs A cirrhosis receiving the 200-mg dose. This suggests that liver fibrosis may affect the pharmacokinetics or distribution of the compound. Finally, *in vitro* enzymatic assays suggested that BILN 2061 was a approximately 50-fold less potent against HCV genotypes 2b and 3a compared to genotype 1. This translated to weaker and more variable efficacy in patients with genotype 2 and 3 HCV treated with 500 mg of BILN 2061 twice daily with only 3/8 patients displaying a >1 log decrease over 2 days and 3/8 displaying no response at all [36].

In contrast, more-gradual decreases were observed in patients with genotype 1 HCV infection, treated with a 2'-methyl nucleoside analogue inhibitor of HCV NS5B polymerase, NM283, for 14 days [37]. Patients receiving 50, 100, 200 and 400 mg once daily (qd) displayed 0.15–0.73 log decreases in HCV RNA after 14 days. Sequential dose escalation from 100 to 800 mg or from 400 to 800 mg was also attempted in combination with an anti-emetic agent to overcome nausea and vomiting observed in some patients at the 400-mg dose. Consistent with IFNα and BILN 2061, a biphasic decline was observed with the first rapid phase lasting for approximately 2 days. The magnitude of the first phase varied with dose. A 0.1–0.2 log first-phase decrease was observed for the 50-, 100- and 200-mg-dose groups and a approximately 0.4 log first-phase decrease was observed for the 400-mg dose compared to the 1–2 log first-phase decrease observed for IFNα and the 2–3 log first phase decrease observed for BILN 2061. The slope of the slower second phase appeared not to vary between the different doses of NM283.

Triphasic clearance kinetics in hepatitis C patients

Recent more-detailed analysis of HCV clearance during therapy, with PEG-IFNα, PEG-IFNα plus ribavirin or standard IFNα plus ribavirin, has revealed a potential third phase of clearance in approximately 50% of patients [38]. The triphasic pattern resembles the staircase pattern described for a proportion of hepatitis B patients [21]. The initial rapid phase of clearance is replaced after approximately 2 days by a plateau, lag or shoulder phase that lasts 3–21 days followed by a phase of accelerated clearance. The third phase appears to occur below a threshold viral load of approximately 100 000 copies/ml and is speculated to be due to enhanced clearance of infected cells following relief of a virally mediated suppression of the immune response once the threshold is reached. A model was developed where the slope of the second 'shoulder' phase was assumed to reflect the rate of clearance of infected cells that existed prior to therapy and the slope of the third phase was assumed to reflect the accelerated rate of clearance of infected cells induced by therapy. The acceleration factor was as high as 100 in 44% and 67% of patients receiving PEG-IFNα or PEG-IFNα plus ribavirin, respectively [31].

Correlation of HCV dynamics parameters with baseline characteristics

Baseline parameters including HCV genotype, ALT, viral load and race have a strong influence on rates of sustained response to therapy. Analysis of the effects of these parameters on the dynamics of HCV clearance has been used to provide some insight into the mechanistic basis for these effects.

HCV genotype has a dominant effect on treatment outcome with sustained response rates as low as 42–46% for genotype 1 and as high as 76–82% for genotypes 2 and 3 in patients receiving PEG-IFNα plus ribavirin [4, 5]. Efficacy of inhibition of viral production by IFNα appears to be the viral-dynamics parameter most sensitive to HCV genotype. Efficacy was reported to be >0.99 in patients with genotype 1 compared to 0.95 in patients with genotype 2 [39]. Therefore patients with genotype 1 HCV have a significantly diminished first phase of clearance. In patients receiving IFNα, a strong positive correlation between baseline ALT and the second phase slope was established [25, 40], consistent with the assumption that the second phase represents the clearance of infected cells. Patients with higher baseline levels of ALT were assumed to have higher intrinsic rates of clearance of infected hepatocytes by cytotoxic T lymphocytes, thereby releasing higher levels of ALT into the plasma. Baseline HCV viral load strongly influences treatment outcome with sustained response rates of 42–53% for patients with viral load >2 000 000 copies/ml and 62–78%% for patients with viral load <2 000 000 copies/ml receiving PEG-IFNα plus ribavirin [4, 5]. In patients treated with IFNα, it was shown that there is a strong negative correlation between baseline viral load and the rate of clearance of infected cells [25]. Subsequently, it was further observed that the slope of the second phase of clearance, in biphasic models, was closely correlated to the viral load at the end of the first phase [41]. This is consistent with the hypothesis discussed above that the rate of clearance of infected cells mediated by the immune system may be governed by a viral-load threshold [38]. Several studies have suggested that sustained response rates to IFNα-based therapies are lower in African–American patients compared to white patients. Analysis of the dynamics of HCV clearance in patients receiving IFNα or IFNα plus ribavirin revealed that efficacy of inhibition of viral production (0.886 *versus* 0.982) and the rate constant for clearance of infected cells (0.13 *versus* 0.20 day^{-1}) were significantly lower in African–American patients [32].

Limitations of HCV viral dynamics analyses

Many of the limitations of viral-dynamics analysis discussed previously for HBV also apply to HCV. However, there are some additional considerations that are specific to hepatitis C that affect the assumptions used to generate the mathematical models. Firstly, the molecular basis for the antiviral effects of both IFNα and ribavirin are less clear than for the direct antiviral agents used to treat hepatitis B. In patients, the effects of both IFNα and ribavirin are likely to be multifactorial. Furthermore the effects of both IFNα and ribavirin are likely to be variable over time due to fluctuating plasma drug concentrations, because of suboptimal dosing schedules and intermittent adherence due to the poor tolerability of these agents, the likely indirect mechanism of both agents and the high propensity for genetic drift in the population of HCV quasispecies. Neither the multifactorial nature nor the potential variability of the effects of these agents are adequately accommodated in the current models. The second consideration is a more practical constraint. Since the clearance of free viral particles appears to be much more rapid for HCV compared to HBV, the first phase of clearance lasts only for 1–2 days compared to 7–10 days for HBV. The rapid clearance kinetics of HCV places an even higher premium on early and frequent sampling during the first few hours and days following initiation of therapy.

Acknowledgements
I would like to thank Avidan U. Neumann for many stimulating discussions about viral dynamics, Shelly Xiong and William E. Delaney IV for critical review of the manuscript, and Lisa Cortopassi and Manuel Tsiang for contributing to the preparation of the manuscript.

References

1 Lai C-L, Chien R-N, Leung NWY, Chang T-T, Guan R, Tai D-I, Ng K-Y, Wu P-C, Dent JC, Barber J et al. (1998) A one year trial of lamivudine for chronic hepatitis B. *N Engl J Med* 339: 61–68
2 Hadziyannis SJ, Tassopoulos NC, Heathcote EJ, Chang T-T, Kitis G, Rizzetto M, Marcellin P, Lim SG, Goodman Z, Wulfsohn MS et al. (2003) Adefovir Dipivoxil for the Treatment of Hepatitis B e Antigen-Negative Chronic Hepatitis B. *N Engl J Med* 348: 800–807
3 Marcellin P, Chang T-T, Lim SG, Tong MJ, Sievert W, Shiffman ML, Jeffers L, Goodman Z, Wulfsohn MS, Xiong S et al. (2003) Adefovir Dipivoxil for the Treatment of Hepatitis B e Antigen-Positive Chronic Hepatitis B. *N Engl J Med* 348: 808–816
4 Manns MP, McHutchison JG, Gordon SC, Rustgi VK, Shiffman M, Reindollar R, Goodman ZD, Koury K, Ling M-H, Albrecht JK et al. (2001) Peginterferon alfa-2b plus ribavirin compared with interferon alfa-2b plus ribavirin for initial treatment of chronic hepatitis C: a randomized trial. *Lancet* 358: 958–965
5 Fried MW, Shiffman ML, Reddy KR, Smith C, Marinos G, Goncales Jr, FL, Häussinger D, Diago M, Carosi G, Dhumeaux D et al. (2002) Peginterferon alfa-2a plus ribavirin for Chronic Hepatitis C virus infection. *N Engl J Med* 347: 975–982
6 Mommeja-Marin H, Mondou E, Blum MR, Rousseau F (2003) Serum HBV DNA as a Marker of Efficacy During Therapy for Chronic HBV Infection: Analysis and Review of the Literature. *Hepatology* 37: 1309–1319
7 Pawlotsky J-M (2003) Diagnostic Testing in Hepatitis C Virus Infection: Viral Kinetics and Genomics. *Semin Liver Dis* 23 (Suppl 1): 3–11
8 Ho DD, Neumann AU, Perelson AS, Chen W, Leonard JM, Markowitz M (1995) Rapid turnover of plasma virions and CD4 lymphocytes in HIV 1 infection. *Nature* 373: 123–126
9 Wei X, Ghosh SK, Taylor ME, Johnson VA, Emini EA, Deutsch P, Lifson JD, Bonhoeffer S, Nowak MA, Hahn BH et al. (1995) Viral dynamics in human immunodeficiency virus type 1 infection. *Nature* 373: 117–122
10 Ribeiro RM, Lo A, Perelson AS (2002) Dynamics of hepatitis B virus infection. *Microbes Infect* 4: 829–835
11 Tsiang M, Gibbs CS (2004) Analysis of hepatitis B virus dynamics and its impact on antiviral development. Methods. *Mol Med* 96: 361–377
12 Herrmann E, Neuman AU, Schmidt JM, Zeuzem S (2000) Hepatitis C virus kinetics. *Antivir Ther* 5: 85–90
13 Layden TJ, Layden JE, Ribeiro RM, Perelson AS (2003) Mathematical modeling of viral kinetics: a tool to understand and optimize therapy. *Clin Liver Dis* 7: 163–178
14 Nowak MA, Bonhoeffer S, Hill AM, Boehme R, Thomas HC, McDade H (1996) Viral dynamics in hepatitis B virus infection. Pro. Natl. *Acad Sci USA* 93: 4398–4402
15 Tsiang M, Rooney JF, Toole JJ, Gibbs CS (1999) Biphasic Clearance Kinetics of Hepatitis B Virus From Patients During Adefovir Dipivoxil Therapy. *Hepatology* 29: 1863–1869
16 Neumann A, Shmailov M, Tsiang M, Kearney B, Currie G, Wulsohn M, Fry J, Brosgart C, Gibbs C (2002) HBV Viral Dynamics in Chronic Hepatitis B Patients Treated with Adefovir Dipivoxil: Effect of Dose and Pretreatment HBV Disease Characteristics. *Hepatology* 36(4; 2): 375A

17 Neumann AU, Eren R, Dagan S (2003) Viral Dynamics Modeling Indicates that the Major Mechanism of Action of Hepex-B Against Hepatitis B Virus is Acceleration of the Free Virus Clearance. *Hepatology* 38(4) Suppl 1: 709A
18 Lau, GKK, Tsiang M, Hou J, Yuen S-T, Carman WF, Zhang L, Gibbs CS, Lam S-K (2000) Combination Therapy with Lamivudine and Famciclovir for Chronic Hepatitis B-Infected Chinese Patients: A Viral Dynamics Study. *Hepatology* 32: 394–399
19 Wolters LMM, Hansen BE, Niesters HGM, Levi-Drummer RS, Neumann AU, Schalm SW, de Man RA (2002) The influence of baseline characteristics on viral dynamic parameters in chronic hepatitis B patients treated with lamivudine. *J Hepatol* 37: 253–258
20 Guidotti LG, Rochford R, Chung J, Shapiro M, Purcell R, Chisari FV (1999) Viral Clearance Without Destruction of Infected Cells During Acute HBV Infection. *Science* 284: 825–829
21 Lewin SR, Ribeiro RM, Walters T, Lau GK, Bowden S, Locarnini S, Perelson AS (2001) Analysis of Hepatitis B Viral Load Decline Under Potent Therapy: Complex Decay Profile Observed. *Hepatology* 34: 1012–1020
22 Neumann A, Havlin Y, Tal R, Tsiang M, Wulfsohn M, Brosgart C, Fry J, Gibbs C (2002) Long Term HBV Kinetics Classification during Treatment with Adefovir Dipivoxil. *J Hepatol* 36 Suppl 1: 121
23 Werle B, Bowden S, Locarnini S, Wursthorn K, Peterson J, Lau G, Trepo C, Marcellin P, Goodman Z, Delaney WE et al. (2004) New Insight on Viral cccDNA Persistence During the Natural History of Chronic Hepatitis B and Adefovir Dipivoxil Therapy. *Gastroenterology* 126: 1750–1758
24 Nguyen TT, Sedghi-Vaziri A, Wilkes LB, Mondala T, Pockros PJ, Lindsay KL, McHutchison JG (1996) Fluctuations in viral load (HCV RNA) are relatively insignificant in untreated patients with chronic HCV infection. *J Viral Hepat* 3: 75–78
25 Neumann AU, Lam NP, Dahari H, Gretch DR, Wiley TE, Layden TJ, Perelson AS (1998) Hepatitis C Viral Dynamics *in Vivo* and the Antiviral Efficacy of Interferon α Therapy. *Science* 282: 103–109
26 Di Bisceglie AM, Conjeevaram HS, Fried MW, Sallie R, Park Y, Yurdaydin C, Swain M, Kleiner DE, Mahaney K, Hoofnagle JH (1995) Ribavirin as therapy for chronic hepatitis C. A randomized, double-blind, placebo-controlled trial. *Ann Intern Med* 123: 897–903
27 DiBisceglie AM, Shindo M, Fong TL, Fried MW, Swain MG, Bergasa NV, Axiotis CA, Waggoner JG, Park Y, Hoofnagle JH (1992) A pilot study of ribavirin therapy for chronic hepatitis C. *Hepatology* 16: 649–654
28 Bodenheimer HC Jr, Lindsay KL, Davis GL, Lewis JH, Thung SN, Seeff LB (1997) Tolerance and efficacy of oral ribavirin treatment of chronic hepatitis C: a multicenter trial. *Hepatology* 26: 473–477
29 Pawlotsky JM, Dahari H, Conrad a, Lonjon I, Hizode C, Germanidis G, Schmid P, Dhumeaux D, Neumann AU (1998) Effect of intermittent interferon (IFN), daily IFN and IFN plus ribavirin induction therapy on hepatitis C virus (HCV) genotype 1b replication kinetics and clearance. *Hepatology* 28: 288A
30 Lee J-H, von Wagner M, Roth WK, Teuber G, Sarrazin C, Zeuzem S (1998) Effect of ribavirin on virus load and quasispecies distribution in patients infected with hepatitis C virus. *J Hepatol* 29: 29–35
31 Herrmann E, Lee J-H, Marinos G, Modi M, Zeuzem S (2003) Effect of Ribavirin on Hepatitis C Viral Kinetics in Patients Treated with Pegylated Interferon. *Hepatology* 37: 1351–1358
32 Layden-Almer JE, Ribeiro RM, Wiley T, Perelson AS, Layden TJ (2003) Viral Dynamics and Response Differences in HCV-Infected African American and White Patients Treated with IFN and Ribavirin. *Hepatology* 37: 1343–1350
33 Hinrichsen H, Benhamou Y, Reiser M, Sentjens R, Wedemeyer H, Calleja JL, Forns X, Crönlein J, Nehmiz G, Steinmann G (2002) First Report on the Antiviral Efficacy of BILN 2061, A Novel Oral HCV Serine Protease Inhibitor, in Patients with Chronic Hepatitis C Genotype 1. *Hepatology* 36(4; 2): 304A
34 Lamarre D, Anderson PC, Bailey M, Beaulieu P, Bolger G, Bonneau P, Bös M, Cameron DR, Cartier M, Cordingley MG et al. (2003) An NS3 protease inhibitor with antiviral effects in humans infected with hepatitis C virus. *Nature* 426: 186–189
35 Herrmann E, Zeuzem S, Sarrazin C, Hinrichsen H, Benhamou H, Wedemeye H, Steinmann G, Nehmiz G (2004) Hepatitis C Viral Kinetics in Chronically Infected Patients Treated with the Serine Protease Inhibitor BILN 2061. *J Hepatol* 40 Suppl 1: 24

36 Reiser M, Hinrichsen H, Benhamou Y, Sentjens R, Wedemeyer H, Calleja L, Forns X, Croenlein J, Yong C, Nehmiz G et al. (2003) Antiviral Effect of BILN 2061, A Novel HCV Serine Protease Inhibitor, After Oral Treatment Over 2 Days in Patients with Chronic Hepatitis C, Non-Genotype 1. *Hepatology* 38(4) Suppl 1: 221A
37 Godofsky E, Afdhal N, Rustgi V, Shick L, Duncan L, Zhou X.J, Chao G, Fang C, Fielman B, Myers M et al. (2004) First Clincal Results for a Noval Antiviral Treatment for Hepatitis C: A Phase I/II Dose Escalation Trial Assessing Tolerance Pharmacokinetics and Antiviral Activity of NM283. *J Hepatol* 40 Suppl 1: 35
38 Bergmann CC, Layden JE, Levy-Drummer RS, Layden TJ, Haagmans BL, Neumann AU (2001) Clinical Implications of a New Tri-Phasic Model for Hepatitis C Viral Kinetics during IFN-α Therapy. *Hepatology* 34(4; 2): 34A
39 Neumann AU, Lam NP, Dahari H, Davidian M, Wiley TE, Mika B, Perelson AS, Layden TJ (2000) Differences in viral dynamics between genotypes 1 and 2 of hepatitis C virus. *J Infect Dis* 182: 28–35
40 Ribeiro RM, Layden-Almer J, Powers KA, Layden TJ, Perelson AS (2003) Dynamics of Alanine Aminotransferase During Hepatitis C Virus Treatment. *Hepatology* 38: 509–517
41 Layden JE, Layden TJ, Reddy KR, Levi-Drummer R, Poulakos J, Neumann AU (2002) First phase viral kinetic parameters as predictors of treatment response and their influence on the second phase viral decline. *J Viral Hepat* 9: 340–345

Hepatitis Prevention and Treatment
Edited by J.M. Colacino and B.A. Heinz

Investigational drugs in clinical development for the treatment of chronic viral hepatitis

Raymund R. Razonable[1,2] and James M. McGill[1,3]

[1] *Bone & Inflammation Research, Lilly Research Laboratories, Indianapolis, IN, USA*
[2] *Division of Infectious Diseases and Internal Medicine, Mayo Clinic College of Medicine, Rochester, MN, USA*
[3] *Division of Gastroenterology and Hepatology, Indiana University School of Medicine, Indianapolis, IN, USA*

Introduction

Chronic infections with the hepatitis B virus (HBV) and hepatitis C virus (HCV) are major health problems worldwide. Chronic hepatitis B (CHB) affects ~350 million people worldwide [1] with the majority of infections occurring in the developing world where preventive HBV vaccination has not been widely implemented. Chronic hepatitis C (CHC) affects ~170 million people worldwide [2]. While extrahepatic disease due to these viruses does occur, the chronic liver disease that results in cirrhosis, end-stage liver disease, and hepatocellular carcinoma are the hallmark for these infections.

Although significant advances have been made during the past decade in the treatment of CHB and CHC, they remain two of the most therapeutically challenging viral diseases. Indeed, the current standard of treatment permits opportunities for further innovation. Lamivudine, a nucleoside analogue that was initially developed for the treatment of human immunodeficiency virus (HIV) infection, was the first antiviral drug (as opposed to an immunomodulator) for the treatment of HBV infection [3]. However, the initial hope that lamivudine would provide a durable cure for CHB was quickly extinguished by lamivudine-resistant HBV, characterized by the YMDD mutation, which rapidly emerged and compromised its efficacy [3]. Recently, adefovir dipivoxil, another nucleotide analogue, has proven effective for treating CHB, including infections due to the lamivudine-resistant YMDD mutant [4]. Nonetheless, longer follow up has demonstrated emergence of adefovir-resistant HBV, albeit at a comparatively lower rate than that for lamivudine therapy [5]. The emergence of drug-resistant HBV during antiviral monotherapy has led to an emerging treatment paradigm, where combination therapy, a strategy that has proven effective in managing HIV infection, is gaining widespread interest.

For CHC, the efficacy of interferon α has improved with PEGylation [treatment with polyethylene glycol (PEG)-attached interferon] and combination with ribavirin. However, the overall clinical response remains modest; virologic relapse often occurs after treatment. Moreover, many patients do not tolerate interferon α because of debilitating adverse events. These factors, together with the many contraindications to interferon α, have resulted in a large number of patients who remain untreated [6].

The limitations to current therapy for CHB and CHC underscore the need for novel compounds and optimized therapeutic strategies. In this chapter, we review the progress in the clinical development of investigational compounds for the treatment of CHB and CHC. These compounds were identified after an extensive search of the MEDLINE and PubMed databases. Proceedings from the meetings of the American Association for the Study of Liver Diseases, the Infectious Diseases Society of America, and the Interscience Conference on Antimicrobial Agents and Chemotherapy were searched. Ongoing clinical trials were further identified at websites such as http://www.clinicaltrials.gov and corresponding company websites. The Investigational Drugs database was also reviewed. Due to space restrictions, we limited our detailed discussion to compounds that possess one or more of the following characteristics: (1) those that have completed phase II clinical evaluation, (2) those that are already commercially available, albeit for other indications, and (3) those that act through novel mechanisms. A brief discussion of representative compounds and novel strategies that are currently undergoing preclinical and early-phase clinical trials is also provided.

Investigational drugs for the treatment of CHB

During the last decade, remarkable progress in antiviral drug discovery research has led to several anti-HBV compounds in different phases of clinical development (Tab. 1 and Fig. 1). These compounds act by different mechanisms and thus have the potential for use in combination anti-HBV therapy. In general, these compounds are classified as either antiviral drugs or immunomodulators. The majority are nucleoside and nucleotide analogues whose primary anti-HBV activity is brought about by their ability to inhibit viral polymerase subsequent to their phosphorylation to the triphosphate equivalent. Most are L-enantiomer nucleoside analogues that exhibit a more potent antiviral activity when compared to D-enantiomers. Because of similarities among viral reverse transcriptases, some compounds have inherent activity against retroviruses. For example, emtricitabine and tenofovir disoproxil fumarate (tenofovir DF), while effective at inhibiting HBV, are also efficacious as HIV therapies. Structural modifications to other compounds have rendered specificity for HBV, and in some cases, have enhanced their anti-HBV potency. New formulations of currently available drugs, such as the adefovir prodrug Hepavir B, have been developed to increase the anti-HBV therapeutic index through a liver-targeting mechanism (i.e. preferential drug distribution in the liver parenchyma) [7]. Novel antiviral mechanisms such as catalytic

Table 1. Investigational drugs in clinical development for the treatment of CHB

Antiviral drugs
Nucleoside and nucleotide analogues
Specific HBV polymerase inhibitors
Entecavir
Telbivudine
Valtorcitabine
β-L-Fd4C
Non-selective HBV polymerase inhibitors
Emtricitabine
Clevudine
Tenofovir disoproxil fumarate
Pre-genomic RNA packaging inbibitor
Alamifovir
Non-nucleoside packaging inhibitors
AT-130
AT-61
Immunomodulators
PEGylated interferons α2a and α2b
Interferon β
Thymosin α-1
HBV therapeutic vaccination
EHT-899 oligonucleotide
Hepex-B monoclonal antibody

ribozymes and RNA inhibition are also being evaluated clinically [8]. Recent investigations that highlighted the critical role of the immune system in controlling HBV infection [1] have also led to the investigational use of therapeutic anti-HBV vaccination as well as the use of oligonucleotides that act through a Toll-like receptor mechanism [9].

Antiviral drugs

Entecavir

The most advanced among the anti-HBV compounds undergoing clinical development is the deoxyguanosine analogue entecavir (BMS-200475, Bristol-Myers Squibb; Figure 2), which selectively and potently inhibits HBV replication by blocking priming, reverse transcription, and DNA-dependent DNA synthesis. Preclinical studies using the woodchuck hepatitis virus (WHV) model demonstrated significant reductions in serum WHV DNA and intrahepatic WHV replicative intermediates and covalently closed circular DNA (cccDNA) during entecavir treatment [10]. The reduction in cccDNA,

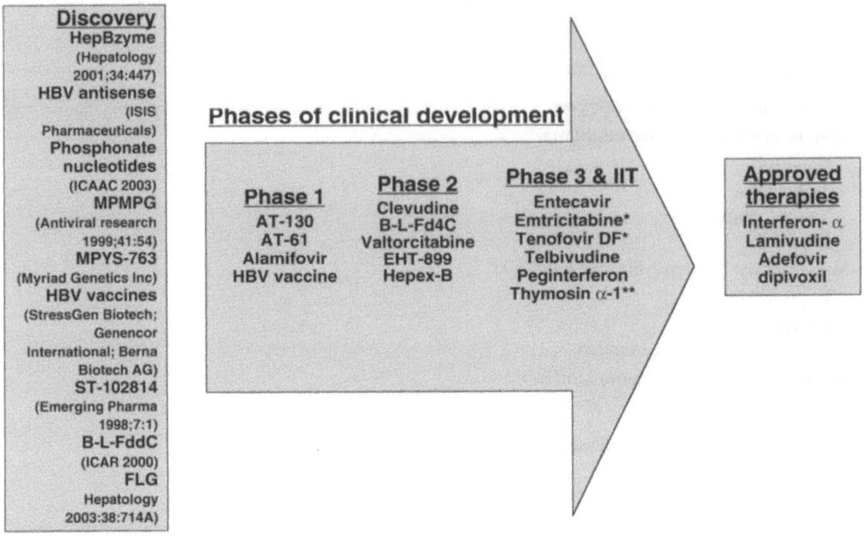

Figure 1. Investigational and approved compounds for the treatment of CHB. References for investigational compounds that were not discussed in the text are indicated in parentheses. If no published reference is available, the company developing the compound is indicated. HBV, hepatitis B virus; ICAAC, Interscience Conference on Antimicrobial Agents and Chemotherapy; ICAR, International Conference on Antiviral Research. *Compounds currently approved in the USA for treatment of HIV virus infection, including those undergoing investigator-initiated trials (IIT) for CHB. **Compounds currently available for CHB treatment in some countries. MPMPG: 9-[2-methylidine-3-(phosphonomethoxy)propyl]guanine.

the major transcriptional template for HBV, is especially promising since cccDNA serves as a reservoir for relapse of infection upon the discontinuation of therapy [8].

Three early-phase clinical trials have demonstrated the superior anti-HBV efficacy of entecavir over placebo or lamivudine in lamivudine-susceptible [11] and -resistant [12] HBV. These studies demonstrated the highest reductions in HBV DNA at entecavir doses of 0.5 and 1.0 mg daily. Depending on the study, the highest magnitude of HBV DNA reductions ranged from $-2.81 \log_{10}$ (among patients who received 0.5 mg of entecavir daily for lamivudine-susceptible HBV) to $-5.11 \log_{10}$ (among patients who received 1.0 mg of entecavir daily for 48 weeks for YMDD mutant infection) [12]. Even in the absence of lamivudine-resistant HBV, the patients who received 0.5 mg entecavir once daily were more likely to attain undetectable HBV DNA than were lamivudine-treated patients (83.7 *versus* 57.5%; $P = 0.008$) [11]. Genotypic analysis of the viral polymerase gene indicated that there were no consistent mutations that emerged after 48 weeks of treatment [13]. Safety analysis of 409 patients in phase II clinical trials demonstrated that entecavir was well tolerated. Adverse events were mostly mild or moderate in intensity and consisted of headache, rhinitis, fatigue, and abdominal pain [14]. Serious

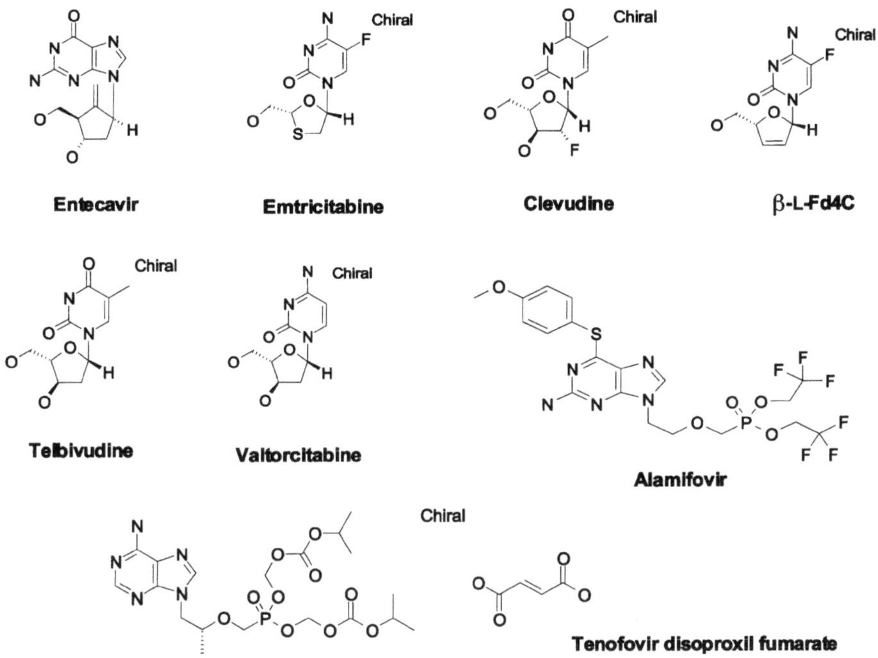

Figure 2. Chemical structures of selected compounds undergoing development for the treatment of CHB.

adverse events (7 and 9% respectively) and treatment discontinuations due to adverse events (3 and 5%) were comparable between entecavir and lamivudine [14]. Currently, entecavir is undergoing phase III clinical trials in nucleoside-naïve and -experienced hepatitis B e antigen (HBeAg)-positive and -negative patients.

Telbivudine

Telbivudine (L-dT, β-L-deoxythymidine, NV-02; Idenex Pharmaceuticals; Figure 2) is one of three β-L-2'-deoxynucleosides that exhibit potent, specific, and selective inhibition of HBV replication through inhibition of viral polymerase [15]; the other related β-L-nucleosides are L-dC [whose prodrug valtorcitabine is also being developed as HBV therapy (see below)] and L-dA (β-L-2'-deoxyadenosine). Analysis of the structure–activity relationship indicates that the presence of the 3'-OH group (in all three related nucleosides) confers HBV-specificity [15, 16]. Preclinical data demonstrated that telbivudine was a potent and selective inhibitor of HBV replication. In woodchucks, the administration of telbivudine, at a dose of 10 mg/kg once daily for 4 weeks, resulted in the reduction of WHV DNA by up to 8 $\log_{10}$, a magnitude of decline that was greater than L-dC (which decreased WHV DNA by 4–6 $\log_{10}$) and lamivudine (which decreased WHV DNA by 0.5–1.0 $\log_{10}$) [15]. Rebound viremia to pre-treatment WHV DNA levels was observed by 8 weeks after drug withdrawal [15].

An international, randomized, blinded, multicenter phase II study demonstrated that telbivudine monotherapy was superior to lamivudine monotherapy and that the addition of a second polymerase inhibitor lamivudine to telbivudine did not produce an additive or synergistic anti-HBV effect [17]. After 52 weeks of treatment, the mean reduction in HBV DNA among patients who received telbivudine (400 or 600 mg) monotherapy (–6.09 $\log_{10}$) was comparable to telbivudine/lamivudine combination therapy (–5.99 $\log_{10}$) and superior to lamivudine monotherapy (–4.57 $\log_{10}$ copies/ml; $P < 0.05$) [17]. Telbivudine-treated patients were also more likely than lamivudine-treated patients to have undetectable HBV DNA (64 *versus* 32%, $P < 0.05$) and normal alanine aminotransferase (ALT; 86 *versus* 63%, $P < 0.05$) [17]. However, HBeAg seroconversion was similar between lamivudine and telbivudine (28 *versus* 33%, respectively). These results reinforce the uncertainty of current efficacy end-points in CHB therapy. A phase III clinical trial that is designed to enroll 1200 patients worldwide is underway.

Emtricitabine

Emtricitabine [(–)-FTC],(–)-β-2',3'-dideoxy-5-fluoro-3'-thiacytidine, Emtriva; Gilead Sciences; Fig. 2] has recently been approved by the US Food and Drug Administration (FDA) for treatment of HIV. Like many nucleoside analogues, emtricitabine reduces HBV replication through inhibition of the viral polymerase [18, 19]. In preclinical studies, emtricitabine was demonstrated initially to inhibit HBV *in vitro* in a HBV-transfected human liver cell line (HepG2 derivative 2.2.15, subclone P5A) [18, 19] and subsequently *in vivo* in a chimeric mouse model [20] and against WHV in naturally infected woodchucks [21]. In a placebo-controlled study using the woodchuck model [22], emtricitabine therapy for 4 weeks caused a dose-dependent decline in WHV DNA levels in the serum and the liver. However, there was no significant effect on cccDNA and intrahepatic WHV RNA, which resulted in rebound viremia within 1–2 weeks after the end of treatment [22].

Early clinical trials demonstrated that emtricitabine causes a dose-dependent reduction in HBV DNA [23]. After 48 weeks of emtricitabine treatment, HBV DNA declined by a median of –2.92 $\log_{10}$ and was undetectable in 55% of patients; HBeAg loss and anti-HBe seroconversion occurred in 50 and 23% respectively [23]. When emtricitabine treatment was extended to 96 weeks, HBV DNA was undetectable in 41%, ALT was normalized in 71%, and HBeAg loss and anti-HBe seroconversion occurred in 51 and 29%, respectively [23]. This study, however, demonstrated the emergence of emtricitabine-resistant HBV, with rates of 6% at 1 year and 19% at 2 years. The emtricitabine-resistant HBV mutant is characterized by L526M ± M550V/I mutations and thus it is likely cross-resistant with lamivudine [23].

Tenofovir DF

Tenofovir DF (9-[R]-2-(phosphonomethoxy) propyl adenine, PMPA, Viread, Gilead Sciences; Figure 2) is a bioavailable prodrug of tenofovir, an acyclic nucleotide analogue with non-specific inhibitory properties against viral polymerase. Like emtricitabine, tenofovir DF is currently approved for treating HIV infection. Preclinical *in vitro* data indicate the potent activity of tenofovir DF against lamivudine-susceptible and -resistant HBV. The clinical data supporting its anti-HBV activity are derived mainly from investigator-initiated observational studies describing the outcomes of the 'off-label' use of tenofovir DF for the treatment of lamivudine-resistant HBV. In HIV-non-infected CHB patients who failed lamivudine treatment, a mean 4.8 $\log_{10}$ reduction in HBV DNA was observed during tenofovir DF treatment [24, 25]. Likewise, in four separate studies that involved 51 HIV- and HBV-coinfected patients, treatment with tenofovir DF for 24 weeks caused a decline in lamivudine-resistant HBV DNA by 3.78–4.30 $\log_{10}$ [26–29]. A similar magnitude of HBV DNA decline was observed when the therapy was extended beyond 24 weeks (~4 $\log_{10}$ at 52 weeks) [30]. In the largest cohort of HIV-infected patients with persistent HBV replication despite lamivudine therapy, 42% of tenofovir DF-treated patients attained undetectable HBV DNA at 9 months; among those with detectable HBV DNA, a 3.6 $\log_{10}$ decline from baseline was observed [31].

Clinical data from much smaller studies than done for adefovir alone also suggest that tenofovir DF is more efficacious than adefovir dipivoxil in the treatment of lamivudine-resistant HBV. At 3 and 6 months after starting treatment, HBV DNA declined by 4.8 and 6.6 $\log_{10}$ respectively, in tenofovir-treated patients, compared to 1.97 and 1.72 $\log_{10}$ in adefovir-treated patients. Tenofovir DF-treated patients were also more likely than adefovir-treated patients to have undetectable HBV DNA at 6 months (57 *versus* 7%) [25].

Antiviral drugs in earlier stages of development

Nucleosides analogues. Several nucleoside analogues are currently in earlier phases of clinical development including β-L-Fd4C (ACH 126.443; Achillion), clevudine (L-FMAU; Gilead Sciences), alamifovir (LY582563, MCC-478; Eli Lilly and Company and Mitsubishi Pharma Corporation), and valtorcitabine (val-L-dC; Idenex Pharmaceuticals) (Fig. 2). β-L-Fd4C (Fig. 2), an L-nucleoside analogue of deoxycytidine, is distinguished by a fluorine atom at 5' position of cytosine, which, together with the longer intracellular retention time of the triphosphorylated metabolite [32], enhances its anti-HBV property [32]. In preclinical studies, the anti-HBV polymerase activity of β-L-Fd4C was observed to be 10-fold greater than lamivudine [32]. Despite this, beta-L-Fd4C could not eliminate cccDNA [33], which likely accounted for the recurrence of viral replication after the discontinuation of therapy in the woodchuck and duck models of HBV [34, 35]. A phase I multicenter, placebo-controlled, and blind study demonstrated that β-L-Fd4C caused a significant decline in HBV DNA (by 1.5–3.0 $\log_{10}$) when compared to placebo [36]. Clevudine (Fig. 2) is a thymidine nucleoside analogue that has also demonstrated potent anti-HBV properties in *in vitro* and animal models of HBV infection [37, 38]. Clevudine therapy resulted in significant but transient reduction in intrahepatic WHV RNA and cccDNA levels in woodchucks [38]. Within 6–12 months of the cessation of clevudine treatment, WHV DNA levels increased towards pre-treatment levels and in some cases the rebound viremia was due to virus strains that carry mutations in the viral DNA polymerase gene [39]. Indeed, it has been demonstrated that lamivudine-associated mutations (M550I, M550V, and L526M) confer cross-resistance between lamivudine and clevudine [40]. In early-phase clinical trials of patients with CHB, a 28-day treatment with 10, 50, and 100 mg of clevudine caused a decline in serum HBV DNA by 2.48, 2.74, and 2.95 $\log_{10}$ respectively [41]. Valtorcitabine (Fig. 2) is unique among the L-nucleosides because it is being developed as a pronucleotide [15, 42]. In preclinical studies using the woodchuck model, a 28-day treatment course with its active component, L-dC, reduced the serum WHV DNA levels by up to 4–6 $\log_{10}$, a magnitude that is superior to lamivudine but less than telbivudine. A preliminary report on the phase I/II dose-escalation study indicated that valtorcitabine causes a consistent and dose-related reduction in serum HBV DNA [43]. Alamifovir (Fig. 2), a 2-amino-6-arylthio-9-phosphonomethoxyethylpurine bis(2,2,2-trifluoroethyl) ester derivative, presents a unique feature among nucleotide analogues because of its ability to influence viral reverse transcription by inhibiting the packaging of pre-genomic RNA into core particles without directly affecting the viral DNA polymerase activity [44]. Preclinical studies using cell-based assays suggested that alamifovir inhibited the replication of wild-type and lamivudine-resistant HBV by up to 80-fold when compared to lamivudine [45]. A phase I clinical trial of 65 HBV-infected patients demonstrated that alamifovir treatment for 28 days caused a significant decline in HBV DNA, with the 20-mg dose inducing a reduction in mean HBV DNA by more than 3 $\log_{10}$ [46, 47].

Non-nucleosides. A series of substituted analogues of phenylpropenamide 6 are also being evaluated for the anti-HBV activity in cell-based assays including AT-61 and AT-130. AT-61 inhibits HBV replication in hepatocyte cell lines [48, 49]. When combined with lamivudine, a synergistic anti-HBV effect was observed, suggesting a complimentary mechanism of action between the two drugs and the potential for AT-61 to be effective against lamivudine-resistant HBV [49]. AT-61 did not suppress the *in vitro* replication of duck hepatitis virus, WHV and HIV type 1, suggesting further that it is a specific inhibitor of HBV [49]. AT-130 is a congener of AT-61 that differs in its rings A and B, which contain *ortho*-methoxy and *para*-nitroso substituents, respectively, and the substitution of the chlorine in AT-61 with a bromine atom. Like AT-61, the anti-HBV activity of AT-130 has also been demonstrated in cell-based assays [48, 49]. In an experimental *in vitro* model using the HBV baculovirus system, AT-130 was observed to inhibit the packaging of the pre-genomic RNA into core particles and thus was able to inhibit viral reverse transcription without affecting the viral DNA polymerase activity or core protein translation (a mechanism of action similar to alamifovir) [50]

Immune modulators

PEGylated interferon

PEGylated interferon α2a. Interferon was the first drug for the treatment of CHB but this was associated with only a modest antiviral response and a variety of adverse effects. The demonstrated efficacy of the PEGylated formulations of interferons for the treatment of CHC has resulted in the current evaluation of these compounds for the treatment of CHB. The superior efficacy of PEGylated interferon α2a (40 kDa) administered over 24 weeks (peginterferon α2a, Pegasys, F. Hoffmann-La Roche) when compared to conventional interferon α (Roferon-A; F. Hoffmann-La Roche) in treating CHB patients was demonstrated in a randomized, multicenter, phase II study of 194 HBeAg and HBV DNA-positive patients infected with genotypes B (32%) and C (66%) [51]. Whereas all three doses of peginterferon α2a (40 kDa) (90, 180 and 270 μg once weekly) were superior to conventional interferon α (4.5 m-i.u. three times weekly), it was demonstrated that the 90 and 180-μg doses were better than the 270 μg dose [51]. At 48 weeks, the proportion of patients with HBV DNA of <500 000 copies/ml, HBeAg loss, and anti-HBe seroconversion was higher among patients who received 90 μg and 180-μg peginterferon α2a (40 kDa) than 270 μg of peginterferon α2a (40 kDa) and conventional interferon α2a [51]. Overall, the combined response rates (HBV DNA suppression, HBeAg loss, anti-HBe seroconversion, and normal ALT) were superior among patients treated with 90 and 180 μg of peginterferon α2a (40 kDa) (27 and 28%, respectively) than among those who received 270 μg of peginterferon α2a (40 kDa) group (19%) and conventional interferon α (12%; $P = 0.036$) [51]. A phase III study comparing the efficacy and safety of peginterferon α2a (40 kDa) with or without lamivudine to lamivudine monotherapy is ongoing.

PEGylated interferon α2b. The anti-HBV efficacy of PEGylated interferon α2b (peginterferon α2b, PEG-Intron, Schering-Plough Corporation) used in combination therapy was evaluated in two studies, where it was directly compared to conventional interferon α2b or lamivudine. In one study [52], patients who received peginterferon α2b (1.5 μg/kg once weekly) plus lamivudine (100 mg daily) were more likely than those who received conventional interferon α2b (5 m-i.u. thrice weekly) plus lamivudine, or lamivudine monotherapy, to have undetectable HBV DNA after 52 weeks of treatment (70, 30, and 25% respectively). In a second study [53] patients who received peginterferon α2b plus lamivudine were more likely than those who received lamivudine monotherapy for 52 weeks to have end-of-treatment virological response (undetectable HBV DNA and HBeAg seroconversion; 75 and 25% respectively; $P = 0.005$), sustained virological response (54 and 10% respectively; $P = 0.08$), and normalization of ALT (95 and 70%, respectively; $P = 0.09$)

Thymosin α-1

Thymosin α-1 (Zadaxin; SciClone Pharmaceuticals), a 28-amino acid thymic peptide with pleiotrophic immunoregulatory effects, has been demonstrated to have anti-HBV properties in several clinical trials that led to its approval for use in many countries. It remains in phase II clinical development in other countries including the United States. Because CHB patients have low endogenous thymosin α-1 levels, a finding associated with CD4+ T cell deficiency [54], it is postulated that thymosin α-1 exerts its antiviral effects by augmenting the maturation of T cells [55]. A recent *in vitro* study also suggested that thymosin α-1 may inhibit HBV directly [56], although the mechanism underlying this finding remains undefined.

A review of clinical studies indicates that the virological and biochemical response after thymosin α-1 therapy is modest, although it is better than placebo or no treatment [57]. Thymosin α-1 treatment for 6 months resulted in undetectable HBV DNA in 14% of patients, compared to 4.5% among untreated patients [57]. It is also suggested that the anti-HBV efficacy of thymosin α-1 is comparable to interferon α. In one study, the end-of-treatment virological response of thymosin α-1 was similar to interferon α [58]. Another study showed that a higher proportion of thymosin α-1-treated patients had normal ALT, undetectable HBV DNA, and HBeAg loss at 12 months (55.6%) compared to interferon α-treated patients (27.3%; $P < 0.05$) [59]. A follow up of this study demonstrated that thymosin α-1-treated patients had significantly higher sustained virological and biochemical responses than interferon α-treated patients [60].

Because of its unique mode of action, thymosin α-1 might provide an additive or synergistic effect when used in combination therapy. The majority of patients who received thymosin α-1 plus interferon α had undetectable HBV DNA and normal ALT at the end of treatment at 52 weeks (18 of 21; 87.7%) and at the end of follow up at 78 weeks (16 of 21; 76.2%) [61]. Patients who received thymosin α-1 plus lamivudine were also more likely to attain undetectable HBV DNA and normal ALT than patients who received interferon α plus lamivudine (53 compared to 28%, $P < 0.05$) [62].

Immune modulators in earlier stages of development

Therapeutic HBV vaccination. A novel concept that is being evaluated currently for the treatment of CHB infection is the use of HBV vaccine in a strategy termed as therapeutic HBV vaccination. The ability of vaccination to overcome the host's immunologic unresponsiveness was initially demonstrated in preclinical studies which showed that repeated immunizations with woodchuck hepatitis surface antigen (WHsAg) resulted in an antibody response in 17 of 18 chronically infected woodchucks [63]. Several lines of clinical evidence have also suggested that HBV vaccination could reduce HBV DNA in

CHB patients. In a pilot study, HBV vaccination reduced HBV replication by inducing HBV-specific CD4+ T lymphocytes to produce high levels of interferon α or γ [64]. In a study of 118 treatment-naïve CHB patients, those who received either five intramuscular injections of a preS2/S (GenHevac B; Pasteur-Merieux; $n = 46$) or an S vaccine (Recombivax; Merck & Co; $n = 34$) were more likely to have undetectable HBV DNA at 3 months than patients who did not receive vaccine ($n = 37$; 16.3 *versus* 2.7%; $P = 0.033$) [65]. These encouraging data led to further animal-model and clinical studies that assessed the therapeutic efficacy of HBV vaccination when combined with antiviral agents (entecavir [66]) and other immunomodulators (interferon [67]).

Nucleic acid-based immunotherapies. Another novel approach for the treatment of CHB is the use of catalytic ribozymes and oligonucleotides that act either by antisense mechanism or by enhancing host immunity. Many of these compounds remain in the preclinical stages of drug development (Fig. 1). EHT-899 (Enzo Biochem Inc) is one such oligonucleotide DNA-based compound that has now reached the clinical phase of development as an immunomodulator for the treatment of CHB. In a preclinical study, the oral administration of EHT-899 suppressed the growth of human liver cancer cells that were experimentally implanted into mice [68]. Early-phase clinical trials have demonstrated that the oral administration of EHT-899 resulted in a reduction in HBV DNA level and serum transaminases in 46% of CHB patients, an improvement in biopsy scores in 58% of patients, and histological improvement in liver necroinflammatory score in 33% of patients [68]

Investigational drugs for the treatment of CHC

The discovery and development of novel drugs to treat CHC was previously hampered by the absence of a stable cell-based system that could support HCV replication and the paucity of effective animal models. The research utility of the chimpanzee model, which is the only well-established animal model that supports HCV replication, is limited by its cost, availability, and ethical issues. Recently, a transgenic mouse model with human hepatocyte xenografts that allow HCV replication was developed. Likewise, a cell-based system that uses human hepatoma cell lines that allow the replication of HCV nonstructural proteins have also been developed for the testing of antiviral drugs. The use of the HCV replicon system is, however, limited by its lack of ability to assess viral infectivity. Nonetheless, the HCV replicon system has allowed the field to undergo substantial progress in recent years.

A number of investigational anti-HCV compounds are now undergoing preclinical and clinical development (Tab. 2 and Fig. 3). Compounds that impair essential HCV-related functions such as enzymes (serine proteases [69], helicase [70], and polymerase [71]) or protein translation [72] are being developed. Strategies that alter the integrity of the HCV genome such as the activa-

Table 2. Investigational drugs in clinical development for the treatment of CHC

Antiviral drugs
Enzyme inhibitors
Serine protease inhibitors
BILN 2061
SCH-6
VX-950
RNA polymerase inhibitors
NM-283
HCV-086 and the VP-50406 series
JTK-003 and JTK-109
Genome-directed strategy
ISIS-14803 and antisense oligonucleotides
Heptazyme and catalytic ribozymes
RNA interference
Ribavirin-like and IMPDH inhibitors
Viramidine
Levovirin
VX-497
Mycophenolate mofetil
p7 inhibitors
UT231B
Immunomodulators
Thymosin α-1
Interferon β
Immunoglobulins
Albuferon
Albuferon β
Civacir
Hepex-C
Therapeutic HCV vaccination
Innovac-C
Chiron HCV vaccine
Intercell HCV vaccine

tion of catalytic ribozymes [73], the use of oligonucleotides that act through antisense mechanisms [74], and short interfering RNA (siRNA) duplexes [75] are also being evaluated. Approaches that enhance HCV-specific humoral and cellular immune responses by the use of peptides, oligonucleotides, immunoglobulins, and vaccines are undergoing clinical development. It is hoped that the collective success of these investigational compounds, alone or

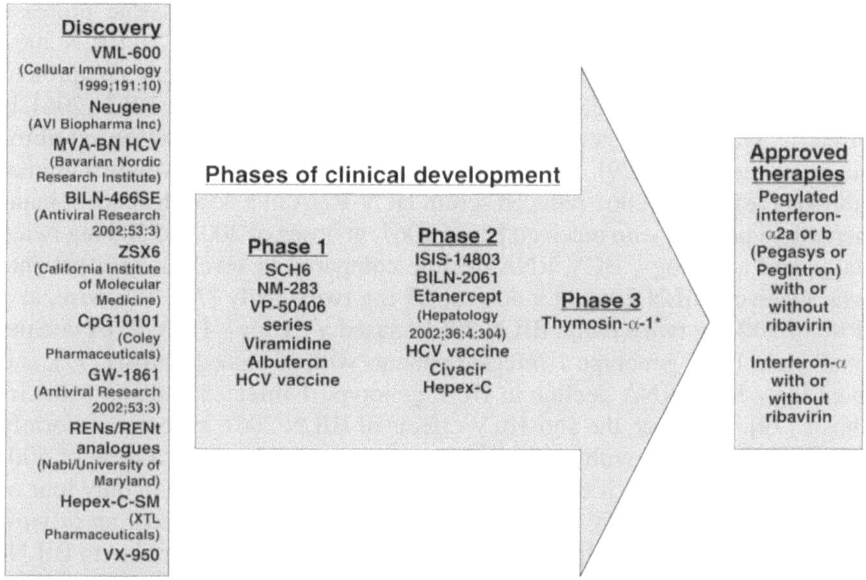

Figure 3. Investigational and approved compounds for the treatment of CHC infection. References for investigational compounds that were not discussed in the text are indicated in parentheses. If no published reference is available, the company developing the compound is indicated. *Compounds currently available for CHC treatment in some countries. RENs/RENt: ring-expanded nucleoside/ ring-expanded nucleotide

hoped that the collective success of these investigational compounds, alone or in combination, will transform CHC from an often-incurable infection into a potentially curable ailment. In contrast to anti-HBV therapies, a distinct advantage of CHC therapy development is the robustness of the clinical end point of undetectable HCV RNA several months following cessation of antiviral therapy. This end point has been proven to predict long-term benefit to patients [2, 6].

Antiviral drugs

The protease and helicase inhibitors

Two HCV proteases—NS2-3 and NS3-4A proteases—are necessary for polyprotein processing. NS2-3 protease catalyzes the cleavage of the HCV polyprotein at the NS2/3 junction while NS3-4A protease cleaves the NS3/NS4A, NS4A/NS4B, NS4B/NS5A, and NS5A/NS5B junctions. The helicase, on the other hand, functions to unwind double-stranded RNA or DNA in preparation for replication or transcription. The essential roles of proteases and helicases in HCV replication have made these enzymes prime targets for several novel anti-HCV compounds [76].

The most advanced in clinical development among the serine protease inhibitors is BILN 2061 (Boehringer Ingelheim Corp; Fig. 4) [77]. BILN 2061 inhibits HCV replication by blocking NS3 protease-dependent polyprotein processing [69]. Three early-phase clinical trials indicate that BILN 2061 is effective against HCV genotype 1 in patients with minimal or advanced fibrosis or cirrhosis [77–79]. In a study of HCV-infected patients with minimal fibrosis [78], BILN 2061 reduced serum HCV RNA in a dose-dependent manner: all 16 patients who received BILN 2061, at doses of 200 and 500 mg twice daily, had a ≥1 $\log_{10}$ HCV RNA decline compared to seven of nine patients who received BILN 2061 at a dose of 25 mg twice daily [78]. Likewise, at a dose of 200-mg twice daily, BILN 2061 caused a ≥1 $\log_{10}$ HCV RNA decline in all eight HCV genotype 1-infected patients with advanced fibrosis [77], and a ≥2 $\log_{10}$ HCV RNA decline in HCV genotype 1-infected patients with cirrhosis [79]. However, the anti-HCV effects of BILN 2061 were not uniformly observed in patients with HCV non-genotype 1 infections. At a dose of 500-mg twice daily, BILN 2061 caused ≥1 $\log_{10}$ HCV RNA decline in only four of eight patients with HCV genotype 2 or 3 infection [80]. Indeed, an *in vitro* assay has demonstrated that HCV non-genotype 1 is less susceptible to BILN-2061 than genotype 1 [81]. This deficiency, combined with recently reported findings of cardiac toxicity, has temporarily halted this compound's clinical development [82].

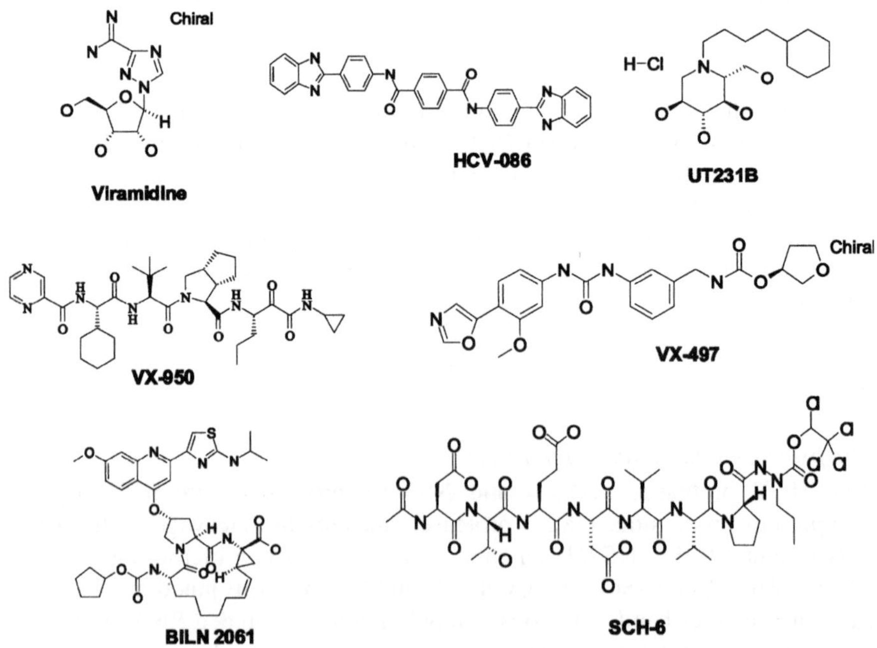

Figure 4. Chemical structures of selected compounds undergoing development for the treatment of CHC.

The other serine proteases in earlier phases of drug development include VX-950 (Vertex Pharmaceuticals; Fig. 4) and SCH-6 (Schering-Plough; Fig. 4). VX-950, a small molecule inhibitor of NS3-4A protease [83], exhibited anti-HCV activity that is more sustained compared to BILN 2061. In addition, VX-950 is effective against viral mutants that were resistant to BILN2061 [83]. SCH-6, a novel ketoamide peptidomimetic inhibitor of NS3-encoded serine protease, has also been demonstrated in the replicon system to exhibit anti-HCV properties and will soon undergo early phases of clinical development [84].

RNA-dependent RNA polymerase inhibitors

The HCV RNA-dependent RNA polymerase NS5B is essential for the synthesis of the negative- and positive-stranded RNA intermediates during viral replication. Thus, the inhibition of this enzyme by nucleoside analogues and non-nucleoside compounds could result in the inhibition of HCV replication. JTK-003 (Akros), a non-nucleoside RNA-dependent RNA polymerase, has been reported to be well-tolerated in phase 1 dose-ranging studies and is being evaluated currently in larger-scale clinical trials [85]. The other non-nucleoside inhibitors of RNA-dependent RNA polymerase that are undergoing preclinical studies include *JTK-109* (Akros) and the benzo-1,2,4-thiazidines (GlaxoSmithKline) [85].

Among the nucleoside inhibitors of RNA-dependent RNA polymerase, *NM-283* (NV-08; Idenix Pharmaceuticals), a prodrug designed to improve the bioavailability of NM-107 [86], is most advanced in its clinical development. In preclinical studies, NM-283 exhibited potent and consistent suppression of HCV [86]. In the primate model, NM-283 rapidly decreased HCV RNA levels by 0.83 and 1.05 $\log_{10}$ after 7 days of treatment with 8.3 and 16.6 mg/kg respectively [87]. In an ongoing phase I/II dose escalation trial, a dose-related decline in HCV RNA was observed at the end of a 14-day treatment course [88]. These encouraging results are anticipated to expand the clinical evaluation of NM-283 [88]. A series of RNA-dependent RNA polymerases collectively known as the VP-50406 series (ViroPharma and Wyeth Pharmaceuticals) is also being developed for the treatment of CHC. Early-phase clinical evaluation of two members of this group of compounds (VP-50406 and HCV-371) demonstrated non-significant reductions in HCV RNA levels and a third member, HCV-086 (Fig. 4), has started phase I clinical evaluation [89].

Nucleic acid-based therapies

The activation of catalytic ribozymes, the use of antisense oligonucleotides, and siRNAs are novel strategies that could alter the integrity of the HCV genome and thus have the potential for the treatment of CHC. The activation of catalytic ribozymes [73], which are RNA or modified RNA polymers that are capable of cleaving specific target RNA sequences, have undergone early-phase clinical evaluation but with discouraging results. The prototype compound of this class, heptazyme (Ribozyme Pharmaceuticals), is a synthetic stabilized 33-mer first-generation ribozyme product that was initially demonstrated to significantly inhibit viral replication in the HCV–poliovirus chimera model [75, 90]. The antiviral effect of hepatazyme was improved *in vitro* if it is combined with interferon α therapy [91]. A pilot clinical trial indicated significant reduction in HCV RNA in approximately 10% of patients receiving heptazyme but toxicities in primate models halted its further clinical development in favor of second-generation anti-HCV ribozymes that are characterized by improved stability, specificity, and therapeutic index. Recent reports, however, indicate that the clinical development of the second-generation heptazyme has also been abandoned in favor of RNA interference [92]. RNA interference is a strategy wherein double-stranded RNAs known as siRNAs are used to down-regulate genes at the post-translational level (termed as gene silencing) [93]. Data from clinical studies are yet to be reported for the siRNA stratregy.

The third nucleic acid-based strategy that is currently undergoing evaluation for the treatment of CHC is the use of antisense oligonucleotides, which are short DNA or RNA sequences that inhibit the HCV translation mechanism by binding to target mRNA, which then results in the (1) disruption of essential secondary or tertiary structure, (2) blockade of the binding of the ribosome to the internal ribosomal entry site (IRES), and (3) cleavage of IRES RNA through the activation of the RNase H, a ubiquitous cellular enzyme that destroys the RNA in a DNA/RNA duplex. The most advanced in clinical development among this group of compounds is ISIS-14803 (Isis Pharmaceuticals), a 20-base 5-methylcytidine phosphorothioate oligodeoxynucleotide, which is complementary to a highly conserved region of HCV IRES that contains the AUG initiation codon, and induces the enzymatic cleavage of IRES [74]. A characteristic effect of the phosphorothioate-based compounds, which include ISIS-14803, is their ability to trigger immune cell stimulation, complement activation, and induction of the coagulation cascade [94]. The 5-methylcytidine modification of the oligonucleotide in ISIS-14803 have rendered it to be less likely to elicit immune stimulation and complement activation, while also increasing its target binding affinity [94]. Early-phase clinical trials of ISIS-14803 indicated modest anti-HCV activity [74, 95, 96]. When CHC patients received ISIS-14803 intravenously or subcutaneously at doses of 0.5–3.0 mg/kg thrice weekly, a dose-dependent response was observed, with $\geq 0.9 \log_{10}$ decline in HCV RNA in some patients who received 2.0 mg/kg [95]. An unexplained ALT elevation was observed in some patients, including levels that exceeded 10 times the upper limit of normal values. Analysis of preliminary data suggested that the ALT elevations were not drug- or dose-related [95], but may be associated with the decline in HCV RNA levels [97]. Higher doses of ISIS-14803 are being evaluated in an ongoing 12-week phase II dose-escalation study. The preliminary data demonstrate that seven of 34 patients who received 6 mg/kg ISIS-14803 once or twice weekly achieved a $\geq 1 \log_{10}$ reduction in HCV RNA [97].

There are, however, certain limitations that have delayed the clinical development of nucleic acid-based strategies, including the logistic issues associated with large-scale production and the anticipated high cost of manufacture. The high specificity of target binding, for example with the ISIS-14803 binding to a specific region in IRES, will likely translate into a reduced or even lack of efficacy against HCV genotypes or quasispecies that contain variation in that specific region of binding. The combination of antisense oligonucleotides that are complementary to different quasispecies may curtail this specific limitation. The degree of dependence of antisense oligonucleotides on RNase H activity may also affect their overall efficacy because of the relatively low level of RNase H in the cytoplasm where HCV replication occurs. There is also the potential for antisense oligonucleotides to inhibit human liver IRES and subsequently affect the expression of human mRNAs [98]. Moreover, since these molecules do not freely cross cell membranes, efforts need to be undertaken to overcome the currently poor delivery systems of antisense oligonucleotides and other gene therapies [99]. The chemical modification of the antisense oligonucleotide backbone and the use of liposomes, synthetic polymers, and non-viral vectors as vehicles for delivery have been suggested as solutions to this limitation [99].

Ribavirin-like drugs and inosine 5'-monophosphate dehydrogenase (IMPDH) inhibitors

Several mechanisms of action have been proposed for ribavirin including the modulation of the host Th1/Th2 cytokine balance, the inhibition of the HCV RNA-dependent RNA polymerase (possibly by acting as a pseudo-substrate) [100], the activity as an RNA viral mutagen [101], and the competitive inhibition of the host IMPDH, an enzyme that catalyzes the essential step in the *de novo* biosynthesis of guanine nucleotides. When used as a monotherapy, ribavirin is ineffective for the treatment of CHC. Its efficacy is demonstrated when used in combination with interferon-based therapy. However, the use of ribavirin is complicated by hemolytic anemia. Viramidine (Ribapharm ; Fig. 4), a carboxamide liver-targeting prodrug that is converted to ribavirin in the liver by the hepatic enzyme adenosine deaminase, is being developed to potentially curtail the current toxicities of ribavirin. Its anti-HCV property is expected to be similar to that of ribavirin. Initial investigations have already indicated that, in contrast to ribavirin, viramidine exhibits a more favorable safety profile. Based on monkey toxicology studies, viramidine is expected to reduce ribavirin-associated hematologic toxicity. A phase I study has demonstrated that humans tolerate viramidine at single doses of up to 1200 mg [102]. In a study that compared viramidine and ribavirin as combination treatment with PEGylated interferon α2a, the ribavirin concentration in red blood cells was significantly lower in subjects receiving the viramidine formulation [103], a finding that correlated with the lower incidence of hemolytic anemia during viramidine therapy [102–104]. A multiple-dose phase II study is ongoing to assess safety, pharmacokinetics, and pharmacodynamics in CHC patients [103, 104].

The non-nucleoside inhibitors of IMPDH that are being evaluated as potential therapies for CHC are VX-497 (Merimepodib; Vertex Pharmaceuticals; Fig. 4) and the immunosuppressive drug mycophenolate mofetil (Cellcept; Roche). VX-497 is a potent reversible noncompetitive inhibitor of IMPDH that is effective against a variety of DNA and RNA viruses [105]. When used as a monotherapy, VX-497 has not been shown to affect HCV RNA levels in humans, but when used in combination with interferon α its anti-HCV effects are comparable to the ribavirin/interferon combination therapy [105]. In the HCV replicon system, the addition of VX-497 to ribavirin has also been shown to result in enhanced antireplicon effect [106], which contrasts to the marginal anti-replicon effect when VX-497 was administered alone. Mycophenolate mofetil is a widely used immunosuppressive drug to prevent allograft rejection in solid organ transplant recipients. Because it inhibits IMPDH noncompetitively, its anti-HCV properties have also been evaluated in clinical trials. In preclinical studies, the combination of mycophenolate mofetil with ribavirin resulted in an enhanced anti-replicon effect [106]. However, preliminary clinical data suggest that combination therapy with interferon α and mycophenolate mofetil was ineffective in improving the virologic response of CHC patients [107]. Moreover, clinical studies have observed poorer outcomes among CHC patients who received mycophenolate mofetil-based immunosuppressive therapy after liver transplantation for CHC [108].

P7 protein inhibitors

A novel structural target that is currently evaluated for treatment of CHC is p7, a 70-amino acid polytopic membrane protein that belongs to a group known as viroporins, which mediate cation permeability across membranes and are important for virion release and maturation. Studies in pestiviruses and bovine viral diarrhea viruses (BVDVs) have suggested that p7 protein does not affect viral replication *per se* but instead may be important in the production of infectious virion particles. The functional deletion of p7 resulted in the production of noninfectious virions. Long-alkyl-chain iminosugar derivatives, which have been demonstrated to exhibit antiviral properties against BVDV (a surrogate model for HCV infection), have also been demonstrated to inhibit HCV p7 ion channels suggesting its potential for the treatment of CHC [109]. UT231B (United Therapeutics; Fig. 4) is the first antiviral iminosugar p7 inhibitor (also known as alkovirs and α-galactosidase inhibitors) that is currently undergoing early-phase clinical trials [109]. The results of these early-phase trials are yet to be reported.

Immune modulators

Thymosin α-1

As discussed above, thymosin α-1 is a synthetic formulation of a 28-amino acid non-glycosylated peptide with multiple biologic effects, including the stimulation of a Th1 cytokine response and the suppression of Th2 response. Several clinical trials have demonstrated the anti-HCV efficacy of thymosin α-1 especially when combined with interferon α [110–112]. While a pilot trial showed that the anti-HCV efficacy of thymosin α-1 was not significantly different than that for placebo [113], at least three other clinical trials and an *in vitro* investigation indicate that thymosin α-1 may be beneficial when used in combination with interferon α. In a randomized, double-blind, placebo-controlled study [111] patients who received thymosin α-1 plus interferon α were more likely than those who received interferon α monotherapy to have undetectable HCV RNA (37.1 *versus* 18.9%) and normalization of ALT (37.1 *versus* 16.2%) at the end of treatment [111]. Currently, thymosin α-1 in combination with PEGylated interferon α is undergoing phase III clinical trials for CHC in the USA [114].

Interferon

Interferon-β. In some countries, a 6–8 week treatment course of interferon β is approved for treatment of CHC, especially in patients with contraindications to interferon α. A recent study reported that interferon β for 6 weeks resulted in a sustained virologic response in four of five elderly patients [115]. A novel recombinant human albumin–interferon β fusion protein (albuferon-β, Human Genome Sciences) is being investigated for the treatment of CHC.

Albuferon. Albuferon (Human Genome Sciences) is a 85.7-kDa fusion protein that consists of a recombinant interferon α and recombinant human albumin, and thus it combines the antiviral properties of interferon α and the long serum half-life of albumin [116]. A phase I, open-label, dose-escalation study is currently ongoing to assess its pharmacokinetics, safety, and immunogenicity in CHC patients who failed interferon α. A preliminary analysis of 63 patients indicates that albuferon increased 2'5'-oligoadenylate synthetase mRNA levels 9-fold, indicating the induction of antiviral protein secretion, which in some patients persisted for 21 days after albuferon administration [117]. A $\geq$0.55 $\log_{10}$ decline in HCV RNA was observed in 52% of patients who received 80 μg of albuferon [117]

Therapeutic vaccination

A novel strategy that is undergoing development as treatment for CHC is the use of a vaccine to induce HCV-specific immunity. Among these vaccines are the structural HCV protein vaccine (Chiron), a vaccine with a 5 T cell epitope peptide (Intercell) and the vaccine containing purified viral E1 protein subtype 1b (InnoVac-C; Innogenetics). InnoVac-C, the most advanced among the vaccines undergoing clinical development, was demonstrated in a phase I trial to induce strong specific Th1-type cellular and humoral immune response in healthy volunteers [118]. In a subsequent phase II, placebo-controlled, clinical trial, CHC patients who received several doses of InnoVac-C vaccine developed anti-E1 antibody and had higher E1-specific T cell proliferation indices compared to patients who received placebo. While HCV RNA remained unchanged following InnoVac-C vaccination, a trend towards a decrease in ALT and improvement in liver histology were noted. The reductions in ALT and Ishak scores correlated with the increase in anti-E1 antibody levels [118]. Long-term studies are now required to confirm the disease-modifying beneficial effects of anti-HCV therapeutic vaccination.

Anti-HCV immunoglobulin

The passive transfer of anti-HCV immunoglobulin is another immunologic approach that is being evaluated for the treatment of CHC [119]. The two members of this class that are undergoing early-phase studies are Hepex-C (XTL Pharmaceuticals) and HCIg (Civacir; Nabi Biopharmaceuticals). Hepex-C, a human monoclonal antibody directed against the HCV envelope protein E2, was generated from peripheral blood B cells of CHC patients [120]. In a phase I study of 15 CHC patients, single doses of Hepex-C of up to 40 mg were well tolerated [120]. In eight of 15 patients there was transient HCV RNA reduction to half of pretreatment levels. The preliminary report of a subsequent multi-dose, phase I study demonstrate a ≥1 log reduction in HCV RNA in 45% of patients who received up to 80 mg of Hepex-C. A phase II pilot study for the prevention of HCV recurrence after liver transplantation for CHC is underway [120]. Civacir has also been demonstrated to shorten the duration and reduce the degree of HCV viremia in the chimpanzee model [121, 122].

Conclusions

With the large number of investigational compounds undergoing clinical development, the landscape of anti-hepatitis virus therapeutics is anticipated to undergo remarkable transformation during the next decade. There will likely be more options for treating CHB and CHC in patients who may have failed current therapies or who possess clinical and virologic characteristics that limit the use of currently approved drugs. Indeed, the dynamic state of anti- hepatitis virus therapeutic research foreshadows an emerging paradigm of combination anti-HBV and -HCV therapy. According to this paradigm, antiviral compounds with complementary mechanisms of action and varying viral targets would be used in what may ultimately become an optimal anti-hepatitis virus-treatment strategy. In reviewing the current pipeline, this may be possible to achieve in the foreseeable future. Obviously this optimism needs to be tempered by the realism that many attractive and promising therapeutic candidates do not survive the rigorous phases of clinical development. Nonetheless, as indicated by the research and clinical progress of investigational drugs discussed herein, the field of anti-hepatitis virus therapeutics is at its most exciting stage ever.

References

1 Mahoney FJ (1999) Update on diagnosis, management, and prevention of hepatitis B virus infection. *Clin Microbiol Rev* 12: 351–366

2 Brown RS Jr, Gaglio PJ (2003) Scope of worldwide hepatitis C problem. *Liver Transpl* 9, S10–S13

3 Lai CL, Chien RN, Leung NW, Chang TT, Guan R, Tai DI, Ng KY, Wu PC, Dent JC, Barber J et al. (1998) A one-year trial of lamivudine for chronic hepatitis B. Asia Hepatitis Lamivudine Study Group. *N Engl J Med* 339: 61–68

4 Hadziyannis SJ, Tassopoulos NC, Heathcote EJ, Chang TT, Kitis G, Rizzetto M, Marcellin P, Lim SG, Goodman Z, Wulfsohn MS et al. (2003) Adefovir dipivoxil for the treatment of hepatitis B e antigen-negative chronic hepatitis B. *N Engl J Med* 348: 800–807

5 Angus P, Vaughan R, Xiong S, Yang H, Delaney W, Gibbs C, Brosgart C, Colledge D, Edwards R, Ayres A et al. (2003) Resistance to adefovir dipivoxil therapy associated with the selection of a novel mutation in the HBV polymerase. *Gastroenterology* 125: 292–297

6 Hadziyannis SJ, Papatheodoridis GV (2003) Peginterferon-alpha2a (40 kDa) for chronic hepatitis C. *Expert Opin Pharmacother* 4: 541–551

7 van Poelje PD, Colby TJ, Fujitaki JM, Reddy KR, Matelich MC, Senekeo KA, Linemeyer DL, Erion MD (2003) MB6866 (Hepavir B), a hepdirect prodrug of adefovir: mechanism of activation and liver targeting [abstract 1143]. *Hepatology* 38: 706A

8 Jackson TE, Kong L, Turnbow C, Hartsough KC, France M, Serebryany V, Adamic J, Radka S, Dickinson B, Zinnen S et al. (2002) Second generation ribozyme development: improved hepatic delivery of a nuclease-resistant ribozyme targeting the hepatitis B virus RNA using galactosamine conjugation [abstract 1927]. *Hepatology* 36: 645A

9 Beckebaum S, Cicinnati VR, Gerken G (2002) DNA-based immunotherapy: potential for treatment of chronic viral hepatitis? *Rev Med Virol* 12: 297–319

10 Colonno RJ, Genovesi EV, Medina I, Lamb L, Durham SK, Huang ML, Corey L, Littlejohn M, Locarnini S, Tennant BC et al. (2001) Long-term entecavir treatment results in sustained antiviral efficacy and prolonged life span in the woodchuck model of chronic hepatitis infection. *J Infect Dis* 184: 1236–1245

11 Lai CL, Rosmawati M, Lao J, Van Vlierberghe H, Anderson FH, Thomas N, Dehertogh D (2002) Entecavir is superior to lamivudine in reducing hepatitis B virus DNA in patients with chronic hepatitis B infection. *Gastroenterology* 123: 1831–1838

12 Chang T, Hadziyannis S, Cianciara J, Rizzeto O, Schiff E, Pastore G, Klesczewski K, Denisky G, DeHertogh D, Hindes R (2002) Sustained viral load and ALT reduction following 48 weeks of entecavir treatment in subjects with chronic hepatitis B who have failed lamivudine [abstract 550]. *Hepatology* 36: 300A

13 Colonno RJ, Rose R (2002) Genotypic analysis following 48 week entecavir treatment of chronic HBV subjects failing lamivudine suggests sustained susceptibility and lack of resistance development [abstract 1925]. *Hepatology* 36: 644A

14 Schiff E, Hindes R, O'Donnell A, DeHertogh D, Kreter B (2002) Summary of phase II clinical and laboratory safety experience with entecavir [abstract 1919]. *Hepatology* 36: 643A

15 Standring DN, Bridges EG, Placidi L, Faraj A, Loi AG, Pierra C, Dukhan D, Gosselin G, Imbach JL, Hernandez B et al. (2001) Antiviral beta-L-nucleosides specific for hepatitis B virus infection. *Antivir Chem Chemother* 12 (Suppl 1): 119–129

16 Bryant ML, Bridges EG, Placidi L, Faraj A, Loi AG, Pierra C, Dukhan D, Gosselin G, Imbach JL, Hernandez B et al. (2001) Antiviral L-nucleosides specific for hepatitis B virus infection. *Antimicrob Agents Chemother* 45: 229–235

17 Lai C-L, Leung NW, Teo EK, Tong MJ, Wong F, Hann W-W, Han S, Poynard T, Myers M, Chao G et al. (2003) A phase IIb comparative trial of LdT, lamivudine, and the combination in hepatitis B patients: Greater antiviral effect with LdT. In: *Program and Abstracts of the 43rd Interscience Conference on Antimicrobial Agents and Chemotherapy*, Chicago, IL. American Society Microbiology Press, Washington

18 Furman PA, Davis M, Liotta DC, Paff M, Frick LW, Nelson DJ, Dornsife RE, Wurster JA, Wilson LJ, Fyfe JA et al. (1992) The anti-hepatitis B virus activities, cytotoxicities, and anabolic profiles of the (–) and (+) enantiomers of cis-5-fluoro-1-[2-(hydroxymethyl)-1,3-oxathiolan-5-yl]cytosine. *Antimicrob Agents Chemother* 36: 2686–2692

19 Schinazi RF, Gosselin G, Faraj A, Korba BE, Liotta DC, Chu CK, Mathe C, Imbach JL,

Sommadossi JP (1994) Pure nucleoside enantiomers of beta-2',3'-dideoxycytidine analogs are selective inhibitors of hepatitis B virus *in vitro*. *Antimicrob Agents Chemother* 38: 2172–2174
20 Condreay LD, Jansen RW, Powdrill TF, Johnson LC, Selleseth DW, Paff MT, Daluge SM, Painter GR, Furman PA, Ellis MN et al. (1994) Evaluation of the potent anti-hepatitis B virus agent (–)-cis-5-fluoro-1-[2-(hydroxymethyl)-1,3-oxathiolan-5-yl]cytosine in a novel *in vivo* model. *Antimicrob Agents Chemother* 38: 616–619
21 Cullen JM, Smith SL, Davis MG, Dunn SE, Botteron C, Cecchi A, Linsey D, Linzey D, Frick L, Paff MT et al. (1997) *In vivo* antiviral activity and pharmacokinetics of (–)-cis-5-fluoro-1-[2-(hydroxymethyl)-1,3-oxathiolan-5-yl]cytosine in woodchuck hepatitis virus-infected woodchucks. *Antimicrob Agents Chemother* 41: 2076–2082
22 Korba BE, Schinazi RF, Cote P, Tennant BC, Gerin JL (2000) Effect of oral administration of emtricitabine on woodchuck hepatitis virus replication in chronically infected woodchucks. *Antimicrob Agents Chemother* 44: 1757–1760
23 Gish RG, Leung NW, Wang C, Corey L, Sacks S, Fried M, Wright TL, Huy T, Chan F, Rousseau F et al. (2002) Antiviral activity, safety, and incidence of resistance in chronically infected hepatitis B patients (CHB) given once daily emtricitabine for 2 years [abstract 838]. *Hepatology* 36: 372A
24 Cecil III BD (2002) Tenofovir 300 mg daily is effective in patients with chronic HBV who fail lamivudine. *Hepatology* 36: 630A
25 van Bommel F, Wunsche T, Schurmann D, Bergk A, Wiedemman B, Hopf U, Berg T (2003) A comparison of adefovir and tenofovir in the treatment of lamivudine-resistant hepatitis B virus (HBV) infection [abstract 246]. *Hepatology* 38: 275A
26 Ristig MB, Crippin J, Aberg JA, Powderly WG, Lisker-Melman M, Kessels L, Tebas P (2002) Tenofovir disoproxil fumarate therapy for chronic hepatitis B in human immunodeficiency virus/hepatitis B virus-coinfected individuals for whom interferon-alpha and lamivudine therapy have failed. *J Infect Dis* 186: 1844–1847
27 Nunez M, Perez-Olmeda M, Diaz B, Rios P, Gonzalez-Lahoz J, Soriano V (2002) Activity of tenofovir on hepatitis B virus replication in HIV-co-infected patients failing or partially responding to lamivudine. *Aids* 16: 2352–2354
28 Benhamou Y, Tubiana R, Thibault V (2003) Tenofovir disoproxil fumarate in patients with HIV and lamivudine-resistant hepatitis B virus. *N Engl J Med* 348: 177–178
29 Park J, Braun J, Kreiswirth S, Goldman D, Mullen M, Dietrich D (2002) Co-infectiion with HIV and HBV: The effect of tenofovir disoproxil fumarate in lamivudine and famciclovir experienced patients [abstract 1911]. *Hepatology* 36: 641A
30 Nelson M, Portsmouth S, Stebbing J, Atkins M, Barr A, Matthews G, Pillay D, Fisher M, Bower M, Gazzard B (2003) An open-label study of tenofovir in HIV-1 and Hepatitis B virus co-infected individuals. *Aids* 17, F7–10
31 Benhamou Y, Piketty C, Katlama C, Rozembaum W, Neau D, Lafeuillade A, Rey D, Yazdanpanah Y, Trimoulet P, Fleury H (2003) Efficacy of tenofovir disoprixil fumarate (TDF) for hepatitis B virus (HBV) in human immunodeficiency virus (HIV) infected patients [Abstract 1155]. *Hepatology* 38: 712A
32 Zhu YL, Dutschman DE, Liu SH, Bridges EG, Cheng YC (1998) Anti-hepatitis B virus activity and metabolism of 2',3'-dideoxy-2',3'-didehydro-beta-L(–)-5-fluorocytidine. *Antimicrob Agents Chemother* 42: 1805–1810
33 Le Guerhier F, Pichoud C, Guerret S, Chevallier M, Jamard C, Hantz O, Li XY, Chen SH, King I, Trepo C et al. (2000) Characterization of the antiviral effect of 2',3'-dideoxy-2',3'-didehydro-beta-L-5-fluorocytidine in the duck hepatitis B virus infection model. *Antimicrob Agents Chemother* 44: 111–122
34 Le Guerhier F, Pichoud C, Jamard C, Guerret S, Chevallier M, Peyrol S, Hantz O, King I, Trepo C, Cheng YC, Zoulim F (2001) Antiviral activity of beta-L-2',3'-dideoxy-2',3'-didehydro-5-fluorocytidine in woodchucks chronically infected with woodchuck hepatitis virus. *Antimicrob Agents Chemother* 45: 1065–1077
35 Zoulim F, Berthillon P, Guerhier FL, Seigneres B, Germon S, Pichoud C, Cheng YC, Trepo C (2002) Animal models for the study of HBV infection and the evaluation of new anti-HBV strategies. *J Gastroenterol Hepatol* 17 (Suppl): S460–S463
36 Afdhal NH, O'Brien C, Oshana SC, Dunkle LM (2002) Rapid, potent anti-HBV activity of ACH-126,443 in a phase 1/2 trial in chronic hepatitis B infection. In: *Programs and Abstracts of the 42nd Interscience Conference on Antimicrobial Agents and Chemotherapy*, San Diego, CA.

American Society Microbiology Press, Washington
37 Aguesse-Germon S, Liu SH, Chevallier M, Pichoud C, Jamard C, Borel C, Chu CK, Trepo C, Cheng YC, Zoulim F (1998) Inhibitory effect of 2'-fluoro-5-methyl-beta-L-arabinofuranosyl-uracil on duck hepatitis B virus replication. *Antimicrob Agents Chemother* 42: 369–376
38 Peek SF, Cote PJ, Jacob JR, Toshkov IA, Hornbuckle WE, Baldwin BH, Wells FV, Chu CK, Gerin JL, Tennant BC, Korba BE (2001) Antiviral activity of clevudine [L-FMAU, (1-(2-fluoro-5-methyl-beta, L-arabinofuranosyl) uracil)] against woodchuck hepatitis virus replication and gene expression in chronically infected woodchucks (*Marmota monax*). *Hepatology* 33: 254–266
39 Zhu Y, Yamamoto T, Cullen J, Saputelli J, Aldrich CE, Miller DS, Litwin S, Furman PA, Jilbert AR, Mason WS (2001) Kinetics of hepadnavirus loss from the liver during inhibition of viral DNA synthesis. *J Virol* 75: 311–322
40 Chin R, Shaw T, Torresi J, Sozzi V, Trautwein C, Bock T, Manns M, Isom H, Furman P, Locarnini S (2001) *In vitro* susceptibilities of wild-type or drug-resistant hepatitis B virus to (–)-beta-D-2,6-diaminopurine dioxolane and 2'-fluoro-5-methyl-beta-L-arabinofuranosyluracil. *Antimicrob Agents Chemother* 45: 2495–2501
41 Lee H, Mommeja-Marin H, Sacks S, Sereni D, Bronowicki J, Conway B, Zarski J, Rousseau F, Mondou E, Snow A et al. (2002) A phase I/II dose escalating trial evaluating safety, tolerability, pharmacokinetics and antiviral activity of clevudine in chronic HBV-infected patients (L-FMAU-102). In: *Programs and Abstracts of the 42nd Interscience Conference on Antimicrobial Agents and Chemotherapy*, San Diego, CA. American Society Microbiology Press, Washington
42 Egron D, Perigaud C, Gosselin G, Aubertin AM, Imbach JL (2001) Synthesis and study of a new series of phosphoramidate derivatives as mononucleotide prodrugs. *Nucleosides Nucleotides Nucleic Acids* 20: 751–754
43 Hodge RA (2004) Telbivudine/Torcitabine Idenix/Novartis. *Curr Opin Invest Drugs* 5: 232–241
44 Kamiya N, Kubota A, Iwase Y, Sekiya K, Ubasawa M, Yuasa S (2002) Antiviral activities of MCC-478, a novel and specific inhibitor of hepatitis B virus. *Antimicrob Agents Chemother* 46: 2872–2877
45 Ono-Nita SK, Kato N, Shiratori Y, Carrilho FJ, Omata M (2002) Novel nucleoside analogue MCC-478 (LY582563) is effective against wild-type or lamivudine-resistant hepatitis B virus. *Antimicrob Agents Chemother* 46: 2602–2605
46 Braun DK, Lowe S, McGill JM, Soon D, Teng CH, Wise S, Yeo KP (2002) Potent antiviral activity of LY582563 (MCC-478) in chronic hepatitis B infected patients. In: *Abstracts of the Digestive Disease Week*, San Francisco, CA, T1369
47 Braun DK, Lowe S, McGill JM, Soon D, Teng CH, Wise SD, Yeo KP (2001) Antiviral activity of a potent novel nucleotide analogue (LY582563) in chronic hepatitis B infected patients. In: *Abstracts of the HEPDART 2001*, Maui, HI, A045
48 Perni RB, Conway SC, Ladner SK, Zaifert K, Otto MJ, King RW (2000) Phenylpropenamide derivatives as inhibitors of hepatitis B virus replication. *Bioorg Med Chem Lett* 10: 2687–2690
49 King RW, Ladner SK, Miller TJ, Zaifert K, Perni RB, Conway SC, Otto MJ (1998) Inhibition of human hepatitis B virus replication by AT-61, a phenylpropenamide derivative, alone and in combination with (–)beta-L-2',3'-dideoxy-3'-thiacytidine. *Antimicrob Agents Chemother* 42: 3179–3186
50 Feld J, Colledge D, Sozzi V, Edwards R, Painter GR, Locarnini SA (2002) The phenylpropenamide derivative AT-130 inhibits HBV replication at viral encapsidation and packaging [abstract 549]. *Hepatology* 36: 300A
51 Cooksley WG, Piratvisuth T, Lee SD, Mahachai V, Chao YC, Tanwandee T, Chutaputti A, Chang WY, Zahm FE, Pluck N (2003) Peginterferon alpha-2a (40 kDa): an advance in the treatment of hepatitis B e antigen-positive chronic hepatitis B. *J Viral Hepat* 10: 298–305
52 Sidiropoulos I, Koliouskas D, Masmanidou M, Mandraveli K, Lazaraki G, Dokas S, Ziakas A (2002) Comparative analysis of peg-interferon alpha-2B and lamivudine in the treatment of chronic hepatitis B patients [abstract 1854]. *Hepatology* 36: 626A
53 Sung JS, Chan HL, Chan F, Wong ML, Hui AY, Leung NW (2002) Prospective randomized study comparing combination of pegylated interferon and lamivudine *versus* lamivudine monotherapy for chronic hepatitis B infection [abstract 1862]. *Hepatology* 36: 628A
54 Sherman KE, Jones CC, Goldstein AL, Naylor PH (1991) Low thymosin alpha-1 concentrations in patients chronically infected with the hepatitis B virus. *Viral Immunol* 4: 195–199
55 Li CL, Zhang T, Saibara T, Nemoto Y, Ono M, Akisawa N, Iwasaki S, Maeda T, Onishi S (2002) Thymosin alpha1 accelerates restoration of T cell-mediated neutralizing antibody response in

immunocompromised hosts. *Int Immunopharmacol* 2: 39–46
56 Tuthill C, Martins E, Rudolph AR (2002) *In vitro* antiviral activity of thymalfasin against hepatitis B [abstract 1904]. *Hepatology* 36: 639A
57 Zavaglia C, Severini R, Tinelli C, Franzone JS, Airoldi A, Tempini S, Bettale G, Ideo G (2000) A randomized, controlled study of thymosin-alpha1 therapy in patients with anti-HBe, HBV-DNA-positive chronic hepatitis B. *Dig Dis Sci* 45: 690–696
58 Andreone P, Cursaro C, Gramenzi A, Zavagliz C, Rezakovic I, Altomare E, Severini R, Franzone JS, Albano O, Ideo G et al. (1996) A randomized controlled trial of thymosin-alpha1 *versus* interferon alfa treatment in patients with hepatitis B e antigen antibody–and hepatitis B virus DNA–positive chronic hepatitis B. *Hepatology* 24: 774–777
59 You J, Zhuang L, Tang BZ, Yang WB, Ding SY, Li W, Wu RX, Zhang HL, Zhang YM, Yan SM, Zhang L (2001) A randomized controlled clinical trial on the treatment of Thymosin a1 *versus* interferon-alpha in patients with hepatitis B. *World. J Gastroenterol* 7: 411–414
60 Zhuang L, You J, Tang BZ, Ding SY, Yan KH, Peng D, Zhang YM, Zhang L. (2001) Preliminary results of Thymosin-a1 *versus* interferon-alpha-treatment in patients with HBeAg negative and serum HBV DNA positive chronic hepatitis B. *World. J Gastroenterol* 7: 407–410
61 Saruc M, Ozden N, Turkel N, Ayhan S, Hock LM, Tuzcuoglu I, Yuceyar H (2003) Long-term outcomes of thymosin-alpha1 and interferon alpha-2b combination therapy in patients with hepatitis B e antigen (HBeAg) negative chronic hepatitis B. *J Pharmacol Sci* 92: 1386–1395
62 Pham Thi TT (2002) Clinical research on the treatment of chronic hepatitis B with thymosin-A1 and lamivudine *versus* interferon and lamivudine [abstract 1843]. *Hepatology* 36: 624A
63 Lu M, Klaes R, Menne S, Gerlich W, Stahl B, Dienes HP, Drebber U, Roggendorf M (2003) Induction of antibodies to the PreS region of surface antigens of woodchuck hepatitis virus (WHV) in chronic carrier woodchucks by immunizations with WHV surface antigens. *J Hepatol* 39: 405–413
64 Couillin I, Pol S, Mancini M, Driss F, Brechot C, Tiollais P, Michel ML (1999) Specific vaccine therapy in chronic hepatitis B: induction of T cell proliferative responses specific for envelope antigens. *J Infect Dis* 180: 15–26
65 Pol S, Nalpas B, Driss F, Michel ML, Tiollais P, Denis J, Brecho C (2001) Efficacy and limitations of a specific immunotherapy in chronic hepatitis B. *J Hepatol* 34: 917–921
66 Foster WK, Miller DS, Marion PL, Colonno RJ, Kotlarski I, Jilbert AR (2003) Entecavir therapy combined with DNA vaccination for persistent duck hepatitis B virus infection. *Antimicrob Agents Chemother* 47: 2624–2635
67 Fruttaldo L, Schettino G, Mongio F (1997) Anti-HBV vaccination before therapy with interferon (IFN) in chronic B hepatitis. *Eur Rev Med Pharmacol Sci* 1: 197–201
68 News Release (2001) Enzo Biochem Announces Plans to Initiate Double-Blind Study of Hepatitis B EHT899 Medicine: Presentation at Dallas Meeting of Liver Specialists Details Successful Phase 2 Study Results Relating to Reduced Key Inflammatory Scores. www.aegis.com/news/bw/2001/BW011110.html (accessed May 8, 2004)
69 Lamarre D, Bailey M, Bolger G, Cameron D, Cartier M, Faucher A, Goudreau N, Kukolj G, Lagace L, Pause A et al. (2002) The discovery of BILN-2061 – an orally bioavailable small molecule inhibitor of the HCV serine protease and a promising antiviral for treatment of hepatitis C [abstract 464]. *Hepatology* 36: 279A
70 Keeney D, Lam AM, Jaffe F, Gallagher R, Frick DN (2002) Defining the DNA binding site on hepatitis C virus NS3 helicase as a potential site for rational drug design [abstract 146]. *Hepatology* 36: 203A
71 Lai VC, Dempsey S, Lau JY, Hong Z, Zhong W (2002) Characterization of HCV replicase complexes and its use for the evaluation of viral polymerase inhibitors [abstract 157]. *Hepatology* 36: 206A
72 Hanecak R, Brown-Driver V, Fox MC, Azad RF, Furusako S, Nozaki C, Ford C, Sasmor H, Anderson KP (1996) Antisense oligonucleotide inhibition of hepatitis C virus gene expression in transformed hepatocytes. *J Virol* 70: 5203–5212
73 Tong MJ, Schiff E, Jensen D, Jacobson IM, Eversen G, McHutchison JG Sr, Aitchison R, Gordon GS, Babcock SA, Enright JH et al. (2002) Preliminary analysis of a phase II study of Heptazyme, a nuclease resistant ribozyme targeting hepatitis C virus (HCV) RNA [abstract 788]. *Hepatology* 36: 360A
74 Witherell GW (2001) *ISIS* 14803 (Isis Pharmaceuticals). *Curr Opin Invest Drugs* 2: 1523–1529
75 Macejak DG, Jensen K, Jamison S, Mokler V, Dickinson B, Zinnen S, McSwiggen J, Pavco P,

Beigelman L (2002) Nuclease resistant siRNAs targeting HCV RNA inhibits replication of a chimeric HCV/poliovirus in Hela cells [abstact 1772]. *Hepatology* 36: 606A
76 Myles DC (2001) Recent advances in the discovery of small molecule therapies for HCV. *Curr Opin Drug Disc Devel* 4: 411–416
77 Benhamou Y, Hinrichsen H, Sentjens R, Reiser M, Manns M, Forns X, Avendano C, Cronlein J, Nehmiz G, Steinmann G (2002) Safety, tolerability and antiviral effect of BILN 2061, a novel HCV serine protease inhibitor, after oral teatment over 2 days in patients with chronic hepatitis C, genotype 1, with advanced liver fibrosis [Abstract 563]. *Hepatology* 36: 304A
78 Hinrichsen H, Benhamou Y, Reiser M, Sentjens R, Wedemeyer H, Calleja JL, Forns X, Cronlein J, Nehmiz G, Steinmann G (2002) First report on the antiviral efficacy of BILN 2061, a novel oral HCV serine protease inhibitor, in patients with chronic hepatitis C genotype 1 [abstract 866]. *Hepatology* 36: 379A
79 Wedemeyer H, Erhardt A, Schmiegel W, Hinrichsen H, Chaves R, Yong CL, Nehmiz G, Steinmann G (2003) Safety and antiviral effect of BILN2061, a novel HCV serine protease inhibitor, after oral treatment over 2 days in patients with chronic hepatitis C genotype 1 and cirrhosis [abstract 294]. *Hepatology* 38: 297A
80 Reiser M, Hinrichsen H, Benhamou Y, Sentjens R, Wedemeyer H, Calleja JL, Forns X, Croenlein J, Yong CL, Nehmiz G, Steinmann G (2003) Antiviral effect of BILN-2061, a novel serine HCV protease inhibitor, after oral treatment over 2 days in patients with chronic hepatitis C, non-genotype 1 [abstract 136]. *Hepatology* 38: 221A
81 Thibeault D, Bousquet C, Gingras R, Lagace L, Maurice R, White PW, Lamarre D (2003) Sensitivity of NS3 proteases from various hepatitis C virus genotypes to the antiviral compound BILN2061 [abstract 300]. *Hepatology* 38: 300A
82 Kronemyer R (2004) Vaccine, new therapies needed for hepatitis C. *Infect Dis News* 17: 45
83 Lin K, Gates CA, Luong Y-P, Perni RB, Kwong AD (2003) VX-950: A tight-binding HCV protease inhibitor with a superior sustained inhibitory response in HCV replicon cells [abstract 137]. *Hepatology* 38: 222A
84 Abid K, Liu R, Rossi C, Bogen SL, Baroudy BM, Negro F (2003) *In vitro* antiviral activity of SCH6, a novel inhibitor of the hepatitis C virus NS3 serine protease [abstract 134]. *Hepatology* 38: 220A
85 Walker MP, Hong Z (2002) HCV RNA-dependent RNA polymerase as a target for antiviral development. *Curr Opin Pharmacol* 2: 534–540
86 Standring DN, Lanford R, Wright TL, Chung RT, Bichko V, Cretton-Scott E, Pan-Zhou X, Bergelson S, Qu L, Tausek M, Bridges EG et al. (2003) NM283 has potent antiviral activity against chronic hepatitis C virus genotype 1 infection in the chimpanzee. In: *Abstracts of the 11th Triennial Symposium on Viral Hepatitis and Liver Diseases*, Sydney, Australia, Australian Center for Hepatitis Virology, WE4
87 Walker MP, Yao N, Hong Z (2003) Promising candidates for the treatment of chronic hepatitis C. *Expert Opin Invest Drugs* 12: 1269–1280
88 Godofsky E, Afdhal NH, Rustgi VK, Shick L, Duncan L, Zhou XJ, Chao G, Fang C, Fielman B, Myers M, Brown NA (2004) A phase I/II dose escalation trial assessing tolerance, pharmacokinetics, and antiviral activity of NM283, a novel antiviral treatment for hepatitis C. In: *Abstracts of the Digestive Disease Week*, New Orleans, LA
89 Press release (2004) Viropharma and Wyeth's HCV-086 enters clinic for HCV. February 25, 2004
90 Macejak DG, Jensen KL, Jamison SF, Domenico K, Roberts EC, Chaudhary N, von Carlowitz I, Bellon L, Tong MJ, Conrad A et al. (2000) Inhibition of hepatitis C virus (HCV)-RNA-dependent translation and replication of a chimeric HCV poliovirus using synthetic stabilized ribozymes. *Hepatology* 31: 769–776
91 Macejak DG, Jensen KL, Pavco PA, Phipps KM, Heinz BA, Colacino JM, Blatt LM (2001) Enhanced antiviral effect in cell culture of type 1 interferon and ribozymes targeting HCV RNA. *J Viral Hepat* 8: 400–405
92 Press release (2003) Sirna therapeutics – setting the plan in motion. May 29, 2003
93 Prabhu R, Vital P, Flemington E, Garry RF, Bastian F, Dash S (2003) Development of small interfering RNA (SIRNAS) against full-length hepatitis C virus 1A strain [abstract 393]. *Hepatology* 38: 349A
94 Brown-Driver V, Eto T, Lesnik E, Anderson KP, Hanecak RC (1999) Inhibition of translation of hepatitis C virus RNA by 2-modified antisense oligonucleotides. *Antisense Nucleic Acid Drug Dev* 9: 145–154

95 McHutchison JG Sr, Pockros PJ, Patel K, Nyberg L, Yu RZ, Kwoh TJ, Dorr FA (2002) A phase Ib dose escalation trial of. *ISIS* 14803, an antisense inhibitor of HCV, in patients with chronic HCV: Final report [abstract 562]. *Hepatology* 36: 303A
96 Soler M, McHutchison JG Sr, Kwoh TJ, Dorr FA, Pawlotsky JM (2002) Treatment of chronic hepatitis C with. *ISIS* 14803, an antisense oligonucleotide inhibitor of HCV. Effect of target region sequence on antiviral efficacy [abstract 469]. *Hepatology* 36: 280A
97 Gordon SC, Bacon BR, Jacobson IM, Shiffman ML, Afdhal NH, Yu RZ, McHutchison JG Sr, Kwoh TJ (2003) Treatment of chronic hepatitis C with. *ISIS* 14803, an antisense inhibitor of HCV, given for 12 weeks [abstract 312]. *Hepatology* 38: 306A
98 Hellen CU, Sarnow P (2001) Internal ribosome entry sites in eukaryotic mRNA molecules. *Genes Dev* 15: 1593–1612
99 Shoji Y, Nakashima H (2004) Current status of delivery systems to improve target efficacy of oligonucleotides. *Curr Pharm Design* 10: 785–796
100 Maag D, Castro C, Hong Z, Cameron CE (2001) Hepatitis C virus RNA-dependent RNA polymerase (NS5B) as a mediator of the antiviral activity of ribavirin. *J Biol Chem* 276: 46094–46098
101 Contreras AM, Hiasa Y, He W, Terella A, Schmidt EV, Chung RT (2002) Viral RNA mutations are region specific and increased by ribavirin in a full-length hepatitis C virus replication system. *J Virol* 76: 8505–8517
102 Arora S, Lau D, Gish RG, Rossi S, Lin C-C, Lau JY, Fang JW (2002) Phase 1 clinical studies of viramidine – a liver-targeting prodrug of ribavirin [abstract 773]. *Hepatology* 36: 356A
103 Arora S, Gish RG, Nelson DR, Hong Z, Lin C-C (2004) Clinical study of viramidine in treatment of hepatitis C supports PBC-sparing mechanism of action. *Gastroenterology* 126(4), supplement 2, 666
104 Gish RG, Nelson DJ, Arora S, Braeckman R, Yu G (2004) Hepatitis C combination therapy with viramidine and peginterferon alpha-2a reduces potential for ribavirin-related hemolytic anemia. *Gastroenterology* 126(4), supplement 2, 681
105 Markland W, McQuaid TJ, Jain J, Kwong AD (2000) Broad-spectrum antiviral activity of the IMP dehydrogenase inhibitor VX-497: a comparison with ribavirin and demonstration of antiviral additivity with alpha interferon. *Antimicrob Agents Chemother* 44: 859–866
106 Zhou S, Liu R, Baroudy BM, Malcolm BA, Reyes GR (2003) The effect of ribavirin and IMPDH inhibitors on hepatitis C virus subgenomic replicon RNA. *Virology* 310: 333–342
107 Cornberg M, Hinrichsen H, Teuber G, Berg T, Naumann U, Falkenberg C, Zeuzem S, Manns MP (2002) Mycophenolate mofetil in combination with recombinant interferon alfa-2a in interferon-nonresponder patients with chronic hepatitis C. *J Hepatol* 37: 843–847
108 Razonable RR, Burak KW, van Cruijsen H, Brown RA, Charlton MR, Smith TF, Espy MJ, Kremers W, Wilson JA, Groettum C et al. (2002) The pathogenesis of hepatitis C virus is influenced by cytomegalovirus. *Clin Infect Dis* 35: 974–981
109 Pavlovic D, Neville DC, Argaud O, Blumberg B, Dwek RA, Fischer WB, Zitzmann N (2003) The hepatitis C virus p7 protein forms an ion channel that is inhibited by long-alkyl-chain iminosugar derivatives. *Proc Natl Acad Sci USA* 100: 6104–6108
110 Rasi G, DiVirgilio D, Mutchnick MG, Colella F, Sinibaldi-Vallebona P, Pierimarchi P, Valli B, Garaci E (1996) Combination thymosin alpha 1 and lymphoblastoid interferon treatment in chronic hepatitis C. *Gut* 39: 679–683
111 Sherman KE, Sjogren M, Creager RL, Damiano MA, Freeman S, Lewey S, Davis D, Root S, Weber FL, Ishak KG, Goodman ZD (1998) Combination therapy with thymosin alpha1 and interferon for the treatment of chronic hepatitis C infection: a randomized, placebo-controlled double-blind trial. *Hepatology* 27: 1128–1135
112 Moscarella S, Buzzelli G, Romanelli RG, Monti M, Giannini C, Careccia G, Marrocchi EM, Zignego AL (1998) Interferon and thymosin combination therapy in naive patients with chronic hepatitis C: preliminary results. *Liver* 18: 366–369
113 Andreone P, Cursaro C, Gramenzi A, Buzzi A, Covarelli MG, Di Giammarino L, Miniero R, Arienti V, Bernardi M, Gasbarrini G (1996) A double-blind, placebo-controlled, pilot trial of thymosin alpha 1 for the treatment of chronic hepatitis C. *Liver* 16: 207–210
114 Billich A (2002) Thymosin alpha1. SciClone Pharmaceuticals. *Curr Opin Invest Drugs* 3: 698–707
115 Komorizono Y, Sako K (2003) Natural beta interferon as an initial treatment for the elderly with chronic hepatitis C. *Am J Gastroenterol* 98: 2103–2105
116 Osborn BL, Olsen HS, Nardelli B, Murray JH, Zhou JX, Garcia A, Moody G, Zaritskaya LS,

Sung C (2002) Pharmacokinetic and pharmacodynamic studies of a human serum albumin-interferon-alpha fusion protein in cynomolgus monkeys. *J Pharmacol Exp Ther* 303: 540–548
117 Balan V, Sulkowski M, Nelson DJ, Everson G, Bambury T, Recta J, Zhong J, Mesghali H, Murray JH, Osborn BL et al. (2003) A phase 1/2 study to evaluate the pharmacokinetics, safety, tolerability, immunogenicity, and pharmacodynamics of albuferon in treatment experienced subjets with chronic hepatitis C [abstract 313]. *Hepatology* 38: 307A
118 Nevens F, Roskams T, Van Vlierberghe H, Horsmans Y, Sprengers D, Elewaut A, Desmet V, Leroux-Roels G, Quinaux E, Depla E et al. (2003) A pilot study of therapeutic vaccination with envelope protein E1 in 35 patients with chronic hepatitis C. *Hepatology* 38: 1289–1296
119 Farci P, Alter HJ, Wong DC, Miller RH, Govindarajan S, Engle R, Shapiro M, Purcell RH (1994) Prevention of hepatitis C virus infection in chimpanzees after antibody-mediated *in vitro* neutralization. *Proc Natl Acad Sci USA* 91: 7792–7796
120 Dagan S, Eren R, Graham N, Nussbaum O, Terkieltaub D, Terrault N, Tur-Kaspa R, Lurie Y, Safadi R, Akerman Z, Galun E (2003) Clinical evaluation of a hman monoclonal antibody against the envelope protein (E2) of HCV for the prevention of HCV infection [abstract 305]. *Hepatology* 38: 302A
121 Krawczynski K, Alter MJ, Tankersley DL, Beach M, Robertson BH, Lambert S, Kuo G, Spelbring JE, Meeks E, Sinha S, Carson DA (1996) Effect of immune globulin on the prevention of experimental hepatitis C virus infection. *J Infect Dis* 173: 822–28
122 Krawczynski K, Fattom A, Spelbring J, Kamili S, Basham L, Purdy M, Sapan C, Naso R, Carson D, Galland G (1999) Experimental immune treatment of HCV infection: passive transfer of anti-HCV in chimpanzees. In: *Abstracts of the XI International Congress of Virology*, Sydney, Australia, 277

Hepatitis Prevention and Treatment
Edited by J.M. Colacino and B.A. Heinz

Problems inherent to antiviral therapy

Tim Shaw and Stephen Locarnini

Victorian Infectious Diseases Reference Laboratory, 10 Wreckyn Street, North Melbourne, Victoria 3051, Australia

Historical background

Twenty years ago, Emeritus Professor Sir Charles Stuart-Harris prefaced a collection of papers from a colloquium on "Problems of Antiviral Therapy" with the following comments:

"Progress towards effective antiviral chemotherapy has proved both tardy and frustrating. In contrast to the giant therapeutic successes achieved against microorganisms ranging from bacteria to *Chlamydiae*, progress against viruses has been minimal. Today, however, antiviral chemotherapy has been achieved and since the first important successes have occurred against differing virus groups, there is a tendency to be over-optimistic. In fact this area of human endeavour has suffered in the past both from too much pessimism and too much optimism. ... But it is particularly important for positive achievement, however limited, to be made in antiviral therapy if only to sustain the optimism needed to further research in this peculiarly difficult field. ... At a fundamental level one may ask whether further progress is likely to result from the mass screening of existing compounds or whether we should look to rational developments in methods of attacking the virus, its enzymes, nucleic acid or proteins. Unfortunately, the quirk of unexpected discovery still exists. ... Probably we should not rule out any way by which new compounds with antiviral properties may be found. This introduction is intended to counter the criticism that progress in antivirals has been too slow, but justification for the delay exists and no apology need be made." [1].

Reports of successful clinical trials of synthetic antiviral drugs did not appear until 1963, although many compounds with *in vitro* antiviral activity had been identified in the preceding 20 years. In 1963, *N*-1-methyl-isatin β-thiosemicarbazone (methisazone) was shown to prevent smallpox in contacts of infected individuals [2] and amantadine was found to prevent induced influenza in volunteers [3]. Coincidentally, 'Australia antigen', later identified as the hepatitis B surface antigen (HBsAg) was discovered the same year [4]. The need for anti-pox virus drugs such as methisazone disappeared with the development of smallpox vaccines and the usefulness of amantadine was limited by its specificity for influenza, poor tolerability and rapid development of resistance. Similarly, the successful clinical use of interferon (IFN) did not occur until about 20 years after its discovery. The phenomenon of viral interference (resistance of infected cells to further viral challenge) had been documented in the early 1940s, but IFN was not isolated until the late 1950s and its successful clinical use (in treating vaccinia keratitis) was not reported until 1962 [5]. It was not until 1976 that IFN was first reported to be efficacious in the treatment of patients with chronic hepatitis B (CHB) [6].

By 1983, nearly 40 years of antiviral research had produced disappointingly few clinically useful drugs. Besides IFN, the drugs that were available were mainly anti-herpes virus agents, including some nucleoside analogues and foscarnet [1]. Viral resistance to existing drugs had already been identified as an emerging problem. Acquired immune deficiency syndrome (AIDS) was being recognized as a clinical entity and its probable viral etiology was suspected but not firmly established; antiretroviral agents were unknown. Combination chemotherapy was acknowledged more as a theoretically sound concept rather than a practical clinical reality [7]. Viral hepatitis was recognized as a challenge and a terminology for distinguishing different types of viral hepatitis was established. Infectious hepatitis and serum hepatitis were recognized as hepatitis A and B respectively; other types of viral hepatitis were referred to as 'non-A, non-B' hepatitis (see Chapter by Kalkeri and Kwong). Attempts to treat hepatitis A and B remained largely empirical and usually unsuccessful [8].

By contrast with the two decades preceding 1983, progress during the two decades since has been rapid, even spectacular in many respects. Perhaps the single most important factor, but certainly the initial impetus, driving antiviral research since 1983, seems to have been the AIDS epidemic in continental USA. By 1983 it was clear that the incidence of AIDS by date of diagnosis had doubled roughly every 6 months during the past 4 years, with a mortality rate exceeding 60% for cases diagnosed more than a year earlier. Significantly, about 80% of AIDS cases were concentrated in six metropolitan areas, primarily in 'high-profile' locations on the east and west coasts [9]. Much progress in antiviral research during the past two decades has been made possible by technological advances directly or indirectly resulting from efforts to control the human immunodeficiency (HIV) epidemic in the USA.

Despite the progress of the past 20 years, many of the fundamental problems identified earlier remain incompletely solved, and each new advance frequently reveals new subsets of problems. Development of clinically useful antiviral drugs, especially treatments for viral hepatitis, has been and remains inherently problematic, as we hope that the remainder of this chapter demonstrates. We have attempted to address the more significant problems and to briefly outline the way in which they have been, are being, or may be approached. Different aspects of hepatitis research exemplify each problem.

Problems that affect development of all antiviral agents

Prevention is better than cure: virucides, chemoprophylaxis and vaccination

Ideally, antiviral chemotherapy could prevent, as well as control, infection. For chemoprophylaxis to be successful, three basic requirements must be fulfilled, namely: (1) it must be possible to administer the prophylactic drug before infection, before the onset of symptoms; (2) the drug must be non-toxic; and (3) it must be active against the specific viral challenge. These requirements turn out to be surprisingly restrictive.

Virucides

Antiviral agents, by analogy with antibacterial agents, can be regarded as either virucidal or virostatic. Virucidal agents are those that are capable of inactivating a virus regardless of its replication state. Since metabolic activity is not required for maintenance of virion infectivity, it may be impossible to find non-toxic virucides, which, by definition, must be able to inactivate extracellular virus. Detergents, aldehydes, heavy metal ions, halogens, and chelating, oxidizing and denaturing agents are amongst compounds that fulfil this criterion, but they are too toxic even for topical application. Exciting claims have been made for a family of synthetic chemical polymers called dendrimers, which resemble biological polymers in terms of size, versatility and physicochemical properties. They appear to possess unlimited potential for future use as virucides and drug carriers [10], but in the immediate future all antiviral drugs for clinical use are likely to be virostatic rather than virucidal, with consequences outlined below.

Chemoprophylaxis

Chemoprophylaxis would be an ideal strategy to prevent epidemics, but would require cheap, stable, broad-spectrum antiviral agents and treatment regimes sufficiently simple to allow treatment of very large at-risk populations. Unfortunately, development of such agents and regimes remains more of a theoretical possibility than a practical achievement. Nearly 40 years ago, amantadine and some related organic amines were fortuitously found to inhibit the transport and uncoating of some RNA viruses, implying that chemoprophylaxis of viral infection is possible [11]. Amantadine and related compounds are weak bases that specifically accumulate in acidic vacuoles in the cell and inhibit the *in vitro* replication of viruses that rely on acid-dependent fusion of virion protein to the vesicular membrane for uncoating. Whether this is their main mechanism of action *in vivo* is not known, but they also have therapeutic activity and recent data indicate other mechanisms of action in infected cells. Administered orally from the onset of symptoms of influenza, amantadine showed prophylactic activity comparable with available vaccines against type A, but not type B, influenza viruses. Although amantadine and related drugs show antiviral activity against several RNA viruses and have been licensed for use against influenza A in some countries, they have been infrequently used for a variety of reasons. These include the inconvenience associated with the need for continuous administration throughout the entire risk period, undesirable side effects, limited demonstrable therapeutic benefit and development of viral resistance. Use of amantadine was also hindered by the lack of facile and reliable diagnostic methods capable of confirming the provisional clinical diagnoses of influenza as well as distinguishing between susceptible and resistant viruses—in other words, identifying cases suitable for treatment, which remains a particular problem for viral hepatitis (see below). Although the need for prophylactic drugs has been reduced by vaccination, they are likely to be the only drugs available to arrest the early spread of pandemics which may otherwise result from infection with new virus strains for which vaccines are not currently available [12]. In this context, reports that compounds that specifically block entry of HIV into cells by inhibiting virus-cell membrane fusion have been effective in early clinical trials is exciting, since it implies the same approach may be successful against other enveloped viruses [13]. IFNs, which can be used prophylactically, are a special case: they have a broad spectrum of action, but their usefulness is severely limited by variety of factors including high cost, limited efficacy and toxicity (see [14] and Chapters by Brown, Boehme and Dev et al. for recent reviews; also see below).

Vaccination

In a few cases, notably polio and smallpox, the need for antiviral agents has been obviated by the development of safe and effective vaccines and by the institution of universal vaccination programs. Although such vaccines against hepatitis B virus (HBV) infection have been available for more than 20 years, economic and logistical obstacles have hampered the global elimination of hepatitis B by universal vaccination. A plasma-derived hepatitis B vaccine became available in 1981, being replaced 5 years later by a recombinant yeast antigen vaccine. These were respectively the world's first licensed viral subunit vaccine and the first licensed recombinant vaccine [15]. By 1999, although at least 10 countries had introduced routine hepatitis B vaccination, it was not feasible in many poorer countries where HBV infection is endemic. Financial backing from the recently formed Global Alliance for Vaccines and Immunization (GAVI) makes it likely that routine hepatitis B vaccination will be possible in most of these countries within the next 5 years [16]. Vaccination cannot benefit the large existing global population of around 400 million individuals who are already infected with HBV and who constitute a reservoir of potentially infectious virus [15]. In this population, antiviral chemotherapy is the only means of controlling active hepatitis B replication, and ameliorating its disastrous clinical sequelae of cirrhosis, liver failure and hepatocellular carcinoma (HCC). However, chemotherapy is not affordable for the majority of those at risk. This, together with the increasing probability of eventual achievement of universal vaccination and rapidly increasing cost of drug development, has become a possible disincentive for pharmaceutical companies to invest in development of new anti-HBV drugs. By contrast, hepatitis C, for which no vaccine is available, is becoming increasingly prevalent in developed nations [17, 18]. Although recombinant vaccines against hepatitis C are being developed, the genetic variability of hepatitis C virus (HCV) (see Chapter by Luo and below) may restrict their effectiveness [19], providing an incentive for funding the quest for other anti-HCV agents.

Limited efficacy: IFN

Theoretically, the broad spectrum of activity of the IFNs suggests that they would be ideal antiviral and antitumour agents, but in practice, attempts to use exogenous IFN for the treatment of human disease has not been the overwhelming success that was hoped for and expected [14]. Until recently, IFNα was the only drug approved for treatment of CHB in most countries [20] and, in its PEGylated form (PEG-IFN, which is covalently linked to polyethylene glycol) and in combination with ribavirin (RBV), is the currently recommended treatment for hepatitis C (see [21, 22] and Chapter by Dev et al.). IFNs have direct antiviral and immunomodulatory effects, the mechanisms of which are extremely complex, involving hundreds of cellular genes [23]. Factors affecting IFN response include genetic polymorphisms in both host and virus, which are only just beginning to be appreciated as expanding genetic databases and technological advances, in particular gene microarray technology, make their investigation feasible [24–27].

The standard course of IFN therapy for CHB (5–10 MU thrice a week for at least 16–20 weeks) produces a sustained virological response (SVR) in less than 40% of highly selected patients [28]; seroconversion is even less frequent (Tab. 1). The proportion of patients with chronic hepatitis C who respond to IFN therapy depends mainly on the viral genotype. The overall success rate after 12 months of therapy with PEGylated IFNα in combination with RBV is about 50%: SVRs may be achieved in almost 80% of patients infected with HCV genotypes other than genotype 1, but they are only about half as frequent (~40%) in genotype 1 infections [29] (Tab. 1). A variety of adverse effects, including psychiatric disturbances, have been associated with IFN-based regimens, for which there are many contraindications, including decompensated cirrhosis and autoimmune disease (Tab. 2). Despite the improvements offered by the longer-acting PEGylated forms of IFN (PEG-IFN), which are replacing conventional IFN [30, 31], low efficacy, high cost, the need for subcutaneous administration and frequent serious adverse side effects continue to limit its usefulness.

Toxicity

Viruses are obligate intracellular parasites that can only replicate after infecting permissive host cells in which cell organelles and enzyme systems are redi-

Table 1. Responses of chronic hepatitis B and hepatitis C to IFN-based therapies

(A) Frequencies of achievement of sustained responses in patients with chronic hepatitis B	
Normalization of serum ALT	26% (18–34%)
Clearance of HBeAg	24% (8–30%)
Sustained loss of HBV DNA	23% (18–29%)
Clearance of HBsAg	5.6% (3.5–7.6%)

(B) Frequencies of achievement of sustained virological response in patients with chronic hepatitis C

<table>
<tr><th>Treatment</th><th>HCV genotype non-1</th><th>HCV genotype 1</th></tr>
<tr><td>Standard IFN alone</td><td colspan="2">12%–19%</td></tr>
<tr><td>PEG-IFN alone</td><td colspan="2">25%–40%</td></tr>
<tr><td>PEG-IFN + RBV lower dose, 24 weeks</td><td>78%</td><td>16–41%</td></tr>
<tr><td>PEG-IFN + RBV higher dose, 24 weeks</td><td>73%</td><td>26–51%</td></tr>
<tr><td>PEG-IFN + RBV lower dose, 48 weeks</td><td>78%</td><td>35–53%</td></tr>
<tr><td>PEG-IFN + RBV higher dose, 48 weeks</td><td>77%</td><td>46–61%</td></tr>
</table>

Data summarized from [20] (HBV) and [45] (HCV) references in text. Lower-dose RBV, 800 mg/day; higher dose, 100–1200 mg/day. ALT, alanine aminotransferase; HBeAg, HBV e antigen.

rected by viral gene products to supporting viral replication. For both HBV and HCV, this is efficiently achieved using minimal genetic resources: each viral genome encodes only a few protein products, the functions of which were unknown when antiviral therapy was first attempted. Early attempts to treat CHB infection were largely empirical: besides miscellaneous compounds including antioxidants, vitamins, intercalating agents, immunosuppressants and immunostimulants, most of the limited number of first-generation antiviral agents that were available were tested unsuccessfully against HBV. The latter included phosphonoformate, ribavirin, adenine arabinoside, ara-AMP, acyclovir and ganciclovir, which were either ineffective or too toxic for long-term use (for review, see [8]). Later, after it was discovered that HBV replication was dependent on reverse transcription, antiretroviral nucleoside analogues were tested as potential anti-HBV drugs, initially with limited success (for review, see [32]). Experience with antiretroviral nucleosides identified inhibition of mitochondrial DNA polymerase (polymerase γ) as the single most important cause of toxicity [33–36]. The seriousness of consequences of inhibition of mitochondrial replication by nucleoside analogues was emphasized by experience with fialuridine, a thymidine analogue that progressed to a phase II clinical trial in patients with CHB. The trial was terminated because of the sudden development of severe multisystem toxicity characterized by hepatic failure, lactic acidosis, and pancreatitis, which resulted in the deaths of five patients [37, 38]. Severe hepatotoxicity from fialuridine was characterized by hepatomegaly with diffuse, predominantly microvesicular steatosis, refractory lactic acidosis hepatocellular glycogen depletion, marked bile ductular

Table 2. Adverse effects of, and contraindications for IFN-based therapies for hepatitis B and C

Adverse effects	Contraindications
Constitutional	
Flu-like illness	Hepatic decompensation
Fever	Portal hypertension
Rigors	
Myalgia	
Arthralgia	
Fatigue	
Haematological	
Leukopenia	Hypersplenism
Thrombocytopenia	
Anaemia*	
Neutropenia*	
Neuropsychiatric	
Depression	History of depression or psychosis
Insomnia	
Irratibility, anxiety	
Concentration loss	
Gastrointestinal	
Nausea*	
Anorexia*	
Diarrhea*	
Autoimmune	
Hypothyroidism	Pre-existing autoimmune disease
Diabetes	Diabetes
Other	
Weight loss	Major system impairment (e.g. renal or cardiopulmonary)
Alopecia	Pregnancy
Ocular disturbances	Continuing intravenous drug abuse
	Conditions that impair long-term compliance

* IFN-RBV combination therapy. Summarized from references in text.

proliferation and cholestasis. Ultrastructural examination revealed intracytoplasmic lipid droplets and mitochondrial abnormalities, consistent with severe mitochondrial defects and the accompanying metabolic abnormalities [39].

Development of assays for mitochondrial toxicity, extensive preclinical screening and, more recently, molecular modeling, have facilitated the identification and development of a new generation of relatively non-toxic nucleosidic drugs (nucleoside and nucleotide analogues and their derivatives) [33, 36, 40–42](see Tab. 3). However, the potential of these drugs to cause unexpected toxicity remains, because (1) nucleosides and nucleotides are involved in almost every aspect of cellular metabolism, and (2) small structural changes frequently result in disproportionately large changes in specificity and resulting biological properties. Isomers represent an extreme case: their pharmacological properties and biological activities are often substantially different. For

Table 3. Relative mitochondrial toxicity of some nucleoside-tide drugs

Drug	Toxicity index	MT_{50} (μM) in HepG2 cells	[mt]% in RPTEC cells
Zalcitabine (ddC)	9000	~1	7
Fialuridine (FIAU)	–	~20	ND
Didanosine (ddI/ddA)	5600	~12	10
Stavudine (d4T)	2000	>100	77
Tenofovir (PMPA)	1.5	>100	109
Lamivudine (LMV)	0.4	>100	ND
Zidovudine (AZT)	0.2	>100	104
Abacavir (CBV)	0.03	>100	ND

Drugs are ranked according to their potential toxicity ('toxicity index') as estimated by Johnson and colleagues [33, 36]. All drugs except FIAU have been used clinically as anti-HIV agents. Didanosine (ddI) is a prodrug for ddA; abacavir is a prodrug for carbovir (CBV) which has demonstrated anti-HBV activity. Toxicity index estimates represent the relative increase in time taken to replicate the mitochondrial genome, based on kinetic measurements of rate of incorporation of the analogue by purified mitochondrial DNA polymerase *versus* the rate of excision by its proofreading exonuclease. FIAU, unlike the other analogues, is not excised and causes mitochondrial toxicity by inducing mutations. MT_{50} values are drug concentrations required to reduce the mitochondrial DNA content of cultured HepG2 cells to 50% of the untreated control value after 7 or 9 days continuous exposure. [mt]% is the amount of mitochondrial DNA (as a percentage of the amount in untreated controls) remaining in primary human renal proximal tubule epithelial cells (RPTEC) after 21 days culture in the presence of high concentrations of each drug. Concentrations were: ddC, 10 μl; ddI, 200 μl, d4T, 200 μl, PMPA, 200 μl; AZT, 200 μl. Data from Birkus and colleagues [41, 42]. Whilst correlation is clear, these data illustrate the difficulty of detecting potential mitochondrial toxicity by means of cell-culture-based assays. ND, no data available.

example, the antiviral activities and cytotoxicity of D-2',3'-dideoxy-3'-thiacytidine are respectively less and greater than lamivudine, its L-enantiomer [43]. Different antiviral and other biological activities have also been shown for isomers of many other nucleoside analogues. At the molecular level, these differences are due to varying abilities of key viral and cellular enzymes to discriminate between isomers [44].

As for hepatitis B, early approaches to hepatitis C treatment were empirical, but by contrast with HBV, HCV was recognized as a member of an established viral taxon (the Flaviviridae) before an intensive search for anti-HCV agents began. Elucidation of the HCV replication strategy and functions of the major HCV gene products was facilitated by existing knowledge of other flaviviruses. Mechanistic details of most of the HCV replication cycle except the early events are now known, and crystal structures of all the HCV non-structural proteins have been solved for several years [45]. Even with the benefit of this accumulation of specific information, the identification of drugs that have a high degree of specificity for HCV has been exceptionally difficult, as attested by the diversity of compounds under investigation as potential anti-HCV agents (for recent reviews, see [45–47] and Chapter by Razonable and McGill). It is not feasible to attempt a discussion of potential toxicities associated with these 'new' agents because of their diversity and because, presently, few data relating to toxicity are available. By contrast, adverse effects of combinations of IFN and RBV have been well documented, although little is known about the molecular mechanisms responsible. Major effects include fatigue, influenza-like symptoms, gastrointestinal disturbances and haematological abnormalities, which may be sufficiently severe to necessitate dose reduction or discontinuation of treatment. Numerous other less frequent adverse effects may also reduce treatment tolerance. PEGylation of IFN, which improves its pharmacokinetic properties, may also influence the nature of side effects experienced. A review of the results of several large phase III clinical trials of different regimes containing standard or PEG-IFN in combination with RBV or placebo showed similar rates of withdrawal (10–14%) due to adverse effects in each group. The need for dose reduction was relatively frequent (32–42%) and mostly due to adverse events, rather than to development of haematological or other laboratory abnormalities [48]. Because most of the side effects are dose-related and reversible, dose-reduction or drug discontinuation are usually sufficient for their management. However, several studies suggest that rigid adherence to prescribed medication regimens maximizes the chance of favourable response, stimulating increasing interest in alternative ways to manage or overcome specific adverse effects [48, 49]. For example, it has been suggested that treatment with cytokines such erythropoietin and granulocyte-macrophage colony-stimulating factor respectively could alleviate haemolytic anaemia and neutropenia, common severe side effects of treatment with RBV and thus enable continued adherence to treatment regimens [49]. An alternative approach is represented by the development of viramidine and levovirin, drugs which are respectively a liver-targeted prodrug of RBV and the latter's L-enantiomer. Both of these drugs share the immunomodulatory properties of RBV, but show reduced toxicity and improved pharmacokinetics [50].

Non-specific symptomology

Unrelated viral pathogens with widely varying replication strategies can cause similar disease symptoms, making it virtually impossible to develop drugs suitable for treatment of common clinical entities like viral diarrhoea, upper respiratory tract infection or even hepatitis. As a corollary, choice of appropriate antiviral drug(s) may frequently require the routine availability of a wide variety of rapid, inexpensive and reliable diagnostic tests. A variety of infectious agents, drugs and toxins can cause hepatitis. The three most important causes of viral hepatitis in humans, hepatitis A, B and C viruses, represent three different viral families (Picornaviridae, Hepadnaviridae and Flaviviridae, respectively) which replicate by radically different strategies, but the clinical presentation of acute hepatitis caused by each is often indistinguishable. A panel of serological tests—most commonly assays for anti-HAV (anti-hepatitis A-virus) IgM, anti-HBc (anti-HBV core antigen) IgM, HBsAg and anti-HCV—is usually required to identify the causative agent(s). Idiosyncracies of serological assays make it desirable to perform confirmatory tests, several of which have only recently become routinely available, in order to avoid misdiagnoses (for review, see [51, 52]). The increasingly apparent need for facile and reliable tests that are able to distinguish between different genotypes of both HBV and HCV and tests that can detect emerging drug resistance is another manifestation of this type of problem (see below). Although such tests have been developed, they are currently available only in a few specialized laboratories [52–55].

Failure to access sites of viral replication: virus sanctuaries

Since viruses replicate intracellularly, effective antiviral agents must either be able to prevent infection and/or inactivate extracellular virions, or be capable of crossing cell membranes and inhibiting active viral replication. As noted above, there are few clinically useful synthetic agents that are capable of inactivating extracellular virions and the usefulness of IFNs and related cytokines that induce intracellular antiviral responses by binding to specific extracellular receptors is likewise limited. Nucleosidic drugs, which currently constitute the largest group of clinically useful antiviral agents, are active only after uptake and entrapment by sequential intracellular phosphorylation, which generates the corresponding deoxynucleoside or ribonucleoside triphosphates [32, 56]. Extracellular nucleotides are too polar to cross cell membranes and are prone to degradation by ubiquitous extracellular phosphatases and nucleases. Uptake of nucleosidic drugs and other xenobiotics generally requires specific transmembrane transporters, but the more lipophilic compounds may enter cells by diffusion. Absence of appropriate transporters may limit the activity of many drugs: as more transmembrane transport systems are identified and characterized, their pharmacological significance, particularly the influence of genetic polymorphisms, is being recognized (for reviews see [57, 58]).

None of the hepatitis viruses encodes a (deoxy)nucleoside kinase: cellular kinases are therefore required to catalyze the sequential intracellular phosphorylation steps that generate the antivirally active (d)NTP analogues from nucleosidic drugs [32, 44, 56]. Activities of all enzymes of nucleoside metabolism are more or less substrate-specific and they are frequently also species-, cell- and tissue-specific, and restricted to cytosol or mitochondria. Their expression may also be dependent on cell-cycle stage and the degree of cell and tissue differentiation (for reviews, see [32, 56, 59]). Cells and tissues that do not express the necessary transmembrane transporters or intracellular kinases, but do possess the necessary receptors for viral entry, have the potential to act as virus sanctuaries during antiviral therapy, whether or not they are capable of supporting complete cycles of viral replication. Similarly, some cell subpopulations that are capable of harbouring virus may be immunologically privileged: in other words, protected from immune attack. In the cases of both HBV and HCV, such cells may include liver cells other than hepatocytes (such as bile duct cells) as well as extrahepatic tissues, in particular lymphoid and epithelial components [60–62]. If a variety of cell types is capable of supporting viral replication, control of viral replication is likely to require treatment with combinations of drugs, since individual antiviral agents may be active only in specific cell or tissue subpopulations [7].

Non-cytopathic infection and disease heterogeneity

Symptoms of many viral diseases are caused by the host's responses to infection, rather than to any direct cytopathic effects of infection. By the time symptoms appear, viral replication, the target of most antiviral drugs, may either be established and stable, declining, or even undetectable. Initiating antiviral therapy at this stage may not alleviate disease. This scenario applies to both hepatitis B and hepatitis C, the clinical course of each being extremely variable and influenced by numerous factors which include viral genotype, host genotype, body mass index, specific genotypic polymorphisms, immune status and age at infection. Environmental factors such as exposure to promutagens and alcohol intake are also influential [20, 21, 63].

The clinical presentation of acute HBV infection ranges from subclinical to acute symptomatic (or, rarely, fulminant) hepatitis. Presentation of chronic infection is similarly variable, ranging from inactive carrier states to chronic active hepatitis, fibrosis, cirrhosis and HCC which occurs in up to 40% of chronic cases and is usually fatal [18, 20]. Initial infection by HCV is commonly asymptomatic or presents with non-specific flu-like prodromal symptoms and acute symptomatic infections are rare and spontaneously resolving infections are probably more frequent than generally supposed because they are rarely detected. Up to 80% of cases in which HCV infection is detected eventually develop CHC, which like CHB, can progress to fibrosis, cirrhosis and HCC [18, 21, 64]. Difficulties in defining the natural histories of chronic hepatitis B and C infections are caused by the indolent course and heterogeneity of associated disease. In both cases liver damage is believed to result mainly from attempts by the host's immune system to eliminate infected cells; chronic infection is established when these attempts fail [65–67].

Lack of experimental systems for study of viral replication, pathogenesis and drug screening

Ethical, economic and other considerations necessitate the use of cell-culture and animal models for controlled studies of viral replication and mechanisms of pathogenesis: such studies are difficult or impossible if suitable systems and convenient laboratory animal models are not readily available. Without knowledge of the natural history and pathogenesis of a 'new' viral disease, it may be difficult to decide whether to initiate treatment, since many viral infections resolve spontaneously. Treatments, if available, invariably entail a degree of risk of known or unknown adverse effects and may not be cost-effective, particularly for drugs with limited efficacy such as IFN. Similarly, inhibition of viral replication in cell culture has traditionally been used for screening compounds for antiviral activity and preclinical screening of drugs for safety and efficacy has traditionally employed animal models. Unfortunately, cell-culture systems have yet to be developed for many viral pathogens and many important viral pathogens of humans are host-specific, having no known counterparts in other animals. Studies of both HBV and HCV have been severely hindered by the lack of small-animal models and cell-culture systems capable of supporting complete cycles of viral infection (reviewed in [68]).

Animal and cell-culture models for hepatitis B

HBV will only infect higher primates, from which host-specific hepadnaviruses closely related to HBV have been isolated [69, 70]. Besides ethical considerations, limited availability, cost and risks associated with housing and handling infected primates prevent their use in routine studies. However, chimpanzees, which are genetically close to humans, have been employed for critical studies related to vaccine development [15]. Ducks infected with duck hepatitis B virus (DHBV) and woodchucks infected with woodchuck hepatitis virus (WHV) have been the most frequently used animal surrogates for hepatitis B, being employed mainly for studies of replication and pathogenesis respectively, although both have been used for *in vivo* and *in vitro* antiviral screening. While the past usefulness of these two models is indisputable, their evolutionary separation and consequent significant differences in drug metabolism and immunology limits their relevance for many studies. For example, results of studies related to host immunity, investigation of drug efficacy, toxicity and resistance and evolution of viral quasispecies may be difficult to interpret if they are based on non-human models ([68, 71] and references therein). Transgenic mouse models have been developed, but suffer similar limitations (for example [72]; also see below). More recently, tree shrews (genus *Tupaia*), which are small rodent-sized animals apparently related to primates, have been found to be susceptible to infection with human HBV [73]. Primary tupaia hepatocytes have been shown to support complete cycles of HBV replication [74, 75], confirming the potential of this model, which is likely to become more widely used for future studies of HBV. Other than primary hepatocytes, several liver-derived cell lines of human or animal origin, either stably or transiently transfected with HBV genomes, have been available for some time, but they have several disadvantages (for review see [68, 71]). Most cells other than primary hepatocytes either fail to support complete cycles of HBV replication or do not produce sufficient virus to make their use feasible for routine studies [68], but the heterogeneity of primary cultures makes standardization of cultures difficult and standardization of antiviral assays virtually impossible. It is possible to standardize experimental conditions by using the HepAD38 cell line, which replicates HBV under conditions that can be controlled by tetracycline [76], or by using systems which permit more efficient and controlled transfection by means of recombinant HBV-baculovirus (or adenovirus) vectors [77]. The major limitation of these systems is that their relevance is limited because the karyotypic and metabolic characteristics of the parent cell lines and their subclones are substantially different both from human hepatocytes and from each other. A recently described human hepatoma cell line named HepaRG, which, it is claimed, retains primary hepatocyte-like morphology and properties, can be infected with HBV and supports complete cycles of replication [78], may overcome these problems.

Animal and cell culture models for hepatitis C

The chimpanzee (*Pan troglodytes*) is the only experimental animal susceptible to infection with HCV. The chimpanzee model of HCV infection was instrumental for early studies on non-A, non-B hepatitis, including observations on the clinical course of infection, characterization of the virus, and eventual cloning of the genome and identification of infectious cDNA clones of HCV [68, 79]. Apart from close genetic identity to humans, the advantages of the chimpanzee model are that it is possible to control the time and route of infection, viral inoculum size and infecting genotype [80]. By contrast, studies of HCV disease in human cohorts usually involve individuals who have been infected for indeterminate periods —often decades. Moreover, selection of human populations is frequently biased towards specific disease status, antibody reactivity, viral genotype or viral load. Chimpanzees are still required, and will continue to be required in the future, to identify specific HCV genome sequences that determine immunogenicity, replication efficiency and pathogenicity as well as for preclinical trials of vaccines when they become available [81, 82]. Recently, DNA microarray technology has been used to investigate changes in cellular gene expression that occur during the course of acute HCV infection of chimpanzees [83] and to investigate gene expression in human liver associated with HCC [84–87]. This is an exciting and rapidly expanding field of study that will doubtless contribute substantially to our understanding of hepatitis B and C, and consequently to the development of new antiviral therapies and strategies.

The disadvantages of being forced to rely on primate model(s) have been noted above and alternative, small-animal and tissue-culture models for the study of HCV have been urgently needed since the virus was first characterized [68, 88, 89]. Despite many attempts to transmit HCV to mammals other than humans and chimpanzees, none has been successful. Preliminary reports that identified tupaias as potential host for HCV remain unconfirmed and controversial [90, 91]. However, GBV-B, a human flavivirus that is closely related to HCV, has been found to infect tamarins and marmosets, raising the possibility that GBV-B-infected tamarins or marmosets could be used as a surrogate models for HCV infection [92, 93]. GBV-B infection of tamarins has been accomplished using either a virus pool injected intravenously or by intrahepatic injection of an infectious RNA clone. The resulting GBV-B infection causes acute hepatitis accompanied by an increase in alanine aminotransferase (ALT) levels. Tamarins clear GBV-B within 14 to 16 weeks of infection and maintain a robust immune response to secondary challenge, suggesting them as a suitable animal model for acute phase infection [92]. It has been suggested that tamarins might also be infected with chimeric GBV-B viruses that depend for replication on the expression of functional HCV gene(s), selective inhibition of which could then be investigated *in vivo*. Unfortunately use of chimeric viruses has several inherent limitations. For example, (1) creation of viable chimeras that cover all aspects of viral replication would be difficult; (2) expression of HCV gene product(s) in isolation may alter their accessibility and/or response to antiviral agents and (3) viral chimeras may not be hepatotrophic.

Mouse models

The mouse offers numerous advantages as a laboratory animal, including high fecundity, low cost, short generation time, ease of manipulation and availability of genetically defined strains. Despite the evolutionary separation, most human genes appear to have mouse homologues. Mice are not susceptible to HCV infection, but creation of transgenic models provides the opportunity to study the effects of individual HCV gene products in a genetically defined background [88]. Unfortunately, different transgenic mouse models that have been used in these studies have often given apparently conflicting results [79]. More recently, Mercer and colleagues [94] generated mice with chimeric mouse–human livers by transplanting normal human hepatocytes into mice with severe combined immune deficiency (SCID), which harboured a plasminogen activator transgene (Alb-uPA). Homozygosity of Alb-uPA was associated with efficient human hepatocyte engraftment, and the mice developed prolonged HCV infections with high viral titers after inoculation with infected human serum. Initial increases of nearly 2000-fold in total viral load were observed, and viral replication was confirmed by detection of negative-strand viral RNA in livers. HCV viral proteins were localized to human hepatocyte nodules, and infection was serially passaged through three generations of mice, confirming both synthesis and release of infectious viral particles. Chimeric mouse systems show the greatest promise for the future, having already been used to evaluate the activities of putative anti-HCV compounds, including a novel genetically engineered inducer of apoptosis which is activated by NS3-mediated cleavage after delivery to hepatocytes by an adenovirus vector [95].

Presently, the chimeric mouse system has two major disadvantages, namely: (1) breeding the required homozygous Alb-uPA mice is difficult because of the toxicity of the transgene and (2) transplantation of primary human hepatocytes into mice a few days after birth is technically challenging. Since the mice are immunodeficient, it will be necessary in future to provide them with components of the human immune system to study issues related to HCV pathogenesis.

HCV replicons

Although continuous hepatoma, B cell and T cell lines, primary hepatocytes and peripheral blood mononuclear cells from humans and chimpanzees have all been reported to support HCV replication, none has been found to be sufficiently robust and reliable to be used for classical virological, biochemical or genetic studies [79]. Thus, to date the mechanism of viral attachment and cellular entry and the precise intracellular steps in HCV RNA replication, virus assembly, and virion release remain largely unknown. The recent development of subgenomic HCV RNA replicons that are capable of successfully replicating in the human hepatoma-derived Huh7 cell line represents a major achievement in the ongoing quest for a reliable cell culture model for HCV [96]. A bicistronic construct containing a neomycin gene upstream of the non-structural HCV genes enabled selection of replicative RNAs in Huh7 cells. Replication was initially observed at low frequency, but adaptive mutations that permitted higher transduction efficiency were found to occur during culture [97]. These adaptive mutations provide an insight into the ways in which HCV may interact with the host cell, since they are often clustered in NS5A, close to the region putatively involved in IFN sensitivity. Adapted replicons initiate RNA replication with high efficiency and produce abundant HCV RNA and proteins (NS3 to NS5B), which should facilitate genetic, biochemical and functional analyses in the future [98, 99]. Meanwhile the replicon system is being used to identify and evaluate new anti-HCV drugs that are targeted at translation or RNA replication and intensive research aims at improving the replicon system so that it produces all the HCV proteins. It is anticipated that a cell-culture system capable of supporting complete cycles of HCV replication will eventually become available and may ultimately lead to the development of a permissive cell line, which would be invaluable for drug screening and studies of viral replication [96]. It is worth noting that even when this is achieved, other means will still be required to study specific aspects of HCV infectivity and pathogenesis.

Difficulty assessing activity of natural products and traditional medicines

As outlined above, the lack of appropriate laboratory assays makes is difficult to screen compounds for antiviral activity, but it is difficult to screen some types of drugs for activity even when assays are available. This applies in particular to natural products and traditional medicines, biological activities of which are invariably difficult to standardize and quantify. The need for additional drugs to control viral hepatitis has generated renewed interest in traditional remedies for liver diseases, which have had a long history of use, particularly in India and Asia. To date, the many attempts to demonstrate the usefulness of these remedies using modern concepts of evidence-based evaluation, dose and formulation standardization and the use of randomized placebo-controlled clinical trials have reached conflicting conclusions. Most of the published results of clinical trials have reported some benefit from the use of antiviral herbal medicines, but since negative findings are rarely published, firm conclusions are not possible. A review of hundreds of clinical trials of herbal remedies by Martin and Ernst [100] found that the majority (~90%) were not adequately controlled. Of those that were, 14 trials described the use of *Phyllanthus* species for treatment of hepatitis B, seven reporting positive and seven reporting negative results. Another recent publication by Thyagarajan and colleagues [101] reviewed the evidence for biological activities of more than 300 traditional preparations used in India for the treatment of jaundice and chronic liver diseases. The preparations were derived from more than 80 different Indian medicinal plants, but acceptable evidence for activity was found for preparations derived from only four species. These were *Sylibum marianum*, which was found to possess anti-oxidant, antifibrotic, anti-inflammatory and immunomodulating activity, *Glycyrrhiza glabra*, which was capable of inducing endogenous IFN, *Picrorhiza kurroa*, which had anti-inflammatory, hepatoprotective and immunomodulatory activities, and *Phyllanthus amarus* which had immunomodulatory and anti-inflammatory properties. All were claimed to be hepatoprotective *in vivo* but evidence of antiviral activity against HBV and HCV was claimed only for *Phyllanthus* species. Unfortunately, neither *in vivo* experiments using the DHBV model [102], nor a recent double-blind placebo-controlled clinical trial over 6 months [103] found any evidence for antiviral efficacy of *Phyllanthus*. On the other hand, another recent randomized double-blind, placebo-controlled multicenter trial involving 216 patients with CHB who were treated with oxymatrine, an alkaloid extracted from Chinese medicinal herb *Sophora alopecuraides*, did show convincing benefit. Decreases in serum HBV DNA, disappearance of HBV e antigen (HBeAg) and increases in the frequency of normalization of serum ALT were observed after 24 weeks of oxymatrine treatment, given either orally or by intraperitoneal injection and the changes were statistically significant compared to changes in a placebo-treated control group [104]. Thus, although presently the laboratory and clinical evidence to support the use of traditional medicines for viral hepatitis is equivocal, there remains a need for larger, more stringently designed, randomized controlled clinical trials to resolve the issue.

Miscellaneous problems related to clinical trials

Optimizing treatment regimes may present problems, particularly if they involve drug combinations. Early clinical trials frequently employed, and continue to employ, selected patient subpopulations, which may not be representative of the general population. For example, clinical trials of IFNs have almost invariably involved selected populations [14]: the clinical anti-HBV potentials of lamivudine, adefovir and tenofovir were first established in patients who were co-infected with HIV [105]. Furthermore, most clinical trials of drug combinations have enrolled predominantly individuals who have previously failed to respond to other treatments [7, 106]. Ethical considerations may preclude placebo-controlled clinical trials of new treatments if an alternative treatment option known to be effective in a majority of cases is already available. This is now the case for both chronic hepatitis B and C for which new treatments must be compared to recognized 'standard' treatments (IFN, lamivudine or adefovir and PEG-IFN + RBV respectively). Finally, geographical isolation, prohibitive cost or sociopolitical obstacles often prevent access to new therapies by populations likely to derive the greatest benefit from them.

Problems with particular significance for hepatitis B and C

Whereas problems outlined above must be overcome in the course of development of all antiviral agents, most have been partly or completely solved or appear potentially soluble in relation to HBV and HCV. A consequence of the relatively recent development of experimental systems for studying HBV and HCV, which has probably been the single most important obstacle to progress, is that resultant advances in our understanding of viral and cellular biology has revealed further challenges. Overcoming these challenges, which arise directly from aspects of the replication strategies employed by HBV and HCV, may prove difficult.

Viral quasispecies

RNA polymerases and reverse transcriptases are inherently error-prone because they lack associated proofreading functions, resulting in mutation rates of the order of 10^{-3} to 10^{-5} misincorporations per nucleotide copied. Low replication fidelity, potential for recombination, short generation time and high replication yields ensure that RNA viruses *in vivo* exist as heterogeneous mixtures of genetic variants termed quasispecies [107, 108]. Because their replication strategy requires reverse transcription, hepadnaviruses also exist *in vivo* as quasispecies, although constraints imposed by overlapping reading frames probably restricts diversity and evolutionary potential [69, 109]. The development of quasispecies is characterized by a continuous process of mutant generation, competition and selection resulting in a genomic composition with a dispersed distribution of low-fidelity copies of the consensus or master sequence, which represents the genome that is most fit for survival in the existing environment [107, 108]. It has been estimated that in infected hosts with viral loads greater than about 10^8 genome copies per ml, the resident quasispecies must include genomes with all possible single point mutations as well as some with two or three point mutations. Mutations cannot occur or be amplified if replication is blocked, but replication requires permissive cell populations, principally hepatocytes in the case of HBV and HCV. The half-life of an infected hepatocyte is between 10 and 100 days, lifespan being inversely related to the severity of liver disease, while the plasma half-lives of HBV and HCV have been estimated to be of the order of hours and days respectively. Independent studies of viral dynamics *in vivo* estimate an HBV production of the order of 10^{11} virions per day during chronic infection, greater than analogous estimates of approximately 10^9 virions per day for HIV and HCV [110]. Such high rates of virion production guarantee that large numbers (in absolute terms) of mutants will be produced. Expansion of particular viral quasispecies is determined by the functional consequences (if any) of individual mutations. The results of so-called 'replication yield' assays have led to the general conclusion that all drug-resistant mutants necessarily have replication defects. However, in the clinical situation, resistant viruses frequently become the dominant quasispecies after only a few months of treatment, indicating that *in vitro* replication does not necessarily correlate with viral fitness *in vivo*. Unfortunately, it is not possible to mimic the *in vivo* situation, which is clearly very complex, using *in vitro* assays. *In vivo*, faster-replicating quasispecies could be expected to predominate in the absence of selection pressure. Although this scenario probably applies during early stages of infection, it is likely to become less important as the available 'replication space' (that is, the number of cells available to establish new infections) is progressively occupied, when resident species would be favored [111].

The quasispecies concept explains the great adaptive potential of many viral pathogens, evidenced by the need for combination antiviral therapy and polyvalent antiviral vaccines to control viral diseases (reviewed in [107, 112]). Various selective pressures, including host immune surveillance, antiviral therapies and virus–cell tropism determine quasispecies composition and the outcome of infection: high quasispecies variability confers rapid but unpredictable evolutionary potential. Since mutations that confer drug resistance generally reduce replication fitness, secondary mutations that compensate for replication deficiencies (which by themselves confer no replicative advantage in the absence of selection pressure) may be co-selected by prolonged selection pressure. Mutations that confer drug resistance frequently increase replication fidelity and decrease processivity, which together would tend to create a bottleneck and limit the potential for further adaptation. On the contrary, compensatory or other mutations that reduced replication fidelity and/or increased processivity would increase the potential for generation of new quasispecies that could continue to increase their 'fitness' by further mutagenesis. The limitation on the latter process is that replication may become so inaccurate that it exceeds the error threshold beyond which viral survival becomes impossible, a situation termed 'error catastrophe' [107, 112, 113]. Quasispecies theory predicts that although viral variants that become less fit due to environmental changes will be retained as minor components of an evolving population (a phenomenon that has been named 'quasispecies memory'), they should eventually be diluted to extinction [114]. However, experimental evidence shows that under conditions of high multiplicity of infection, complementation by wild type virus allows retention of quasispecies memory (i.e. less-fit variants) for longer than predicted [114]. These phenomena have important practical implications, as Briones and colleagues [108] showed recently for HIV. In addition to demonstrating retention of quasispecies replicative memory *in vivo*, they show that HIV quasispecies also retain 'cellular' memory; that is, minor quasispecies that have been 'hidden' in cellular or anatomical compartments for long periods of time.

Impact of viral quasispecies in hepatitis B

Several new drugs that have been shown to possess *in vitro* anti-HBV activity have recently progressed to phase II/III clinical trials. They include nucleosidic drugs that (presumably) inhibit the viral reverse transcriptase (NRTIs), as well as structurally diverse inhibitors that block other steps in the viral replication cycle (for review, see [115] and Chapter by Razonable and McGill). The most promising of the latter group appear to act as inhibitors of virion assembly and packaging [116]. Quasispecies theory predicts that the development of resistance to all these compounds is inevitable once they become available for clinical use [107] and that combination chemotherapy analogous to highly active antiretroviral therapy will be required to suppress HBV replication *in vivo* [7, 106]. To date detailed information about HBV drug resistance is available only for lamiudine and adefovir, and more information—especially information on cross-resistance patterns—will be required to facilitate the design of optimal combination chemotherapy.

Treatment of CHB with lamivudine results in a rapid decrease in viremia and reversal or arrest of existing liver disease in the majority of cases (for recent review, see [117] and Chapter by Brown and Boehme). Unfortunately, these improvements are rarely sustained in the longer term due to development of drug resistance. HBV resistance to lamivudine has been well characterized, both molecularly and clinically (e.g. [118, 119]). Mutations that confer lamivudine resistance decrease *in vitro* sensitivity to lamivudine from 20- to greater than 100-fold. The incidence of lamivudine resistance in patients with CHB rises progressively at rates of between 14 and 32% annually, approaching 100% after 48 months of treatment [118, 120]. This is a relatively slow rate of development of resistance compared to HIV, probably a reflection of constraints imposed by overlapping reading frames in the HBV genome and by reduced opportunities for complementation and recombination imposed by genetic haploidy [69]. Factors known to increase the risk of development of resistance to lamivudine include high pre-therapy serum HBV DNA and ALT levels, incomplete suppression of viral replication, longer duration of therapy, infection with particular HBV serotypes and genotypes and high body mass index (for review, see [20, 118, 120]).

Specific mutations which result in replacement of methionine in the tyrosine-methionine-aspartate-aspartate (YMDD) catalytic site motif of the viral reverse transcriptase (rt) by valine, isoleucine or serine are necessary and sufficient to confer resistance to lamivudine. These changes are designated rtM204V, rtM204I and rtM204S respectively using a recently adopted genotype-independent nomenclature [121]. The rtM204I substitution has been detected in isolation, but rtM204V, and rtM204S are found only in association with other changes. Other secondary changes in the rt sequence, some of which may be compensatory, commonly occur in conjunction with rtM204V/I/S [122, 123]. Lamivudine-resistant HBV mutants have been screened for cross-resistance to several other NRTIs in a variety of *in vitro* assay systems [111, 122]: results are summarized in Table 4.

Adefovir dipivoxil, an orally available prodrug for the acyclic dAMP analogue adefovir, gained US Food and Drug Administration approval in September 2002. Large-scale multicenter clinical trials showed that the efficacy of adefovir was comparable to that of lamivudine [124]. Viral resistance was not observed during 60 weeks of treatment with adefovir; on the contrary, increasing suppression of viremia was observed as treatment progressed [125, 126]. However, an HBV mutant that exhibited a 6–15-fold reduction in sensitivity to adefovir *in vitro* was isolated from two of 124 HBeAg-negative

Table 4. HBV polymerase mutants: resistance to lamivudine and other nucleosidic drugs

Drug	Polymerase change and associated change in $\log_{10}$ IC_{50} relative to wild type				
	None (wt)	rtL180M	rtM204I	rtM204V	rt180M+rtM204V
Lamivudine	0	0.2	>2	>1	>2
Adefovir	0	–0.3	–0.2	0.2	0.5
Tenofovir	0	0	–0.2	0	0.5
Penciclovir	0	1	>2	NA	>2
Clevudine (L-FMAU)	0	0.7	>2	0.3	>3
Entecavir	0	1	>2	NA	>2
Coviracil (FTC)	0	1	NA	NA	>1
Telbivudine (L-dT)	0	0.3	>2	NA	>2
L-Fd4C	0	0.5	NA	NA	>2
DAPD	0	–0.1	–0.1	–0.4	–0.1
FLG	0	0	<1	<1	<1

Relative sensitivity of lamivudine-resistant HBV mutants to other nucleosidic inhibitors of replication. Data are $\log_{10}$ values for the ratio mutant IC_{50}/wt IC_{50}, rounded to the nearest integer except when the log-transformed ratio is less than 1. For example, for no increase in IC_{50}, the ratio is 1, giving a $\log_{10}$ value of 0; $\log_{10}$ values of 0.7, 1 and 2 correspond to 5-, 10- and 100-fold increases in IC_{50} respectively. NA, data not available because of inefficient replication of mutant. Penciclovir is the active component of famciclovir; DAPD, diaminopurine dioxolane; L-Fd4C, L-dideoxy-2,3-didehydro-β-L-fluorocytidine; LG 3'-fluoro, 2,3-dideoxyguanosine. Modified from [111]. wt, wild type; rt, reverse transcriptase.

patients who had been treated with adefovir dipivoxil for 96 weeks [127]. Thus, HBV resistance to adefovir occurs less frequently than does resistance to lamivudine and does not confer as great a reduction in viral susceptibility. Adefovir resistance is conferred by substitution of threonine for asparagine 236 (rtN236T) [127, 128].

There are several molecular mechanisms for resistance, which are not mutually exclusive (see [111] for summary). They include (1) steric hindrance, which appears to be the main mechanism for lamivudine resistance [129], (2) reduced catalytic efficiency which also contributes to lamivudine resistance and (3) excision of chain-terminating nucleotides by pyrophosphorolysis or pyrophosphate exchange, which are essentially reversals of the polymerization reaction. The latter has not yet been demonstrated for anti-HBV drugs. Steric hindrance describes the situation where the viral polymerase is unable to accommodate analogues such as lamivudine because of the reduced size of a binding site [129]. The mechanism of adefovir resistance has not been determined, but it is unlikely to involve steric hindrance, since threonine is smaller but more hydrophilic than serine. Therefore, both the mutations that confer lamivudine resistance and the mechanisms by which they confer resistance are different from those that confer adefovir resistance. Consistent with this, cross-resistance between adefovir and lamivudine has not been observed, and adefovir has been found to be effective against lamivudine-resistant HBV [130–133]. These observations support the case for use of adefovir and lamivudine in combination [106, 130, 133].

Impact of viral quasispecies in hepatitis C

Quasispecies theory also predicts that drug resistance and vaccine escape are likely to be major problems that may restrict the use of newly developed anti-HCV drugs and vaccines when they became available for clinical use. Indeed, the demonstration that specific mutations in the HCV genome confer resistance to putative antiviral drugs has come to be regarded as evidence for specific antiviral activity and used to elucidate mechanisms of action. The apparent involvement of HCV gene products in resistance to innate immunity and endogenous IFN are special cases that are discussed in detail in earlier chapters and elsewhere [134, 135]. In many cases, mutations that confer drug resistance have been identified in the course of identification and preclinical development new drugs, well in advance of clinical trials. This applies to all classes of potential anti-HCV drugs, including inhibitors of the viral helicase, polymerase and proteases and inhibitors that act by binding to viral regulatory elements [136–138].

RBV, a guanosine analogue that has broad-spectrum antiviral activity, serves as a specific example. Although RBV has been used as an antiviral agent for more than 30 years, its mechanisms of action are still poorly understood [139]. Recently, Crotty et al. [113] showed that it could act as an RNA mutagen and induce error catastrophe in poliovirus *in vitro*; these authors proposed this as its primary mechanism of action against RNA viruses *in vivo*.

RBV also has the potential to cause mutations in HCV *in vitro* [140–142]. After intracellular phosphorylations, which generate RBV triphosphate, it can be incorporated internally, albeit with very low efficiency, into nascent viral RNA by the HCV RNA polymerase, which incorporates it opposite a pyrimidine (C or U), but not purine (A or G) residues. Conversely, HCV polymerase will incorporate a pyrimidine (but neither purine nor RBV) monophosphate opposite internal RBV residues, again with low catalytic efficiency compared to the efficiencies of incorporation templated by A or G residues. Consequently, efficiency of RNA polymerization on RBV-containing templates is very low, indicating that RBV is not only mutagenic for HCV, but also inhibits viral RNA synthesis [140]. To test whether these mechanisms operate *in vivo*, Young and colleagues [142] studied the evolution of the nucleotide sequences of HCV RNA in the nonstructural NS5B region in patients receiving RBV, placebo, or IFNα monotherapy. The rate of HCV quasispecies change was found to be faster in the RBV-treated group than in either the IFNα or placebo group. RBV caused preferentially A-to-G and U-to-A mutations. In addition, a mutation which caused a phenylalanine-to-tyrosine substitution at amino acid residue 415 of NS5B (F415Y) was observed in all (five of five) patients infected with HCV genotype 1a during treatment with RBV. Furthermore, the parental (415F) strain re-emerged in some patients after RBV treatment was stopped. Assays using an HCV subgenomic replicon system demonstrated that RBV caused dose-dependent inhibiton of viral RNA synthesis by the B415F replicon, but not the B415Y replicon, confirming that F415Y confers RBV resistance. Three-dimensional modeling revealed that the NS5B residue 415 is located in the 'thumb' subdomain of the viral polymerase, which has been proposed to be involved in binding the HCV template–primer duplex. These results not only confirm that RBV can act as a weak viral mutagen *in vivo* in HCV-infected patients, but also confirm that RBV inhibits the HCV polymerase more directly. Confirmation that RBV can act as a viral mutagen *in vivo* raises an intriguing question: how will this affect evolution of drug resistance when RBV is used in combination with other antiviral drugs? Would it drive the virus towards error catastrophe or hasten the development of drug resistance?

The key to HBV persistence: cccDNA and minichromosomes

The double-stranded covalently closed circular (ccc) form of the HBV genome is found in hepatocyte nuclei (see Chapter by Condreay and Harris and [143]). Since HBV infection is not cytopathic, viral cccDNA may persist in the nuclei as long as the host cells survive. The mechanisms by which viral cccDNA is generated and maintained provides some explanations for the recalcitrance of CHB to conventional antiviral therapy, for the 'rebound' phenomenon which occurs almost inevitably when antiviral therapy is stopped [144], and for the consequent need for long-term administration of treatment.

Viral cccDNA is produced from the relaxed circular (rc) DNA found in viral nucleocapsids. These may originate either from infecting virions following *de novo* infection or from newly assembled cytoplasmic nucleocapsids after infection is established. Conversion of rc DNA to cccDNA requires its import into the cell nucleus. In the nucleus, the single-stranded gap region is replicated ('repaired'), the capped oligoribonucleotide and the HBV polymerase protein bound to the 5' ends of the strands are removed, the short terminally redundant sequence of the minus strand is removed and both strands are covalently ligated. Viral cccDNA associates with cellular histones and other nuclear proteins to form minichromosomes (see below). Host RNA polymerase II transcribes all the viral mRNA species, including the 3'- polyadenylated, terminally redundant primary transcript, the pregenomic (pg) RNA, from minichromosomal cccDNA.

Early in infection, newly synthesized nucleocapsids are transported to the cell nucleus where their genomes are converted to the cccDNA form ('intracellular conversion') [145, 146], a process that cannot be prevented by conventional antiviral agents [147]. Nuclear cccDNA accumulates until there are between 5 and 50 genome copies per infected cell nucleus. Once the nuclear cccDNA pool is established, nucleocapsids containing newly synthesized genomes are enveloped and exported from the cell rather than being recycled to the nucleus. The fate of newly synthesized nucleocapsids is determined by the availability of viral preS envelope protein, which increases as more nuclear cccDNA templates become available for transcription. Presumably depletion of the cytoplasmic envelope protein pool resulting from damage to or decay of viral cccDNA template would trigger its replenishment by the intracellular conversion pathway.

Complete inhibition of HBV polymerase activity should prevent replenishment of the cccDNA pool by blocking synthesis of its rc precursor. The recalcitrance of CHB infection to treatment with nucleoside and nucleotide analogues such as lamivudine and adefovir (both of which are potent inhibitors of the RNA- and DNA-dependent activities of the HBV polymerase) indicates that the cccDNA pool in infected cells is relatively stable. If production of new virions could be completely blocked, cccDNA should eventually be eliminated as a result of either death of infected hepatocytes or intracellular decay. The lifespan of infected human hepatocytes appears to be variable, depending on the severity of liver disease, and intracellular turnover of cccDNA is probably similarly variable. Some individuals can eliminate HBV cccDNA and resolve HBV infection without treatment. The mechanisms by which this is achieved are poorly understood, but the average half-life of cccDNA must clearly be shorter during spontaneously resolving acute infection than during CHB infection. Two pathways have been proposed to account for the elimination of cccDNA: (1) cytokine-mediated non-cytopathic clearance of intracellular virus [148], and (2) T cell-mediated destruction of infected cells [149]. Recent studies indicate that these mechanisms are not mutually exclusive. Guidotti and colleagues [148] studied the response of chimpanzees to acute HBV infection.

They found that most HBV DNA was cleared from liver and blood long before the peak of T cell infiltration and occurrence of severe liver damage. These observations indicate that both HBV replicative intermediates and cccDNA can be eliminated from infected cells by non-cytopathic mechanisms. Further studies in chimpanzees confirmed this observation but also revealed that elimination of HBV cccDNA is a biphasic process [150]. During the early phase, elimination of the HBV cccDNA was faster than elimination of HBsAg-positive hepatocytes and coincided with an influx of small numbers of IFNγ-producing CD8+ T cells into the liver. Clearance of residual cccDNA was associated with the peak of liver disease and hepatocellular turnover and coincided with a rapid influx of large numbers of CD8+ T cells into the liver. These results were interpreted as indicating that HBV cccDNA clearance is a biphasic process mediated by the cellular immune response. They suggested the pool of cccDNA molecules is reduced noncytolytically, probably by eliminating their relaxed circular DNA precursors, perhaps by destabilizing them, during the first phase, and that the the process is completed by destruction of infected hepatocytes during the second phase. Surprisingly, despite this robust antiviral response, traces of viral cccDNA persisted indefinitely in the chimpanzees' livers, suggesting that it may confer lifelong immunity to reinfection by providing a continuous low-level antigenic stimulus. This is consistent with recent clinical observations which indicate that occult HBV infection is much more prevalent than formerly supposed [151–153]. Similarly, results of extensive studies of the woodchuck model by Michalak and colleagues have demonstrated that WHV infection can be maintained indefinitely, irrespective of whether it is symptomatic or silent [61, 154, 155]; (see Chapter by Michalak).

Other studies in woodchucks indicate that a large fraction of the infected hepatocyte population is killed and replaced by cell division during resolution of both transient and chronic WHV infections [156, 157]. During WHV clearance induced by antiviral chemotherapy, the amount of cccDNA was reduced by 20–100-fold, while the amount of integrated viral DNA remained constant, indicating that the regenerated cell population is derived mainly by cell division of virus-infected cells, rather than uninfected progenitors. In addition, these studies revealed that frequency of integrated WHV DNA in chronically infected woodchucks was 1 or 2 orders of magnitude higher than that in transiently infected woodchucks. This suggests that the potential for genetic damage and deregulation due to integration of viral genome fragments increases over the duration of infection and remains even after infection is cleared, supporting a case for early antiviral intervention.

Regardless of the mechanisms by which cccDNA is maintained and cleared, these processes must be influenced by multiple factors. For example, cell-cycle status has been shown to influence amplification and maintenance of the cccDNA pool *in vitro*, but how it affects HBV cccDNA *in vivo* is unclear. Inhibition of nuclear DNA synthesis by aphidicolin in HBV containing HepG2 cells has been reported to induce amplification of viral cccDNA [158], but, paradoxically, blockage of the cell cycle by *n*-butyrate was found to inhibit amplification of duck HBV cccDNA in primary duck hepatocytes [159].

Viral cccDNA may be intrinsically unstable, as suggested by the relatively short half-life estimate of 3–5 days for DHBV cccDNA in cultures of primary duck hepatocytes, which are mainly quiescent [160]. This estimate is inconsistent with longer *in vivo* estimates of 35–57 days [161] suggesting that gradual dedifferentiation under non-physiological conditions may reduce primary hepatocytes' ability to maintain cccDNA [162]. Conversely, the presence of cell subpopulations that can act as virus sanctuaries during antiviral therapy may bias *in vivo* estimates. Several studies of the kinetics of HBV clearance during antiviral therapy indicate that the half-life of HBV cccDNA during chronic infection is determined mainly by the rate of turnover of infected cells. These studies show an initial rapid decline of serum HBV due to blockage of viral replication followed by a much slower phase, interpreted as being due to clearance of infected cells [110, 163]. Indeed, recent studies indicate that the second phase may continue indefinitely, with viral genomes being detectable —if sufficiently sensitive assays are used—in liver and/or lymphoid tissue in both immunologically detectable and occult persistent infection [151–155].

Viral minichromosome structure

Genomes of other small DNA viruses besides HBV also exist as minichromosomes in nuclei of infected cells. These include simian virus 40, polyoma virus, human papilloma virus, bovine papilloma virus and the cauliflower mosaic virus [164]. While little is known about the enzymology and mechanics of HBV minichromosome formation, there is a rapidly accumulating body of information describing the structure and function of chromatin. Nuclear genomes of all eukaryotic cells are packaged into chromatin. Chromatin structure plays a major role in all aspects of DNA metabolism including transcription, replication and repair by forming a physical barrier that protects the DNA from damage and controls access of enzymes and regulatory proteins [165]. Chromatin structure is dynamic, changing in response to a variety of cellular stimuli. The fundamental structural units of chromatin are nucleosomes. Each nucleosome consists of 147 bp of DNA wrapped in a left-handed superhelix around a core histone octamer containing 2 molecules each of histones H2A, H2B, H3 and H4. Variable lengths (up to about 100 bp) of nuclease-sensitive linker DNA associated with histones H1 and H5 link nucleosomal DNA, giving a characteristic 'beads on a string' appearance in scanning electron micrographs. Nucleosome spacing (the repeat length) varies with cell type, species and stage of development, ranging from about 156 to 250 bp, but there is a general trend indicating that short repeat lengths identify transcriptionally active chromatin. The repeating nucleosome-linker units are folded into arrays that form fibres 30 nm in diameter and these are organized into higher-order structure [165]. The existence of hepadnaviral minichromosomes was first demonstrated by Bock and colleagues [164], who isolated HBV minichromosomes from the nuclei of a human hepatoblastoma cell line and found that they contained host histone and non-histone proteins, as well as HBV core antigen associated with viral cccDNA. Microcococcal nuclease digestion and electron microscopy showed an average of 18 nucleosomes per HBV minichromosome, corresponding to repeat length of 180 bp, shorter than the 200 bp for host cell chromatin. Analogous findings were reported for DHBV, which was found to form minichromosomes containing up to 20 nucleosomes, corresponding to a repeat length of 151 bp, again shorter than host-cell repeat length of 200 bp [143]. The population of DHBV minichromosomes was found to be heterogeneous, consisting mainly of approximately equal amounts of fully chromatinized and half-chromatinized species. These authors postulated that these species may exist in dynamic equilibrium and may perform different functions as a result of exposure of different promotors. Specific proteins associated with DHBV chromatin were not identified. Further investigation of the structure of HBV minichromosomes confirmed that HBV core antigen was a specific component of HBV chromatin and that by preferentially binding to HBV cccDNA, it reduced nucleosomal spacing [166]. Together with earlier observations that demonstrated a correlation between the amount of core detectable in hepatocyte nuclei and active HBV replication, these findings support the notion that HBV core antigen (HBcAg) may regulate viral transcription by altering the structure of viral chromatin. Packaging of viral cccDNA into chromatin implies that it is also subject to regulation by cellular factors.

DNA incorporated into chromatin must be able to function as a template for both transcription and replication and be accessible to enzymes that repair and modify it. Changes in chromatin stucture that control these activities are accomplished by covalent modification of histone and non-histone proteins and by changes in nucleosome positioning. Post-translational modifications include acetylation, methylation and phosphorylation [167, 168]. The main targets for modification are the N-terminal sequences of histones, although non-histone proteins can also be modified [169–171].

Histone acetyltransferases (HATs) alter histone-binding properties by acetylating positively charged lysine residues, thus neutralizing their charge; histone deacetylases catalyse the reverse reaction [169–171]. Hyper-acetylation leads to an increase in expression of target genes, while hypo-acetylation results in the opposite effect. Methylation of histones creates transcriptionally inert heterochromatin [172]. Gene expression can also be controlled by methylation of cytosine residues in DNA, a process crucial for silencing expression of specific genes during development and differentiation and for preventing the uncontrolled amplification of repetitive DNA [172, 173]. Methylation produces more thermodynamically stable products than acetylation and phosphorylation, and targets for methylation, once methylated, may persist in that state [173]. Other chromatin-modifying enzymes include the nucleosome-remodelling complexes that alter nucleosome position in relation to specific DNA sequences, an energy-dependent process that is usually driven by ATP hydrolysis [174, 175].

Can HBV minichromosomes be eliminated by therapy?
The presence of functional HBV minichromosomes in hepatocyte nuclei is essential for viral replication and their complete elimination is necessary to effect a lasting cure. Because it is replicated via an RNA intermediate, HBV cccDNA is not directly affected by inhibitors of DNA polymerase activity. Although these drugs do inhibit production of cccDNA precursors, the long *in vivo* half-life of cccDNA during chronic infection and low viral replication fidelity makes it likely that drug-resistant virus will emerge before cccDNA is completely depleted. Several alternative strategies aimed at destabilizing this key replicative intermediate have been proposed, but few have been tested experimentally.

One approach has been to attempt to destabilize cccDNA and prevent its formation by treatment with compounds that either interact directly with viral cccDNA or that inhibit specific enzymes required for its formation and processing. The first group includes intercalators and compounds that bind to DNA; the second are mostly inhibitors of topoisomerases, the enzymes that catalyse interconversion of DNA topoisomers by introducing transient single- or double-strand breaks via formation of covalent enzyme–DNA intermediates. Stabilization of topoisomerase–DNA 'cleavable complexes' by inhibitors and cross-linking agents blocks progress of polymerases, ultimately causing permanent strand breakage [176]. This approach has been shown to work in principle in primary duck hepatocytes infected with DHBV. Agents such as these that have demonstrated antiviral activity include adriamycin, tenoposide and camptothecin, all of which have been used clinically as anticancer drugs, but which are clearly too non-specific and too toxic to be used long term. However, logical extensions to this approach may include design of antisense RNAs, silencing RNAs, or peptides that specifically bind to viral cccDNA to block transcription. An alternative approach might involve the use of a new generation of antineoplastic drugs that interfere with histone and DNA modification [177–179]. Whether or not these compounds have clinically useful antiviral activity awaits evaluation, but their use in studies *in vitro* should generate useful information about the structure and function of hepadnaviral minichromosomes.

Conclusion and future directions

Development of clinically useful antiviral chemotherapy, particularly in the case of viral hepatitis, is inherently difficult for many reasons. Nearly three decades after its use against HBV was reported, IFN, despite its limited efficacy, remains one of only three treatment options for CHB, and the only option (in combination with RBV) for chronic HCV. There now seems to be a general consensus on two issues: (1) the severity of liver disease can be reduced and its progression delayed by antiviral therapy, especially if started early; and (2) long-term combination chemotherapy will be required to control both hepatitis B and hepatitis C infection. However, consensus has yet to be reached regarding optimal therapies in both cases. The major problems that have hampered development of antiviral therapies for chronic hepatitis B and C are essentially similar, namely: (1) diagnostic problems caused by the heterogeneity and delayed manifestation of signs and symptoms of disease; (2) a lack of safe and efficacious antiviral agents and variability in individual response to the few available; (3) paucity of information about viral replication strategies, potential virus-specific targets and the natural history of disease; (4) the lack of convenient small-animal and cell-culture systems which would otherwise provide the means to tackle (2) and (3). It must be acknowledged that despite these problems, recent progress has been spectacular. Technological advances have made it possible to solve many diagnostic problems, and should in future permit easy distinction between viral genotypes and rapid recognition of developing drug resistance. Several new anti-HBV drugs have progressed to clinical trials and more are being developed; identification of virus-specific targets for anti-HCV drugs led to development of a host of potential anti-HCV agents that are currently under investigation. The recently developed cell-culture systems and small-animal models for studies of both HBV and HCV have already been useful for studies viral replication and pathogenesis and should greatly facilitate screening of new drugs. Rapid expansion of gene data bases in conjunction with new technologies—notably gene microarrays—that allows the analysis of gene expression and regulation on a scale previously inconceivable will permit newly available resources to be used with maximum efficiency. To paraphrase Stuart-Harrris [1], we are now fortunate to be in the position where we are able "to undertake mass screening of existing compounds *and* to look to rational developments in methods of attacking the virus, its enzymes, nucleic acid or proteins". But with each new development, we "should (still) not rule out any way by which new compounds with antiviral properties may be found". In the immediate future, ongoing experience with newer anti-HBV agents and accumulating clinical and molecular data should allow development of rational strategies designed to arrest HBV replication, prevent disease progression and minimize the risk of drug resistance. For HCV, important short-term aims will be to extend the potential of the chimeric mouse system and to develop cell-culture systems capable of supporting complete cycles of HCV replication. These aims now seem realistic and achievable. Their eventual realization will represent an important milestone on the long and difficult path to the control and ultimate elimination, by means of antiviral chemotherapy, of liver disease caused by HBV and HCV.

References

1 Stuart-Harris CH, Oxford J (eds) (1983) *Problems of antiviral chemotherapy*. Academic Press, London
2 Bauer DJ (1963) The chemotherapy of ectromelia infection with isatin beta-dialkylthiosemicar-

bazones. *Br J Exp Pathol* 44: 233–242
3 Jackson GG, Muldoon RL, Akers LW (1963) Serological Evidence for Prevention of Influenzal Infection in Volunteers by an Anti-Influenzal Drug Adamantanamine Hydrochloride. *Antimicrob Agents Chemother* 161: 703–707
4 Blumberg BS (2002) The discovery of the hepatitis B virus and the invention of the vaccine: a scientific memoir. *J Gastroenterol Hepatol* 17 Suppl: S502–503
5 Jones BR, Galbraith JE, Al-Hussaini MK (1962) Vaccinial keratitis treated with interferon. *Lancet* 1: 875–879
6 Greenberg HB, Pollard RB, Lutwick LI, Gregory PB, Robinson WS, Merigan TC (1976) Effect of human leukocyte interferon on hepatitis B virus infection in patients with chronic active hepatitis. *N Engl J Med* 295: 517–522
7 Shaw T, Locarnini S (2000) Combination chemotherapy for hepatitis B virus: the path forward? *Drugs* 60: 517–531
8 Thomas HC, Lok AS (1983) Approaches to the treatment of chronic hepatitis B viral infection. *Prog Clin Biol Res* 143: 379–393
9 MMWR (1982) *Morb Mortal Weekly Report*, Sept 24
10 Boas U, Heegaard PM (2004) Dendrimers in drug research. *Chem Soc Rev* 33: 43–63
11 Davies WL, Grunert RR, Haff RF, McGahen JW, Neumayer EM, Paulshock M, Watts JC, Wood TR, Hermann EC, Hoffmann CE (1964) Antiviral Activity of 1-Adamantanamine (Amantadine). *Science* 144: 862–863
12 Monto AS (2003) The role of antivirals in the control of influenza. *Vaccine* 21: 1796–1800
13 Moore JP, Doms RW (2003) The entry of entry inhibitors: a fusion of science and medicine. *Proc Natl Acad Sci USA* 100: 10598–10602
14 Parmar S, Platanias LC (2003) Interferons: mechanisms of action and clinical applications. *Curr Opin Oncol* 15: 431–439
15 Hilleman MR (2003) Critical overview and outlook: pathogenesis, prevention, and treatment of hepatitis and hepatocarcinoma caused by hepatitis B virus. *Vaccine* 21: 4626–4649
16 Kane MA, Brooks A (2002) New immunization initiatives and progress toward the global control of hepatitis B. *Curr Opin Infect Dis* 15: 465–469
17 Davis GL, Albright JE, Cook SF, Rosenberg DM (2003) Projecting future complications of chronic hepatitis C in the United States. *Liver Transpl* 9: 331–338
18 Llovet JM, Burroughs A, Bruix J (2003) Hepatocellular carcinoma. *Lancet* 362: 1907–1917
19 Koff RS (2003) Hepatitis vaccines: recent advances. *Int J Parasitol* 33: 517–523
20 de Franchis R, Hadengue A, Lau G, Lavanchy D, Lok A, McIntyre N, Mele A, Paumgartner G, Pietrangelo A, Rodes J et al. (2003) EASL International Consensus Conference on Hepatitis B. 13–14 September, 2002 Geneva, Switzerland. Consensus statement (long version). *J Hepatol* 39 (Suppl 1): S3–S25
21 NIH (2002) NIH Consensus Statement on Management of Hepatitis C: 2002. *NIH Consens State Sci Statements* 19: 1–46
22 Di Bisceglie AM, Hoofnagle JH (2002) Optimal therapy of hepatitis C. *Hepatology* 36: S121–S127
23 Samuel CE (2001) Antiviral actions of interferons. *Clin Microbiol Rev* 14: 778–809
24 Ben-Ari Z, Mor E, Papo O, Kfir B, Sulkes J, Tambur AR, Tur-Kaspa R, Klein T (2003) Cytokine gene polymorphisms in patients infected with hepatitis B virus. *Am J Gastroenterol* 98: 144–150
25 Gale M Jr (2003) Effector genes of interferon action against hepatitis C virus. *Hepatology* 37: 975–978
26 Ji X, Cheung R, Cooper S, Li Q, Greenberg HB, He XS (2003) Interferon alfa regulated gene expression in patients initiating interferon treatment for chronic hepatitis C. *Hepatology* 37: 610–621
27 Geiss GK, Carter VS, He Y, Kwieciszewski BK, Holzman T, Korth MJ, Lazaro CA, Fausto N, Bumgarner RE, Katze MG (2003) Gene expression profiling of the cellular transcriptional network regulated by alpha/beta interferon and its partial attenuation by the hepatitis C virus nonstructural 5A protein. *J Virol* 77: 6367–6375
28 Manns MP (2002) Current state of interferon therapy in the treatment of chronic hepatitis B. *Semin Liver Dis* 22 Suppl 1: 7–13
29 Cornberg M, Wedemeyer H, Manns MP (2002) Treatment of chronic hepatitis C with PEGylated interferon and ribavirin. *Curr Gastroenterol Rep* 4: 23–30
30 Craxi A, Cooksley WG (2003) Pegylated interferons for chronic hepatitis B. *Antivir Res* 60: 87–89

31 Fried MW, Shiffman ML, Reddy KR, Smith C, Marinos G, Goncales FL Jr, Haussinger D, Diago M, Carosi G, Dhumeaux D et al. (2002) Peginterferon alfa-2a plus ribavirin for chronic hepatitis C virus infection. *N Engl J Med* 347: 975–982
32 Shaw T, Locarnini SA (1995) Hepatic purine and pyrimidine metabolism: implications for antiviral chemotherapy of viral hepatitis. *Liver* 15: 169–184
33 Johnson AA, Ray AS, Hanes J, Suo Z, Colacino JM, Anderson KS, Johnson KA (2001) Toxicity of antiviral nucleoside analogs and the human mitochondrial DNA polymerase. *J Biol Chem* 276: 40847–40857
34 Lewis W, Day BJ, Copeland WC (2003) Mitochondrial toxicity of NRTI antiviral drugs: an integrated cellular perspective. *Nat Rev Drug Discov* 2: 812–822
35 Brinkman K, Kakuda TN (2000) Mitochondrial toxicity of nucleoside analogue reverse transcriptase inhibitors: a looming obstacle for long-term antiretroviral therapy? *Curr Opin Infect Dis* 13: 5–11
36 Lee H, Hanes J, Johnson KA (2003) Toxicity of nucleoside analogues used to treat AIDS and the selectivity of the mitochondrial DNA polymerase. *Biochemistry* 42: 14711–14719
37 Colacino JM (1996) Mechanisms for the anti-hepatitis B virus activity and mitochondrial toxicity of fialuridine (FIAU). *Antivir Res* 29: 125–139
38 Honkoop P, Scholte HR, de Man RA, Schalm SW (1997) Mitochondrial injury. Lessons from the fialuridine trial. *Drug Safety* 17: 1–7
39 Lewis W, Levine ES, Griniuviene B, Tankersley KO, Colacino JM, Sommadossi JP, Watanabe KA, Perrino FW (1996) Fialuridine and its metabolites inhibit DNA polymerase gamma at sites of multiple adjacent analog incorporation, decrease mtDNA abundance, and cause mitochondrial structural defects in cultured hepatoblasts. *Proc Natl Acad Sci USA* 93: 3592–3597
40 Lim SE, Ponamarev MV, Longley MJ, Copeland WC (2003) Structural determinants in human DNA polymerase gamma account for mitochondrial toxicity from nucleoside analogs. *J Mol Biol* 329: 45–57
41 Birkus G, Gibbs CS, Cihlar T (2003) Comparative effects of adefovir and selected nucleoside inhibitors of hepatitis B virus DNA polymerase on mitochondrial DNA in liver and skeletal muscle cells. *J Viral Hepat* 10: 50–54
42 Birkus G, Hitchcock MJ, Cihlar T (2002) Assessment of mitochondrial toxicity in human cells treated with tenofovir: comparison with other nucleoside reverse transcriptase inhibitors. *Antimicrob Agents Chemother* 46: 716–723
43 Chang CN, Doong SL, Zhou JH, Beach JW, Jeong LS, Chu CK, Tsai CH, Cheng YC, Liotta D, Schinazi R (1992) Deoxycytidine deaminase-resistant stereoisomer is the active form of (+/–)-2',3'-dideoxy-3'-thiacytidine in the inhibition of hepatitis B virus replication. *J Biol Chem* 267: 13938–13942
44 Focher F, Spadari S, Maga G (2003) Antivirals at the mirror: the lack of stereospecificity of some viral and human enzymes offers novel opportunities in antiviral drug development. *Curr Drug Targets Infect Disord* 3: 41–53
45 Hugle T, Cerny A (2003) Current therapy and new molecular approaches to antiviral treatment and prevention of hepatitis C. *Rev Med Virol* 13: 361–371
46 Isaguliants MG (2003) Hepatitis C virus clearance: the enigma of failure despite an impeccable survival strategy. *Curr Pharm Biotechnol* 4: 169–183
47 McHutchison JG, Patel K (2002) Future therapy of hepatitis C. *Hepatology* 36: S245–252
48 McHutchison JG, Fried MW (2003) Current therapy for hepatitis C: pegylated interferon and ribavirin. *Clin Liver Dis* 7: 149–161
49 Dieterich DT, Spivak JL (2003) Hematologic disorders associated with hepatitis C virus infection and their management. *Clin Infect Dis* 37: 533–541
50 Watson J (2002) Prospects for hepatitis C virus therapeutics: levovirin and viramidine as improved derivatives of ribavirin. *Curr Opin Invest Drugs* 3: 680–683
51 Harris KR, Dighe AS (2002) Laboratory testing for viral hepatitis. *Am J Clin Pathol* 118 Suppl: S18–S25
52 Pawlotsky JM (2002) Molecular diagnosis of viral hepatitis. *Gastroenterology* 122: 1554–1568
53 Comanor L, Elkin C, Leung K, Krajden M, Kronquist K, Nicolas K, Horansky E, deMedina M, Kittichai P, Sablon E et al. (2003) Successful HCV genotyping of previously failed and low viral load specimens using an HCV RNA qualitative assay based on transcription-mediated amplification in conjunction with the line probe assay. *J Clin Virol* 28: 14–26
54 Sablon E, Shapiro F, Zoulim F (2003) Early detection of hepatitis B drug resistance: implications

for patient management. *Expert Rev Mol Diagn* 3: 535–547
55 Gardner SN, Kuczmarski TA, Vitalis EA, Slezak TR (2003) Limitations of TaqMan PCR for detecting divergent viral pathogens illustrated by hepatitis A, B, C, and E viruses and human immunodeficiency virus. *J Clin Microbiol* 41: 2417–2427
56 Van Rompay AR, Johansson M, Karlsson A (2003) Substrate specificity and phosphorylation of antiviral and anticancer nucleoside analogues by human deoxyribonucleoside kinases and ribonucleoside kinases. *Pharmacol Ther* 100: 119–139
57 Kong W, Engel K, Wang J (2004) Mammalian nucleoside transporters. *Curr Drug MeTab.* 5: 63–84
58 Gerloff T (2004) Impact of genetic polymorphisms in transmembrane carrier-systems on drug and xenobiotic distribution. *Naunyn Schmiedebergs Arch Pharmacol* 369: 69–77
59 Mazzon C, Rampazzo C, Scaini MC, Gallinaro L, Karlsson A, Meier C, Balzarini J, Reichard P, Bianchi V (2003) Cytosolic and mitochondrial deoxyribonucleotidases: activity with substrate analogs, inhibitors and implications for therapy. *Biochem Pharmacol* 66: 471–479
60 Lee JY, Culvenor JG, Angus P, Smallwood R, Nicoll A, Locarnini S (2001) Duck hepatitis B virus replication in primary bile duct epithelial cells. *J Virol* 75: 7651–7661
61 Michalak TI (2000) Occult persistence and lymphotropism of hepadnaviral infection: insights from the woodchuck viral hepatitis model. *Immunol Rev* 174: 98–111
62 Zhu Q, Guo JT, Seeger C (2003) Replication of hepatitis C virus subgenomes in nonhepatic epithelial and mouse hepatoma cells. *J Virol* 77: 9204–9210
63 Fattovich G (2003) Natural history and prognosis of hepatitis B. *Semin Liver Dis* 23: 47–58
64 Poynard T, Yuen MF, Ratziu V, Lai CL (2003) Viral hepatitis C. *Lancet* 362: 2095–2100
65 Bertoletti A, Ferrari C (2003) Kinetics of the immune response during HBV and HCV infection. *Hepatology* 38: 4–13
66 Nakamoto Y, Kaneko S (2003) Mechanisms of viral hepatitis induced liver injury. *Curr Mol Med* 3: 537–544
67 Wodarz D (2003) Hepatitis C virus dynamics and pathology: the role of CTL and antibody responses. *J Gen Virol* 84: 1743–1750
68 Olive DS, Konishi M, Wu GY (2004) Cell culture and animal models for human viral hepatitis. *Hepatol Res* 28: 61–67
69 Simmonds P (2001) The origin and evolution of hepatitis viruses in humans. *J Gen Virol* 82: 693–712
70 Starkman SE, MacDonald DM, Lewis JC, Holmes EC, Simmonds P (2003) Geographic and species association of hepatitis B virus genotypes in non-human primates. *Virology* 314: 381–393
71 Delaney WE 3rd, Locarnini S, Shaw T (2001) Resistance of hepatitis B virus to antiviral drugs: current aspects and directions for future investigation. *Antivir Chem Chemother* 12: 1–35
72 Singh M, Kumar V (2003) Transgenic mouse models of hepatitis B virus-associated hepatocellular carcinoma. *Rev Med Virol* 13: 243–253
73 Cao J, Yang EB, Su JJ, Li Y, Chow P (2003) The tree shrews: adjuncts and alternatives to primates as models for biomedical research. *J Med Primatol* 32: 123–130
74 Ren S, Nassal M (2001) Hepatitis B virus (HBV) virion and covalently closed circular DNA formation in primary tupaia hepatocytes and human hepatoma cell lines upon HBV genome transduction with replication-defective adenovirus vectors. *J Virol* 75: 1104–1116
75 Glebe D, Aliakbari M, Krass P, Knoop EV, Valerius KP, Gerlich WH (2003) Pre-s1 antigen-dependent infection of Tupaia hepatocyte cultures with human hepatitis B virus. *J Virol* 77: 9511–9521
76 Ladner SK, Otto MJ, Barker CS, Zaifert K, Wang GH, Guo JT, Seeger C, King RW (1997) Inducible expression of human hepatitis B virus (HBV) in stably transfected hepatoblastoma cells: a novel system for screening potential inhibitors of HBV replication. *Antimicrob Agents Chemother* 41: 1715–1720
77 Delaney WE 3rd, Isom HC (1998) Hepatitis B virus replication in human HepG2 cells mediated by hepatitis B virus recombinant baculovirus. *Hepatology* 28: 1134–1146
78 Gripon P, Rumin S, Urban S, Le Seyec J, Glaise D, Cannie I, Guyomard C, Lucas J, Trepo C, Guguen-Guillouzo C (2002) Infection of a human hepatoma cell line by hepatitis B virus. *Proc Natl Acad Sci USA* 99: 15655–15660
79 Grakoui A, Hanson HL, Rice CM (2001) Bad time for Bonzo? Experimental models of hepatitis C virus infection, replication, and pathogenesis. *Hepatology* 33: 489–495
80 Lanford RE, Bigger C, Bassett S, Klimpel G (2001) The chimpanzee model of hepatitis C virus

infections. *Ilar J* 42: 117–126
81 Thimme R, Bukh J, Spangenberg HC, Wieland S, Pemberton J, Steiger C, Govindarajan S, Purcell RH, Chisari FV (2002) Viral and immunological determinants of hepatitis C virus clearance, persistence, and disease. *Proc Natl Acad Sci USA* 99: 15661–15668
82 Thomson M, Nascimbeni M, Havert MB, Major M, Gonzales S, Alter H, Feinstone SM, Murthy KK, Rehermann B, Liang TJ (2003) The clearance of hepatitis C virus infection in chimpanzees may not necessarily correlate with the appearance of acquired immunity. *J Virol* 77: 862–870
83 Bigger CB, Brasky KM, Lanford RE (2001) DNA microarray analysis of chimpanzee liver during acute resolving hepatitis C virus infection. *J Virol* 75: 7059–7066
84 Iizuka N, Oka M, Yamada-Okabe H, Mori N, Tamesa T, Okada T, Takemoto N, Hashimoto K, Tangoku A, Hamada K et al. (2003) Differential gene expression in distinct virologic types of hepatocellular carcinoma: association with liver cirrhosis. *Oncogene* 22: 3007–3014
85 Smith MW, Yue ZN, Korth MJ, Do HA, Boix L, Fausto N, Bruix J, Carithers RL Jr, Katze MG (2003) Hepatitis C virus and liver disease: global transcriptional profiling and identification of potential markers. *Hepatology* 38: 1458–1467
86 Smith MW, Yue ZN, Geiss GK, Sadovnikova NY, Carter VS, Boix L, Lazaro CA, Rosenberg GB, Bumgarner RE, Fausto N et al. (2003) Identification of novel tumor markers in hepatitis C virus-associated hepatocellular carcinoma. *Cancer Res* 63: 859–864
87 Otsuka M, Aizaki H, Kato N, Suzuki T, Miyamura T, Omata M, Seki N (2003) Differential cellular gene expression induced by hepatitis B and C viruses. *Biochem Biophys Res Commun* 300: 443–447
88 Fimia GM, Tripodi M, Alonzi T (2003) Transgenic models for Hepatitis C virus pathogenesis. *Cell Death Differ* 10 Suppl 1: S16–S18
89 Pietschmann T, Bartenschlager R (2003) Tissue culture and animal models for hepatitis C virus. *Clin Liver Dis* 7: 23–43
90 Xie ZC, Riezu-Boj JI, Lasarte JJ, Guillen J, Su JH, Civeira MP, Prieto J (1998) Transmission of hepatitis C virus infection to tree shrews. *Virology* 244: 513–520
91 Zhao X, Tang ZY, Klumpp B, Wolff-Vorbeck G, Barth H, Levy S, von Weizsacker F, Blum HE, Baumert TF (2002) Primary hepatocytes of Tupaia belangeri as a potential model for hepatitis C virus infection. *J Clin Invest* 109: 221–232
92 Martin A, Bodola F, Sangar DV, Goettge K, Popov V, Rijnbrand R, Lanford RE, Lemon SM (2003) Chronic hepatitis associated with GB virus B persistence in a tamarin after intrahepatic inoculation of synthetic viral RNA. *Proc Natl Acad Sci USA* 100: 9962–9967
93 Bright H, Carroll AR, Watts PA, Fenton RJ (2004) Development of a GB virus B marmoset model and its validation with a novel series of hepatitis C virus NS3 protease inhibitors. *J Virol* 78: 2062–2071
94 Mercer DF, Schiller DE, Elliott JF, Douglas DN, Hao C, Rinfret A, Addison WR, Fischer KP, Churchill TA, Lakey JR et al. (2001) Hepatitis C virus replication in mice with chimeric human livers. *Nat Med* 7: 927–933
95 Hsu EC, Hsi B, Hirota-Tsuchihara M, Ruland J, Iorio C, Sarangi F, Diao J, Migliaccio G, Tyrrell DL, Kneteman N et al. (2003) Modified apoptotic molecule (BID) reduces hepatitis C virus infection in mice with chimeric human livers. *Nat Biotechnol* 21: 519–525
96 Bartenschlager R (2002) Hepatitis C virus replicons: potential role for drug development. *Nat Rev Drug Disc* 1: 911–916
97 Bukh J, Pietschmann T, Lohmann V, Krieger N, Faulk K, Engle RE, Govindarajan S, Shapiro M, St Claire M, Bartenschlager R (2002) Mutations that permit efficient replication of hepatitis C virus RNA in Huh-7 cells prevent productive replication in chimpanzees. *Proc Natl Acad Sci USA* 99: 14416–14421
98 Blight KJ, McKeating JA, Marcotrigiano J, Rice CM (2003) Efficient replication of hepatitis C virus genotype 1a RNAs in cell culture. *J Virol* 77: 3181–3190
99 Blight KJ, McKeating JA, Rice CM (2002) Highly permissive cell lines for subgenomic and genomic hepatitis C virus RNA replication. *J Virol* 76: 13001–13014
100 Martin KW, Ernst E (2003) Antiviral agents from plants and herbs: a systematic review. *Antivir Ther* 8: 77–90
101 Thyagarajan S, Jayaram S, Gopalakrishnan V, Hari R, Jeyakumar P, Sripathi M (2002) Herbal medicines for liver diseases in India. *J Gastroenterol Hepatol* 17 (Suppl 3): S370–S376
102 Niu JZ, Wang YY, Qiao M, Gowans E, Edwards P, Thyagarajan SP, Gust I, Locarnini S (1990) Effect of Phyllanthus amarus on duck hepatitis B virus replication *in vivo*. *J Med Virol* 32:

212–218
103 Chan HL, Sung JJ, Fong WF, Chim AM, Yung PP, Hui AY, Fung KP, Leung PC (2003) Double-blinded placebo-controlled study of Phyllanthus urinaris for the treatment of chronic hepatitis B. *Aliment Pharmacol Ther* 18: 339–345
104 Lu LG, Zeng MD, Mao YM, Li JQ, Wan MB, Li CZ, Chen CW, Fu QC, Wang JY, She WM et al. (2003) Oxymatrine therapy for chronic hepatitis B: a randomized double-blind and placebo-controlled multi-center trial. *World. J Gastroenterol* 9: 2480–2483
105 Benhamou Y (2004) Antiretroviral therapy and HIV/hepatitis B virus coinfection. *Clin Infect Dis* 38 Suppl 2: S98–103
106 Shaw T, Bowden S, Locarnini S (2004) Rescue therapy for drug resistant hepatitis B: another argument for combination chemotherapy? *Gastroenterology* 126: 343–347
107 Domingo E (2003) Quasispecies and the development of new antiviral strategies. *Prog Drug Res* 60: 133–158
108 Briones C, Domingo E, Molina-Paris C (2003) Memory in retroviral quasispecies: experimental evidence and theoretical model for human immunodeficiency virus. *J Mol Biol* 331: 213–229
109 Park SG, Kim Y, Park E, Ryu HM, Jung G (2003) Fidelity of hepatitis B virus polymerase. *Eur J Biochem* 270: 2929–2936
110 Lewin S, Walters T, Locarnini S (2002) Hepatitis B treatment: rational combination chemotherapy based on viral kinetic and animal model studies. *Antivir Res* 55: 381–396
111 Locarnini S (2003) Hepatitis B viral resistance: mechanisms and diagnosis. *J Hepatol* 39 Suppl 1: S124–132
112 Eigen M (2002) Error catastrophe and antiviral strategy. *Proc Natl Acad Sci USA* 99: 13374–13376
113 Crotty S, Andino R (2002) Implications of high RNA virus mutation rates: lethal mutagenesis and the antiviral drug ribavirin. *Microbes Infect* 4: 1301–1307
114 Wilke CO, Novella IS (2003) Phenotypic mixing and hiding may contribute to memory in viral quasispecies. *BMC Microbiol* 3: 11
115 Feld J, Locarnini S (2002) Antiviral therapy for hepatitis B virus infections: new targets and technical challenges. *J Clin Virol* 25: 267–283
116 Deres K, Schroder CH, Paessens A, Goldmann S, Hacker HJ, Weber O, Kramer T, Niewohner U, Pleiss U, Stoltefuss J et al. (2003) Inhibition of hepatitis B virus replication by drug-induced depletion of nucleocapsids. *Science* 299: 893–896
117 Dienstag JL, Goldin RD, Heathcote EJ, Hann HW, Woessner M, Stephenson SL, Gardner S, Gray DF, Schiff ER (2003) Histological outcome during long-term lamivudine therapy. *Gastroenterology* 124: 105–117
118 Lai CL, Dienstag J, Schiff E, Leung NW, Atkins M, Hunt C, Brown N, Woessner M, Boehme R, Condreay L (2003) Prevalence and clinical correlates of YMDD variants during lamivudine therapy for patients with chronic hepatitis B. *Clin Infect Dis* 36: 687–696
119 Liu X, Schinazi RF (2002) Hepatitis B virus resistance to lamivudine and its clinical implications. *Antivir Chem Chemother* 13: 143–155
120 Leung N (2002) Clinical experience with lamivudine. *Semin Liver Dis* 22 Suppl 1: 15–21
121 Stuyver LJ, Locarnini SA, Lok A, Richman DD, Carman WF, Dienstag JL, Schinazi RF (2001) Nomenclature for antiviral-resistant human hepatitis B virus mutations in the polymerase region. *Hepatology* 33: 751–757
122 Delaney WE 3rd, Edwards R, Colledge D, Shaw T, Torresi J, Miller TG, Isom HC, Bock CT, Manns MP, Trautwein C et al. (2001) Cross-resistance testing of antihepadnaviral compounds using novel recombinant baculoviruses which encode drug-resistant strains of hepatitis B virus. *Antimicrob Agents Chemother* 45: 1705–1713
123 Delaney WE 3rd, Yang H, Westland CE, Das K, Arnold E, Gibbs CS, Miller MD, Xiong S (2003) The hepatitis B virus polymerase mutation rtV173L is selected during lamivudine therapy and enhances viral replication *in vitro*. *J Virol* 77: 11833–11841
124 Dando T, Plosker G (2003) Adefovir dipivoxil: a review of its use in chronic hepatitis B. *Drugs* 63: 2215–2234
125 Marcellin P, Chang TT, Lim SG, Tong MJ, Sievert W, Shiffman ML, Jeffers L, Goodman Z, Wulfsohn MS, Xiong S et al. (2003) Adefovir dipivoxil for the treatment of hepatitis B e antigen-positive chronic hepatitis B. *N Engl J Med* 348: 808–816
126 Hadziyannis SJ, Tassopoulos NC, Heathcote EJ, Chang TT, Kitis G, Rizzetto M, Marcellin P, Lim SG, Goodman Z, Wulfsohn MS et al. (2003) Adefovir dipivoxil for the treatment of hepatitis B e

antigen-negative chronic hepatitis B. *N Engl J Med* 348: 800–807
127 Angus P, Vaughan R, Xiong S, Yang H, Delaney W, Gibbs C, Brosgart C, Colledge D, Edwards R, Ayres A et al. (2003) Resistance to adefovir dipivoxil therapy associated with the selection of a novel mutation in the HBV polymerase. *Gastroenterology* 125: 292–297
128 Villeneuve JP, Durantel D, Durantel S, Westland C, Xiong S, Brosgart CL, Gibbs CS, Parvaz P, Werle B, Trepo C et al. (2003) Selection of a hepatitis B virus strain resistant to adefovir in a liver transplantation patient. *J Hepatol* 39: 1085–1089
129 Das K, Xiong X, Yang H, Westland CE, Gibbs CS, Sarafianos SG, Arnold E (2001) Molecular modeling and biochemical characterization reveal the mechanism of hepatitis B virus polymerase resistance to lamivudine (3TC) and emtricitabine (FTC). *J Virol* 75: 4771–4779
130 Perrillo R, Hann HW, Mutimer D, Willems B, Leung N, Lee WM, Moorat A, Gardner S, Woessner M, Bourne E et al. (2004) Adefovir dipivoxil added to ongoing lamivudine in chronic hepatitis B with YMDD mutant hepatitis B virus. *Gastroenterology* 126: 81–90
131 Schiff ER, Lai CL, Hadziyannis S, Neuhaus P, Terrault N, Colombo M, Tillmann HL, Samuel D, Zeuzem S, Lilly L et al. (2003) Adefovir dipivoxil therapy for lamivudine-resistant hepatitis B in pre- and post-liver transplantation patients. *Hepatology* 38: 1419–1427
132 Barcena Marugan R, Cid Gomez L, Lopez Serrano P (2003) Use of adefovir in the treatment of the chronic hepatitis B virus infection with resistance to lamivudine. *Transplant Proc* 35: 1841–1843
133 Peters MG, Hann Hw H, Martin P, Heathcote EJ, Buggisch P, Rubin R, Bourliere M, Kowdley K, Trepo C, Gray Df D et al. (2004) Adefovir dipivoxil alone or in combination with lamivudine in patients with lamivudine-resistant chronic hepatitis B. *Gastroenterology* 126: 91–101
134 Foy E, Li K, Wang C, Sumpter R Jr, Ikeda M, Lemon SM, Gale M Jr (2003) Regulation of interferon regulatory factor-3 by the hepatitis C virus serine protease. *Science* 300: 1145–1148
135 Pawlotsky JM (2003) The nature of interferon-alpha resistance in hepatitis C virus infection. *Curr Opin Infect Dis* 16: 587–592
136 Migliaccio G, Tomassini JE, Carroll SS, Tomei L, Altamura S, Bhat B, Bartholomew L, Bosserman MR, Ceccacci A, Colwell LF et al. (2003) Characterization of resistance to non-obligate chain-terminating ribonucleoside analogs that inhibit hepatitis C virus replication *in vitro*. *J Biol Chem* 278: 49164–49170
137 Tomei L, Altamura S, Bartholomew L, Bisbocci M, Bailey C, Bosserman M, Cellucci A, Forte E, Incitti I, Orsatti L et al. (2004) Characterization of the inhibition of hepatitis C virus RNA replication by nonnucleosides. *J Virol* 78: 938–946
138 Trozzi C, Bartholomew L, Ceccacci A, Biasiol G, Pacini L, Altamura S, Narjes F, Muraglia E, Paonessa G, Koch U et al. (2003) *In vitro* selection and characterization of hepatitis C virus serine protease variants resistant to an active-site peptide inhibitor. *J Virol* 77: 3669–3679
139 Hong Z, Cameron CE (2002) Pleiotropic mechanisms of ribavirin antiviral activities. *Prog Drug Res* 59: 41–69
140 Vo NV, Young KC, Lai MM (2003) Mutagenic and inhibitory effects of ribavirin on hepatitis C virus RNA polymerase. *Biochemistry* 42: 10462–10471
141 Bougie I, Bisaillon M (2003) Initial binding of the broad spectrum antiviral nucleoside ribavirin to the hepatitis C virus RNA polymerase. *J Biol Chem* 278: 52471–52478
142 Young KC, Lindsay KL, Lee KJ, Liu WC, He JW, Milstein SL, Lai MM (2003) Identification of a ribavirin-resistant NS5B mutation of hepatitis C virus during ribavirin monotherapy. *Hepatology* 38: 869–878
143 Newbold JE, Xin H, Tencza M, Sherman G, Dean J, Bowden S, Locarnini S (1995) The covalently closed duplex form of the hepadnavirus genome exists *in situ* as a heterogeneous population of viral minichromosomes. *J Virol* 69: 3350–3357
144 Dean J, Bowden S, Locarnini S (1995) Reversion of duck hepatitis B virus DNA replication *in vivo* following cessation of treatment with the nucleoside analogue ganciclovir. *Antivir Res* 27: 171–178
145 Wu TT, Coates L, Aldrich CE, Summers J, Mason WS (1990) In hepatocytes infected with duck hepatitis B virus, the template for viral RNA synthesis is amplified by an intracellular pathway. *Virology* 175: 255–261
146 Tuttleman JS, Pourcel C, Summers J (1986) Formation of the pool of covalently closed circular viral DNA in hepadnavirus-infected cells. *Cell* 47: 451–460
147 Kock J, Schlicht HJ (1993) Analysis of the earliest steps of hepadnavirus replication: genome repair after infectious entry into hepatocytes does not depend on viral polymerase activity. *J Virol*

67: 4867–4874
148 Guidotti LG, Rochford R, Chung J, Shapiro M, Purcell R, Chisari FV (1999) Viral clearance without destruction of infected cells during acute HBV infection. *Science* 284: 825–829
149 Guo JT, Zhou H, Liu C, Aldrich C, Saputelli J, Whitaker T, Barrasa MI, Mason WS, Seeger C (2000) Apoptosis and regeneration of hepatocytes during recovery from transient hepadnavirus infections. *J Virol* 74: 1495–1505
150 Wieland SF, Spangenberg HC, Thimme R, Purcell RH, Chisari FV (2004) Expansion and contraction of the hepatitis B virus transcriptional template in infected chimpanzees. *Proc Natl Acad Sci USA* 101: 2129–2134
151 Hu KQ (2002) Occult hepatitis B virus infection and its clinical implications. *J Viral Hepat* 9: 243–257
152 Torbenson M, Thomas DL (2002) Occult hepatitis B. *Lancet Infect Dis* 2: 479–486
153 Roche B, Feray C, Gigou M, Roque-Afonso AM, Arulnaden JL, Delvart V, Dussaix E, Guettier C, Bismuth H, Samuel D (2003) HBV DNA persistence 10 years after liver transplantation despite successful anti-HBS passive immunoprophylaxis. *Hepatology* 38: 86–95
154 Mulrooney PM, Michalak TI (2003) Quantitative detection of hepadnavirus-infected lymphoid cells by *in situ* PCR combined with flow cytometry: implications for the study of occult virus persistence. *J Virol* 77: 970–979
155 Michalak TI, Mulrooney PM, Coffin CS (2004) Low doses of hepadnavirus induce infection of the lymphatic system that does not engage the liver. *J Virol* 78: 1730–1738
156 Summers J, Jilbert AR, Yang W, Aldrich CE, Saputelli J, Litwin S, Toll E, Mason WS (2003) Hepatocyte turnover during resolution of a transient hepadnaviral infection. *Proc Natl Acad Sci USA* 100: 11652–11659
157 Summers J, Mason WS (2004) Residual integrated viral DNA after hepadnavirus clearance by nucleoside analog therapy. *Proc Natl Acad Sci USA* 101: 638–640
158 Yeh CT, Chiu HT, Chu CM, Liaw YF (1998) G1 phase dependent nuclear localization of relaxed-circular hepatitis B virus DNA and aphidicolin-induced accumulation of covalently closed circular DNA. *J Med Virol* 55: 42–50
159 Turin F, Borel C, Benchaib M, Kay A, Jamard C, Guguen-Guillouzo C, Trepo C, Hantz O (1996) n-Butyrate, a cell cycle blocker, inhibits early amplification of duck hepatitis B virus covalently closed circular DNA after *in vitro* infection of duck hepatocytes. *J Virol* 70: 2691–2696
160 Civitico GM, Locarnini SA (1994) The half-life of duck hepatitis B virus supercoiled DNA in congenitally infected primary hepatocyte cultures. *Virology* 203: 81–89
161 Addison WR, Walters KA, Wong WW, Wilson JS, Madej D, Jewell LD, Tyrrell DL (2002) Half-life of the duck hepatitis B virus covalently closed circular DNA pool *in vivo* following inhibition of viral replication. *J Virol* 76: 6356–6363
162 Galle PR, Schlicht HJ, Kuhn C, Schaller H (1989) Replication of duck hepatitis B virus in primary duck hepatocytes and its dependence on the state of differentiation of the host cell. *Hepatology* 10: 459–465
163 Tsiang M, Rooney JF, Toole JJ, Gibbs CS (1999) Biphasic clearance kinetics of hepatitis B virus from patients during adefovir dipivoxil therapy. *Hepatology* 29: 1863–1869
164 Bock CT, Schranz P, Schroder CH, Zentgraf H (1994) Hepatitis B virus genome is organized into nucleosomes in the nucleus of the infected cell. *Virus Genes* 8: 215–229
165 Widom J (1998) Structure, dynamics, and function of chromatin *in vitro*. *Annu Rev Biophys Biomol Struct* 27: 285–327
166 Bock CT, Schwinn S, Locarnini S, Fyfe J, Manns MP, Trautwein C, Zentgraf H (2001) Structural organization of the hepatitis B virus minichromosome. *J Mol Biol* 307: 183–196
167 Cunliffe VT (2003) Memory by modification: the influence of chromatin structure on gene expression during vertebrate development. *Gene* 305: 141–150
168 Ito T (2003) Nucleosome assembly and remodeling. *Curr Top Microbiol Immunol* 274: 1–22
169 Turner BM (2002) Cellular memory and the histone code. *Cell* 111: 285–291
170 Imhof A, Becker PB (2001) Modifications of the histone N-terminal domains. Evidence for an "epigenetic code"? *Mol Biotechnol* 17: 1–13
171 Neely KE, Workman JL (2002) Histone acetylation and chromatin remodeling: which comes first? *Mol Genet MeTab.* 76: 1–5
172 Geiman TM, Robertson KD (2002) Chromatin remodeling, histone modifications, and DNA methylation-how does it all fit together? *J Cell Biochem* 87: 117–125
173 Reik A, Gregory PD, Urnov FD (2002) Biotechnologies and therapeutics: chromatin as a target.

Curr Opin Genet Develop 12: 233–242
174 Becker PB, Horz W (2002) ATP-dependent nucleosome remodeling. *Annu Rev Biochem* 71: 247–273
175 Langst G, Becker PB (2004) Nucleosome remodeling: one mechanism, many phenomena? *Biochim Biophys Acta* 1677: 58–63
176 Civitico G, Wang YY, Luscombe C, Bishop N, Tachedjian G, Gust I, Locarnini S (1990) Antiviral strategies in chronic hepatitis B virus infection: II. Inhibition of duck hepatitis B virus *in vitro* using conventional antiviral agents and supercoiled-DNA active compounds. *J Med Virol* 31: 90–97
177 Kelly WK, O'Connor OA, Marks PA (2002) Histone deacetylase inhibitors: from target to clinical trials. *Expert Opin Invest Drugs* 11: 1695–1713
178 Szyf M (2000) The DNA methylation machinery as a therapeutic target. *Curr Drug Targets* 1: 101–118
179 Gabrielli BG, Johnstone RW, Saunders NA (2002) Identifying molecular targets mediating the anticancer activity of histone deacetylase inhibitors: a work in progress. *Curr Cancer Drug Targets* 2: 337–353

Index

MIX
Papier aus verantwortungsvollen Quellen
Paper from responsible sources
FSC® C105338

If you have any concerns about our products,
you can contact us on
ProductSafety@springernature.com

In case Publisher is established outside the EU,
the EU authorized representative is:
Springer Nature Customer Service Center GmbH
Europaplatz 3, 69115 Heidelberg, Germany

Printed by Libri Plureos GmbH
in Hamburg, Germany